ANIMAL INNOVATION

ANIMAL INNOVATION

Edited by

SIMON M. READER

Department of Biology
McGill University
Montréal
Canada

and

KEVIN N. LALAND

Centre for Social Learning and Cognitive Evolution
School of Biology
University of St Andrews
Scotland
UK

OXFORD
UNIVERSITY PRESS

OXFORD

UNIVERSITY PRESS

Great Clarendon Street, Oxford OX2 6DP

Oxford University Press is a department of the University of Oxford.
It furthers the University's objective of excellence in research, scholarship,
and education by publishing worldwide in

Oxford New York

Auckland Bangkok Buenos Aires Cape Town Chennai
Dar es Salaam Delhi Hong Kong Istanbul Karachi Kolkata
Kuala Lumpur Madrid Melbourne Mexico City Mumbai Nairobi
São Paulo Shanghai Taipei Tokyo Toronto

A catalogue record for this title is available from the British Library

Library of Congress Cataloging in Publication Data

Data available

ISBN 0 19 852621 0 (Hbk)
0 19 852622 9 (Pbk)

10 9 8 7 6 5 4 3 2 1

Typeset by Newgen Imaging Systems (P) Ltd., Chennai, India
Printed in Great Britain
on acid-free paper by Biddles Ltd, Guildford and King's Lynn

PREFACE

In most instances where learned behaviour patterns spread through animal populations a single individual will have triggered the diffusion, by devising a novel means of exploiting the environment. This capacity for 'innovation' is a component of behavioural flexibility, important to the survival of many animals, particularly those with generalist or opportunistic lifestyles. It is potentially of considerable significance to those endangered or threatened species forced to adjust to changed or impoverished environments. Moreover, there is growing evidence that innovation may have played a critical role in avian and primate brain evolution. Yet, until recently, animal innovation has been subject to almost complete neglect by behavioural biologists, psychologists, social learning researchers, and conservation-minded biologists. Such inattention is particularly striking when one considers that animal innovation seemingly overlaps with several research areas (the animal mind, animal culture, brain evolution, social intelligence) that are currently *en vogue* and subject to unprecedented attention from scientists and non-specialists alike. *Animal Innovation* is designed to redress this imbalance.

This edited volume grew out of a symposium on *Animal Innovation* held at the recent International Ethological Congress in Tuebingen, Germany (August 2001), at which many of the contributors to this volume presented papers. The popularity of this symposium, together with the exponential growth in the number of papers dedicated to this topic published in scientific journals, led us to the view that an edited volume would be both valuable and timely. This collection of essays draws together researchers from a wide variety of disciplines to focus attention for the first time on, which individuals invent new behaviour patterns, what ecological variables influence innovation, what processes underlie the invention of new behaviour, and how innovation may drive the evolutionary process. The contributors represent some of the world's leading scientific authorities on animal and human innovation.

We would like to thank the following people for commenting on one or more chapters and for discussing the material in the book: Robert Aunger, Robert Barton, Johan Bolhuis, Robert Boyd, Gillian Brown, Josep Call, Nicky Clayton, Isabelle Coolen, Rachel Day, Todd Freeberg, Jeff Galef, Luc-Alain Giraldeau, Jean-Guy Godin, Celia Heyes, Louis Lefebvre, Jan Lindstrom, Bill McGrew, Steve Nowicki, John Odling-Smee, Henry Plotkin, Denis Réale, Robert Seyfarth, David Sherry, Richard Sibly, Charles T. Snowdon, Daniel Sol, Yfke van Bergen, Carel van Schaik, Bill Vickery, Elisabetta Visalberghi, and Tim Wright. We would also like to express our thanks to the authors of this volume, both for their contributions and for their encouragement throughout *Animal Innovation*'s progress.

SMR and KNL
March 2003

CONTENTS

Human innovation

Discussion

CONTRIBUTORS

Johan J. Bolhuis   Department of Behavioural Biology, Utrecht University, Padualaan 14, PO Box 80086, 3508 TB Utrecht, The Netherlands

Hilary O. Box   School of Psychology, University of Reading, Whiteknights, Reading RG6 6AL, UK

Richard W. Byrne   Scottish Primate Research Group, School of Psychology, University of St Andrews, St Andrews, Fife KY19 9JU, Scotland, UK

Bennett G. Galef, Jr.   Department of Psychology, McMaster University, 1280 Main Street West, Hamilton, Ontario L8S 4K1, Canada

Jane Goodall   The Jane Goodall Institute, PO Box 14890, Silver Spring, MD 20911, USA

Russell Greenberg   Smithsonian Migratory Bird Centre, National Zoological Park, Washington, DC 20008, USA

Marc D. Hauser   Department of Psychology and Program in Neurosciences, 33 Kirkland Street, Harvard University, Cambridge, MA 02138, USA

Hans Kummer   Grundrebenstr.119, CH-8932 Mettmenstetten, Switzerland

Robert F. Lachlan   IEEW, University of Leiden, Leiden 2311 GP, The Netherlands

Kevin N. Laland   Centre for Social Learning and Cognitive Evolution, School of Biology, University of St Andrews, Bute Medical Building, Queen's Terrace, St Andrews, Fife KY16 9TS, Scotland, UK

Phyllis C. Lee   Department of Biological Anthropology, University of Cambridge, Downing Street, Cambridge CB2 3DZ, UK

Louis Lefebvre   Department of Biology, McGill University, 1205, avenue Docteur Penfield, Montréal, Québec, Canada H3A 1B1

Katharine MacDonald   Department of Archaeology, University of Southampton, SO17 1BJ, UK

Simon M. Reader   Department of Behavioural Biology, Utrecht University, Padualaan 14, PO Box 80086, 3508 TB Utrecht, The Netherlands

Anne E. Russon   Department of Psychology, Glendon College of York University, 2275 Bayview Avenue, Toronto, ON, Canada M4N 3M6

Dean Keith Simonton   Department of Psychology, One Shields Avenue, University of California at Davis, Davis, CA 95616–8686, USA

Peter J. B. Slater School of Biology, Bute Building, University of St Andrews, Fife KY16 9TS, Scotland, UK

Daniel Sol Department of Biology, McGill University, 1205 avenue Docteur Penfield, Montréal, Québec, Canada H3A 1B1

Yfke van Bergen Sub-Department of Animal Behaviour, University of Cambridge, High Street, Madingley, Cambridge CB3 8AA, UK

DEFINITIONS AND KEY QUESTIONS

CHAPTER 1

ANIMAL INNOVATION: AN INTRODUCTION

SIMON M. READER AND KEVIN N. LALAND

Do animals innovate?

In 1953 a young female Japanese macaque called Imo began washing sweet potatoes in water before eating them, presumably to remove dirt and sand grains (Figure 1.1, Kawai, 1965). Soon other monkeys had adopted this behaviour, and potato-washing gradually spread throughout the troop. When, 3 years after her first invention, Imo devised a second novel foraging behaviour, that of separating wheat from sand by throwing mixed handfuls into water and scooping out the floating grains (Kawai, 1965), she was almost instantly heralded around the world as a 'monkey genius' (Wilson, 1975).

Imo is probably the most celebrated of animal innovators. In fact, many animals invent new behaviour patterns, modify existing behaviour to a novel context, or respond to social and ecological stresses in an appropriate and novel manner (Kummer and Goodall, 1985; Lefebvre *et al.*, 1997; Reader and Laland, 2001). We submit that such behaviour can sensibly be termed innovation, and discuss below where the boundaries demarcating innovative behaviour should be drawn. However, research into animal innovation is at a formative stage, and there are many unanswered questions. It is not yet clear whether innovation

Figure 1.1 Sweet potato-washing (Kawai, 1965). Reprinted with permission.

should be treated as a unitary phenomenon, or even on what basis this can be decided, and conceivable that a variety of alternative psychological processes underlie innovation. If we accept that animals do innovate, there are still open questions as to what roles innovation plays in facilitating the flexibility of individual development, survival, reproductive success, and evolution of animal species. Is the ability to innovate an adaptation or a by-product of selection on some other character? Under what circumstances is innovation adaptive, and what is its functional significance? Is innovation intrinsically important to processes such as brain evolution or is it simply a covariate of more fundamental characteristics (say, sociality or foraging mode) or an indicator of more general abilities (such as behavioural flexibility)?

This book is about the psychological bases, natural ecology, evolution, and adaptive function of behavioural innovation in animals. We draw together researchers from a wide variety of disciplines to focus attention on questions such as

- Which individuals invent new behaviour patterns?
- What ecological variables influence innovation?
- What psychological processes underlie the invention of new behaviour?
- Can innovation drive the evolutionary process?

In this chapter, we argue that a strong case can be made for the assertions that many animals—not just humans—innovate, that innovation can be regarded as qualitatively distinct from related processes such as exploration and learning, and that innovation is likely to play important roles in the lives and evolution of many animals. The goal of this chapter is to provide definitions, identify key questions related to animal innovation, and present an overview of the current trends in innovation-related research.

On the significance of behavioural innovation in animals

Behavioural scientists have long observed that innovation is phylogenetically widespread, but have also noted species differences in innovative tendency (Thorpe, 1956; Cousteau, 1958; Cambefort, 1981; Lefebvre et al., 1997), with greatest innovation associated with 'higher organisms' (Lloyd Morgan, 1912; McDougall, 1936) or animals with larger relative brain sizes (Wyles et al., 1983; Lefebvre et al., 1997). For instance, as long ago as 1912 in his book *Instinct and Experience* Conwy Lloyd Morgan speculated that the behaviour of animals may be composed of a repetitive component that has occurred before, and a smaller proportion of novel behaviour that can be regarded as a creative departure from routine. For Lloyd Morgan the 'higher' organisms have a larger measure of this latter creative component. While there are long traditions of research into related topics, such as neophilia, exploration, and insight learning in animals (discussed later), and while innovation in humans has been subject to considerable investigation, it is only in the last 20 years that animal innovation has begun to receive attention.

A landmark paper in this respect was Hans Kummer and Jane Goodall's review of primate innovation (Kummer and Goodall, 1985; Chapter 10), which almost for the first time stimulated general interest in the topic of animal innovation. They suggested that some innovations derive from the ability of the individual to profit from an accidental happening, while others result from the ability of the higher primates to use existing behaviour patterns for new purposes. For instance, chimpanzees frequently used leaves to wipe their soiled bodies, but one female used leaves to wipe away ants and bees, and one male used leaves to wipe the inside of a baboon skull to get at the brain. A third kind of innovation is the performance of a completely new pattern, such as 'wrist-shaking', a novel gesture adopted by a juvenile female chimpanzee for use in aggressive contexts. They also suggested that innovations may be occasioned by a sudden change in the environment such as when wild populations are provisioned or during the outbreak of disease, or during periods of environmental stability or plenty that result in an excess of leisure and energy. This is exemplified in some captive situations, a hypothesis we discuss subsequently.

Kummer and Goodall were also among the first to draw a link between innovation and tactical deception, a theme developed further by Byrne (Chapter 11) in his examination of innovation in social contexts. They observe

> A subordinate individual *A*, to attain a particular goal in spite of the inhibiting proximity of a social superior *B* must either form a coalition with a third individual *C*... or follow a more devious route, such as moving out of *B*'s sight, or persuading *B* to move away, or distracting *B*'s attention. Some of the solutions to social problems of this sort represent novel ones on the part of the individual concerned. (p. 207, italics added)

Taking a more functional approach, Lee (1991; Chapter 12) suggested that innovation may originate as a solution to a specific ecological problem. She argues that the classic examples of innovation, behaviour patterns concerned with the extraction, preparation, and processing of food, demonstrate how the intake of food is either enhanced or made energetically more efficient when that food is exploited in a novel way. Examples include the washing of wheat and potatoes by Japanese macaques, novel tool-use in primates, particularly chimpanzees *Pan troglodytes* (Goodall, 1964; Beck, 1980; McGrew, 1994), and milk-bottle-top opening by British titmice, *Parus* spp. (Figure 1.2, Fisher and Hinde, 1949; Hinde and Fisher, 1951). Innovation may provide benefits but may also carry costs, such as increasing the risk of predation or of consuming hazardous foods.

Other examples of innovation function in different domains. Reader and Laland (2001, 2002) surveyed primate innovation by searching four leading primate behaviour journals, plus other relevant literature, recording 533 cases of innovation in 41 separate species. Nearly half these innovations occurred in a foraging context, but innovations were recorded across a broad range of contexts. Non-foraging examples included aggressive behaviour, such as the first reported use of a club to attack a snake by a capuchin monkey (Boinski, 1988), apparent play behaviour like snowball rolling by macaques (Figure 1.3; Eaton, 1976) and communicative behaviour patterns such as the creation of a new gesture by a Hamadryas baboon, inviting an infant to be carried (Kummer and Goodall, 1985), and the invention by a Japanese macaque of a rattlesnake alarm call (Rowe, 1996). Social innovations are also commonly reported (see Chapters 11 and 12). For instance, Goodall (1986) describes a male chimpanzee, Mike, that augmented his threat display by banging together empty kerosene cans, this behaviour coinciding with a dramatic rise in his dominance status.

Figure 1.2 Milk bottle top opening. See Fisher and Hinde (1949). © Brian Bevan, Ardea London.

Another important paper in the study of animal innovation was Lefebvre *et al.*'s (1997) investigation of the relationship between feeding innovations and forebrain size in birds. These researchers collated data from avian behaviour journals on the incidence of complex foraging traits associated with behavioural flexibility by using keywords such as 'novel' or 'never seen before' to classify behaviour patterns. Lefebvre and co-workers (Chapter 2 and references therein) have now gathered over 2200 examples of feeding innovation in birds, representing the most complete existing survey of animal innovation. In all such examples, as a result of the innovation a novel food source is utilized, or an established food resource is exploited more efficiently. Lefebvre *et al.* (1997) found that absolute and relative innovation frequency per order were positively correlated with measures of forebrain size. Their pioneering work has inspired further analyses, in birds and in primates, which explore the relationship between innovative tendency and a variety of ecological and cognitive traits (see Chapters 2–4). This kind of study, which relies on statistical analysis of a sample of innovative behaviour from the field, is vulnerable to a number of biases: for instance, some species, sexes or ages may be studied more than others, and field workers may be more inclined to describe as innovation the behaviour of large-brained species such as apes, compared with smaller-brained species, such as prosimians. Nevertheless, Lefebvre and colleagues have devised statistical methods for evaluating and counteracting these biases, and there are reasonable grounds for confidence in the reliability of these findings

Figure 1.3 Snowball rolling in macaques. See Eaton (1976). © M. Iwago, Foto Natura.

(see Chapters 2–4). It is largely due to the excitement generated by these analyses that the topic of animal innovation has begun to receive attention by biologists.

When a novel learned behaviour spreads through an animal population, as individuals learn from one another, typically, a single individual will have initiated the process. Such a diffusion requires two processes: there has to be the initial inception of the behavioural variant, which is what we describe as innovation, and the novel trait must spread between individuals, which is known as social transmission. In the 50 years since Imo was first observed washing sweet potatoes, great strides have been made in understanding the social learning processes that underpin diffusion, as can be seen by the 'explosion of interest' in animal social learning (Galef and Giraldeau, 2001; Shettleworth, 2001), exemplified by a proliferation of books, conferences and papers dedicated to the topic (Zentall and Galef, 1988; Heyes and Galef, 1996; Box and Gibson, 1999; Fragaszy and Perry, in press). In comparison, innovation has received very little attention.

Apart from its significance to researchers interested in social learning and the evolution of culture, innovation might be expected to be central to a number of other disciplines. For instance, innovation is an important component of behavioural flexibility, vital to the survival of individuals in species with generalist or opportunistic lifestyles (Lefebvre and Bolhuis, Chapter 2). Innovation might also be of critical importance to those endangered or threatened species forced to adjust to changed or impoverished environments (Greenberg and Mettke-Hofman, 2001; Sol, Chapter 3; Box, Chapter 9).

Moreover, innovation is central to a variety of hypotheses concerning brain evolution. For instance, Berkeley biochemist Allan Wilson noted a crude correlation between brain size and rate of evolution among vertebrates. To account for this, Wilson (1985) proposed a 'behavioural drive' hypothesis, which maintained that episodes of innovation and cultural transmission are more frequent in large-brained than small-brained species, leading these animals to exploit the environment in new ways and so exposing them to novel selection pressures (see also Wyles *et al.*, 1983). Such processes may increase the rate of evolution in big-brained species. Only in the last few years has evidence begun to accumulate among birds and primates that supports the assumptions of behavioural drive (Lefebvre *et al.*, 1997; Reader and Laland, 2002; Sol *et al.*, 2002), and innovation rate has recently been shown to correlate with avian species richness (Nicolakakis *et al.*, 2003). Hence the topic of animal innovation overlaps with several research areas (the animal mind, social evolution, brain evolution, social intelligence) that have been *en vogue* for over two decades.

In sum, despite being of importance to a variety of research traditions, the topic of animal innovation has been subject to almost complete neglect by behavioural biologists, psychologists, social learning researchers, and conservation-minded biologists alike. At this point of time there is not even a well-formulated definition of innovation. Moreover, there has been virtually no contact between researchers interested in animal and human innovation. Even a cursory look through citation indexes or bookstores using the keyword 'innovation' reveals the immense interest of human innovation to economists and business people (Figure 1.4). Human scientists too have long been interested in human innovation

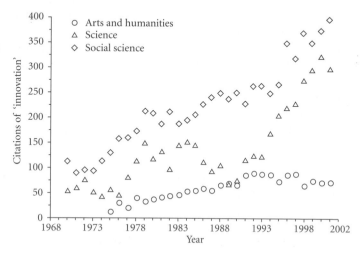

Figure 1.4 Number of articles published with the word 'innovation' in the title from 1970 to 2001. Data were taken from three Institute for Scientific Information citation indexes: Science, Social Science, and Arts and Humanities. The social science citation index includes economics, business and finance journals.

and creativity (e.g. Eysenck, 1995; Rogers, 1995; see Simonton, Chapter 14). Galef (Chapter 6) discusses why innovation in animals has only recently begun to receive the attention it deserves. We hope this book will address the neglected status of animal innovation and provide a bridge between disciplines.

Defining innovation

Perhaps now is the time to pause and discuss definitional issues. Without a clear definition of innovation and underlying concerns, evaluation of the utility of the innovation concept to each field of study will remain difficult. According to Kummer and Goodall (1985; Chapter 10):

> An innovation can be: a solution to a novel problem, or a novel solution to an old one; a communication signal not observed in other individuals in the group (at least at that time) or an existing signal used for a new purpose; a new ecological discovery such as a food item not previously part of the diet of the group (p. 205).

Note that in Kummer and Goodall's definition, innovation refers to the end *product* of an inceptive or creative process, that is, innovation is a novel pattern of behaviour. In contrast, Lee (1991) and Wyles *et al.* (1983) regard innovation as a *process*:

> Innovation refers to the introduction of a novel mode of coping with the environment. The behaviour can be either one not previously known, or it can be an existing behaviour, or a combination of behaviours that are applied in a novel context (Lee, 1991, p. 46).

> Behavioural innovation refers to...the origin of a new skill in a particular individual, leading it to exploit the environment in a new way (Wyles *et al.*, 1983, p. 4396).

The human innovation literature also appears to describe innovation as both product and process. For instance, Rogers (1995) describes innovation as 'an idea, practice, or object that is perceived as new by an individual' (p. 11), while Simonton (Chapter 14) treats innovation as a creative process. A failure to recognize this product–process distinction lies at the heart of much of the terminological confusion surrounding innovation. For example, a new behaviour may be introduced into a population through immigration of an individual, social learning from an out-group member or invention by a group member, but only the latter would be a manifestation of a process of innovation. Note also that Rogers's broad definition allows innovations to be either (1) ideational, (2) behavioural phenomena, or (3) objects or constructions.

Both Kummer and Goodall's and Lee's definitions regard innovation as a behaviour pattern not previously found in the population or a process that generates such a novel variant. In contrast, for Rogers (1995): 'It matters little...whether or not an idea is objectively new.... If the idea seems new to the individual, it is an innovation' (p. 11).

In part, this disparity is due to Rogers' focus on the spread of innovations, with the critical variable of interest being the adoption, rather than the creation, of an innovation. Thus, for Rogers (1995) 'invention' perhaps approaches more closely what other authors would term innovation: 'Invention is the process by which a new idea is discovered or created. In contrast, innovation...occurs when a new idea is adopted or rejected' (footnote, p. 135). Rogers describes 'the diffusion of innovations' as the spread of technological advances, such as hybrid corn, methods of water purification and horseback riding, first invented elsewhere but introduced into a population through migration or social learning.

The birdsong and animal vocal learning literatures appear to be using the term innovation in a much more specific manner. Slater and Lachlan (Chapter 5; Janik and Slater, 2000) distinguish between *innovation*, where new behaviour patterns are derived by modifications of pre-existing ones, *invention* which is the appearance of a behaviour pattern that is totally novel, *improvisation* which represents spontaneous, transient modifications and inventions, and *immigration* where new behaviour is introduced through the movement of individuals. The human innovation literature, at least as exemplified by Rogers (1995), does not dwell on these distinctions. It seems likely that in practice the distinctions would be easier to make when the subject is vocal learning (where repertoires and novelty can more easily be quantified) as compared to the learning of other forms of behaviour.

Let us consider the above definitions, try to crystallize any core components and make policy decisions about points of apparent disagreement. We make the following proposals.

Proposal 1

Innovation is either regarded as a *new or modified behaviour pattern* (henceforth innovation *sensu* product) or as a *process that results in new or modified behaviour* (henceforth innovation *sensu* process).

It is clear that few researchers have made the distinction between innovation as product and process, that these two uses of innovation are both commonly employed, and that terminological confusion abounds partly as a consequence. To make our meaning absolutely clear we distinguish exhaustively between the product ('an innovation') and the innovatory process in the following paragraphs.

Proposal 2

Innovations (*sensu* product) are *learned* behaviour patterns while innovation (*sensu* process) requires *learning*.

Sometimes evolutionary biologists refer to newly evolved anatomical and behavioural characters as 'innovations'. Conversely, our use of the term requires that the novel behaviour should not be a regular or standard feature of the species' behavioural repertoire, and that its emergence is contingent on appropriate experience. However, here too there is some confusion, with some researchers describing any novel behaviour as an innovation, irrespective of whether or not it is learned, and others insisting that learning is part of the definition. It might be argued that the essence of innovation is the discovery of novel information, in which case it seems counterintuitive to incorporate learning as an important definitional component. Our proposal is based on the view that, without learning as an essential characteristic of innovation, it would be difficult to distinguish innovation from exploration, or from idiosyncratic or accidental behaviour (or indeed, from unlearned behaviour). Neither the apparent complexity nor the supposed goal of a behaviour pattern are usable defining features of innovation. However, we are not suggesting that learning is the only important feature of innovation, and other processes such as exploration, the generation of novelty, and insight may be critical in particular cases.[1]

In practice it may often be difficult to determine whether, or at what point, a novel behaviour has been learned, and there is a danger that a strict adherence to the need to demonstrate learning might suppress much interesting data. While we regard learning as a central component of innovation, we recognize that in practice, at least at this embryonic stage of investigation, it is sometimes likely to be useful to interpret this definition loosely. For instance, a significant proportion of the reports of innovation in birds used by Lefebvre *et al.* (1997) were observed only once, with little indication that they are an established feature of the innovator's behaviour. However, presumably, bird enthusiasts would not bother to write up their observations and journal editors would not publish them, if they believed that the innovation was a trivial and idiosyncratic one-off. Lefebvre *et al.* (1997) emphasize this point when they term innovations as 'novel *opportunistic* behaviours'. It is because innovations are deemed to have become a stable component of the behavioural repertoire of an individual, a component that is a significant and interesting departure from established behaviour, and that potentially can spread to other group members and form the basis of between-population diversity, that they are regarded as important.

We envisage that in the future, when data sets are richer, researchers will be able to circumvent these problems by, for instance, only describing as innovations individually learned novel behaviour patterns. It may not be particularly onerous to determine whether a behaviour pattern is part of the learned behavioural repertoire of an individual. For example, simply observing an individual performing the same novel, unusual behaviour pattern several times may give us reasonable confidence that the behaviour is not a chance one-off, especially if there seems to be a clear payoff to the behaviour pattern.[2] Presumably, we would not want to describe a lemur's tail accidentally falling into a pond as an innovation, but if we repeatedly observed a lemur tail-dipping, this might suggest that something

---

[1] An important issue that we suspect lies at the heart of these differences of opinion is the need to operationalize the innovation concept. Our emphasis on an operational definition reflects a belief that 'innovation' must first and foremost be a useful and usable concept. The definitions below are designed to provide a practical tool to facilitate the identification of behaviour patterns that we can be reasonably sure are innovation, even if this means excluding some apparently compelling examples. Definitions of imitation have come across similar problems and have incorporated additional qualifications in order to determine with certainty that imitation has actually taken place. It seems perfectly possible to imitate an act already known to the imitator, but many definitions of imitation include some requirement that the act imitated be novel, since without such provisos it is difficult to distinguish imitation from other processes such as social facilitation (Whiten and Ham, 1992).

[2] A case can be made that, including learning in our definition of innovation allows us to determine what novel discoveries the animal itself finds relevant. For instance, an animal may be observed making a novel discovery, such as exploiting a nutritious food source, but if the animal does not repeat this incident the discovery is likely to be of little importance, not least because it might suggest that the event is of little significance to the animal itself. For instance, Nishida (1994) observed a male chimpanzee placing a probe in its nostril, which was followed by the chimpanzee sneezing, and interprets this as functional tool use designed to remove a nasal plug. A sceptical observer might assume that what they had witnessed had merely been a random or idiosyncratic behaviour on the part of the chimpanzee. If the functional interpretation is correct and if this behaviour is of utility to the chimpanzee we would expect it to repeat this behaviour. Nishida reports that the same chimpanzee has been seen to perform this behaviour on four separate occasions.

more interesting is going on, especially if we subsequently observed lemurs drinking from their tails and so using them to access water sources (Figure 1.5, Hosey *et al.*, 1997). Indeed, many reported avian foraging innovations fall into this more interesting category of repeatedly observed, apparently useful behaviours (Lefebvre, personal communication, e.g. Taylor, 1972; Thompson *et al.*, 1996; Lefebvre *et al.*, 2001; Reader *et al.*, 2002). The observability of repeatedly performed innovations will also be greater than that of one-offs. In practice the application of an operational definition of innovation, particularly to natural populations, might require researchers to focus on the 'durability' of novel behaviour rather than directly on 'learning', on the assumption that the former is an indication of the latter. Further discussion of definitional issues and some pragmatic solutions can be found in Chapters 4 and 7.

Proposal 3

While an innovation (*sensu* product) can be learned from others, and as a consequence it may increase in frequency in a population, *the social learning of a novel behaviour in its complete form is not innovation (sensu process)*. While immigration of individuals or information is not an innovatory process it can introduce an innovation (*sensu* product) into the population for the first time.

Among non-human animals, the process of innovation as generally conceived is not heavily dependent on observation or interaction with other individuals or their products. Innovation is usually regarded as the result of a single individual's creativity. To give an example, if a juvenile male bird learns its entire song from a tutor, then few people would want to describe this learning as innovation; rather, it is a case of social learning. However, there are some interesting exceptions worthy of consideration. What if, for instance, a bird incorporates a novel element into its song through learning from a conspecific (see Slater and Lachlan, Chapter 5)? Despite the fact that the element is learned socially, this might still be regarded as an example of innovation—although the element was not novel, it has been incorporated to form a novel song. Incorporation of more unusual elements into a song (e.g. man-made noises such as mobile phone rings) would seem to be a clearer example of innovation, yet only differ from the previous example in the source of the novel song element. Clearly, it will be an important challenge for researchers investigating innovation to come up with clear and non-arbitrary definitions of 'novel behaviour'.[3]

In other instances, an innovation appears to result from the activities of more than one individual. Experimental studies of social learning in rats, *Rattus norvegicus*, and common chimpanzees reveal that it is possible for an innovation to be created through the activities of several animals (Laland and Plotkin, 1990; Paquette, 1992). Novel foraging behaviour emerged as a result of the accumulated efforts of several individuals without one particular creative individual to initiate it. Moreover, at least among humans, cooperative innovation

---

[3] 'Novelty' is well-known to be a problematic, amorphous concept. Few behaviour patterns can be considered completely novel, since virtually all will share at least the most basic properties with some previously performed behaviour pattern. Similarly, novel stimuli cannot be distinguished from familiar stimuli by their physical properties (Berlyne, 1950, 1960; Birke and Archer, 1983). We accept that there may be differences of opinion between researchers in what is regarded as 'novel enough' to qualify as innovation.

(a)

(b)

Figure 1.5 Tail-dipping in lemurs: (a) tail-dipping, (b) drinking from their tails. See Hosey *et al.* (1997). Reprinted with permission.

(*sensu* process) on the part of more than one individual is possible, but this would only be regarded as innovation if it results in a novel invention.

To the above three definitional considerations we make the following operational proposal.

Proposal 4

An innovation (*sensu* product) is a behaviour pattern *not previously found in the population*, while innovation (*sensu* process) *introduces novel behavioural variants* into a population's repertoire.

Clearly, in reality, two individuals in a population could independently come up with the same novel behaviour. We envisage that in the wild it would frequently be extremely difficult to establish whether the two really had acquired the novel variant independently. Moreover, if all independent invention is described as innovation then the distinction between innovation and asocial learning becomes blurred.

If innovation as a process is to be defined, in part, by its product (the introduction of a novel behavioural variant into a population) is the product/process distinction meaningful? We believe yes, because the process of innovation is not the only process that can result in the introduction of a novel behavioural variant into a population. As intimated above, the immigration into a population of an individual exhibiting a new behaviour, or the social learning of behaviour from a source external to the population, can both introduce innovations (*sensu* product) into the population. An understanding of the innovatory process would require an investigation into the biological and psychological processes that underpin asocial learning, individual creativity, and behavioural flexibility. Conversely, an understanding of innovation as a product would include an investigation of the social processes that can result in its introduction and spread.

The above considerations lead us to the following two *operational* definitions.

Innovation sensu product.   An innovation is a new or modified learned behaviour not previously found in the population.

Innovation sensu process.   Innovation is a process that results in new or modified learned behaviour and that introduces novel behavioural variants into a population's repertoire.[4]

We view as similar phenomena, and treat as examples of innovation, novel behaviour patterns such as the exploitation of novel resources and foods, the development of new food-processing methods, the use of an established behaviour in a novel context, and learning to use novel tools or technologies. We make no distinction between totally novel behaviour and modifications of existing behaviour, as has been the practice among researchers studying vocal learning. While we recognize that further distinctions and narrower definitions may be of utility in particular fields or at some point in the future, we see problems in applying these distinctions to non-vocal innovative phenomena. Our

---

[4] Our use of the phrase 'population's repertoire' is not meant to imply that all individuals in the population will necessarily acquire the novel behaviour, but rather that at least one individual in the population will behave in a manner not previously seen.

definitions are deliberately broad, applying equally to the learning of the route through a maze by a fish, invention of a new song element by a bird, or a new food processing technique in a monkey.

Some objectors might prefer to reserve the term 'innovation' for qualitatively new or cognitively demanding tasks or processes. In our view this would be a mistake. A broad definition is justified, given the primitive state of knowledge of animal innovation. The key characteristic of innovation is the introduction of a novel behaviour pattern into a population's repertoire, and it would be foolhardy to insist that in the process the innovator must express a previously unobserved motor pattern, or exhibit some unusual level of intelligence. For instance, there is no reason to believe that the food-washing macaques moved their bodies in ways they had never moved before, while there is reason to believe that food washing is common in macaques (Suzuki, 1965; Visalberghi and Fragaszy, 1990; Galef, 1992), which means that the innovation shown by Imo involved the application of a familiar behaviour pattern to a novel food source. Moreover, subjective judgements of intelligence are vulnerable to be prejudiced by assumptions based on phylogenetic proximity to humans. Making premature distinctions potentially jeopardizes the ability to see genuine relationships between different kinds of novel behaviour.

The processes underlying innovation

In spite of the fact that, outside of humans, innovation has been subject to comparatively little direct research, a number of seemingly related topics, such as animal exploration, curiosity, and insight learning have been the focus of investigation. An important question is to what extent innovation can be treated separately from these other processes that are likely to play a role in the creation of innovative behaviour. This question will have implications for the study of the causal mechanisms, evolution, development, and function of innovation; all four of Tinbergen's questions of ethology (Tinbergen, 1963). In this section, we discuss the processes underlying innovation and clarify what innovation is and what it is not. Naturally, many authors in this volume will discuss these processes more fully and we can only hope to give an overview here. Candidate processes that may underlie innovation are summarized in Table 1.1.

Neophilia and neophobia

Attraction to and avoidance of novelty (neophilia and neophobia, respectively) are likely to be critical predictors of innovative behaviour. Before an animal utilizes or learns about an unfamiliar resource, the resource must be approached, explored, and sampled (Greenberg and Mettke-Hofman, 2001). Greenberg (Chapter 8) suggests that 'the concept of innovation is inexorably wed to that of newness', arguing that the response to novelty in animals plays a pivotal role in the probability that innovative behaviours will develop and spread. Greenberg discusses the preconditions for innovative foraging behaviour in birds, and notes that novelty responses are not necessarily associated with the complex information processing often supposed to accompany innovation. This implies that selection acting on novelty responses may influence innovative propensities without the requirement for changes in cognitive processing. Lefebvre (2000) anticipates a negative relationship

Table 1.1 Candidate processes underlying or influencing innovation.

Process	Possible contribution to innovation
Novelty responses (neophilia and neophobia)	Attraction to or avoidance of novelty will determine which resources or situations are approached, explored, sampled, or learned about (See Greenberg, Chapter 8)
Exploration	Exploration of the environment and investigation of objects ('object exploration') may increase the encounter rate of novel situations and objects, and allow the discovery of novel environmental affordances. (See Laland and van Bergen, Greenberg, Chapters 7 and 8).
Asocial (individual) learning	Learning the affordances of an object or situation during innovation, or learning the innovation itself. (See Lefebvre and Bolhuis, Simonton, Chapters 2 and 14).
Insight	Understanding of a problem without recourse to trial and error learning. (See Simonton, Hauser, Chapters 14 and 15).
Creativity	Generation of novel combinations of ideas or behaviour (See Simonton, Russon, Hauser, Chapters 13–15).
Behavioural flexibility, such as inhibition of existing responses	Inhibition of a pre-existing response, perhaps a previously successful one, to try a novel strategy. (See Lefebvre and Bolhuis, Sol, Reader and Macdonald, Hauser, Chapters 2–4, 15).
Social processes	In what may be quite limited circumstances, innovation may result as the combined efforts of several individuals (e.g. cooperative innovation, ratchet cultural evolution). The social milieu will also be an important influence on the costs, benefits, and likelihood of innovation (See Galef, Chapter 6).

Note: some of these processes may be context-specific (Coleman and Wilson, 1998; Wilson, 1998). This would mean the fact an animal was innovative in one context (such as consuming novel foods) may say little about the probability of innovation in other contexts (such as anti-predator responses).

between neophobia and innovation across species and individuals, an association observed in callitrichid monkeys and columbid and passerine birds (Seferta *et al.*, 2001; Webster and Lefebvre, 2001; Bouchard, 2002; Day *et al.*, 2003). However, as Greenberg (Chapter 8) explains, the relationship between novelty responses and behavioural flexibility is a complex one, for which there are many competing hypotheses and some seemingly conflicting data.

Exploration and curiosity

Researchers from both the comparative psychology and ethological traditions have noted an inherent tendency for exploration in animals. It has been well established that many animals will explore their environment, investigate or manipulate novel objects, and inspect predators or unfamiliar conspecifics in the absence of any direct reward ('intrinsic' exploration, Harlow *et al.*, 1950; Thorpe, 1956; McFarland, 1981; Archer and Birke, 1983). Darwin famously observed monkeys in London Zoo repeatedly opening a bag to examine a snake inside, unable to 'resist taking a momentary peep', and leaping back in fear (Darwin, 1897, p. 72). Pavlov referred to a response to novelty by 'curiosity' as 'the investigatory reflex' (Boakes, 1984), Berlyne (1950) suggested that novel stimuli elicit a 'curiosity' response leading to exploration and learning, and psychologist William McDougall (1936) described an 'instinct of curiosity' which 'determines fuller perception and so leads on to appropriate action' (p. 143). Seemingly accepting Lloyd Morgan's argument that big-brained animals were more innovative than their smaller-brained counterparts (Lloyd Morgan, 1912), McDougall noted that

> The omnivorous monkeys...display this instinct in great strength and versatility; and this feature of their constitution was probably of the first importance in leading them on in the scale of intellectual development (p. 144)

This reference to a natural tendency on the part of animals to investigate their environment was further developed by ethologist William Thorpe in his classic 1956 text *Learning and Instinct*. Thorpe suggested that this was exploration for its own sake, rather than the result of trial and error learning. In support of this assertion, Thorpe pointed out that in rhesus monkeys pure manipulation of a mechanical puzzle is an adequate motivation for the learning of new discriminations and the acquisition of new responses, and moreover visual exploration alone will serve as a reward (Harlow, 1953). However, for Thorpe behavioural innovation was likely to result from a combination of exploration and learning:

> In 'higher' vertebrates...complex behaviour patterns are to a greater extent built up by a process of 'taking notice of' or 'investigating' a large variety of environmental factors which have not any specific valence for the animal to start with. This building up, no doubt, proceeds by all types of learning, but probably much of it is due to something very like that which, when seen in a rat in a maze, we call latent learning and which is based on a tendency to 'explore the environment' (Thorpe 1956, p. 96).

The fact that animals may learn during this intrinsic exploration supports the hypothesis that the function of such behaviour is to gather information (Glickman and Sroges, 1966; Renner, 1987, 1988). Thus exploration may facilitate the development of new innovative behaviour patterns that permit exploitation of the environment in new ways (Russell, 1983). Exploration is one area relevant to innovation where developmental and motivational issues have been well studied, typically focusing on the influence of enriched over impoverished environments during rearing, or of factors such as hunger level (Barnett, 1970; Renner, 1988; Nicol and Guildford, 1991; Heinrich, 1995; Mettke-Hofmann, 2000). A particularly interesting research program on exploration demonstrates consistent individual differences in exploratory tendencies in birds that are linked with other behavioural contexts such as aggressive behaviour and social learning (Verbeek *et al.*, 1994, 1996, 1999; Drent, 1997; Marchetti and Drent, 2000). These differences are heritable and current research is examining the fitness payoffs of exploratory differences in wild-living birds (Dingemanse *et al.*, 2002; Drent *et al.*, 2003).

Asocial learning and problem solving

We discussed earlier our reasons for placing learning as a central core of an operational definition of innovation. Is innovation then simply a special type of asocial learning? We would argue no. Innovation involves learning, including, importantly, learning the affordances of an object, tool or novel situation prior to the innovative solution being first discovered. However, it is likely that other processes (e.g. exploration, curiosity, motivational factors, mental simulation) will frequently contribute to the innovation process and hence asocial learning is perhaps best considered as an important component of innovation. It is an empirical question as to whether the kind of learning that results in innovation is psychologically distinctive from the kind that does not, but at this point there is little clear evidence either way.

Insight

Perhaps because, in humans, we commonly associate innovation with intelligence and moments of inspiration, there has been a historical connection between innovation and 'insight learning'. Between 1913 and 1917 the German psychologist Wolfgang Köhler conducted observations and experiments on the intelligence of chimpanzees at a field station in north-west Africa (Köhler, 1925). In one study a male chimpanzee, Sultan, solved the problem of getting a suspended banana, by moving a box and climbing on it (Figure 1.6; Köhler, 1925). Chimpanzees were also observed to use a short stick to get a long stick to get the bananas, fit two sticks together to rake in a banana, and untying a rope to let down a suspended basket of bananas. Köhler contrasted this 'insightful' behaviour with other forms of learning that develop gradually as a result of reinforcement. Since Köhler's original study, numerous examples of apparent insight in animals have been reported, particularly in primates and birds (Pitocchelli, 1985; Heinrich, 2000). Apart from chimpanzees, a number of non-human primates (gorillas, orang-utans, gibbons) seem capable of similar problem solving (Davey, 1981). Moreover, insight has been repeatedly invoked to explain the behaviour of various birds in reaching food on a string (Bierens de Haan, 1933; Thorpe, 1943) in a research tradition that continues to this day (Heinrich, 2000). However, the interpretation of the problem-solving abilities of apes and birds as manifestations of 'insight learning' remains controversial (Gould and Gould, 1994). For instance, Schiller (1952) found that chimpanzees will fit sticks together, or stack boxes for fun, even if there is no problem to be solved. Thus components of problem-solving behaviour may occur by chance in an experimental situation. At this stage it is not clear to what extent animal innovation is reliant on insight-like processes. In fact, the same reservations apply to our own species. Sudden flashes of inspiration, 'eureka' experiences, obscure the fact that they were typically preceded by a body of work essential to the moment of insight (Eysenck, 1995). Moreover, in the absence of a good understanding of the psychological mechanisms that underlie insight, it is not clear what is to be gained by describing an instance of innovation as 'insight'.

Creativity

In essence, innovation would seem to require at least a degree of creativity, for instance, a novel behaviour or tool must be devised, or prior knowledge must be applied to a novel

Figure 1.6 A chimpanzee box-stacking (Köhler, 1925). Reprinted with permission from Kegan Paul.

problem. Before an innovation can become a learned component of an individual's repertoire, novel combinations of ideas or behaviour must be generated. Russon (Chapter 13) reports on a large number of discoveries, particularly foraging techniques, made by rehabilitant orang-utans after their release into natural habitat. She argues that orang-utan innovation is a natural and quite common feature of their development, resulting from a generative cognitive system in which low-level elements are combined into higher-level cognitive structures. Hauser (Chapter 15) notes that plasticity will frequently depend upon the ability to inhibit a pre-existing response, perhaps one that was previously successful, and to try a novel strategy.

How novel behavioural variants arise has also been the subject of attention by psychologists researching human creativity (Sternberg, 1999; Simonton, Chapter 14). Simonton (Chapter 14) argues that creativity is the result of a Darwinian process in which ideas are generated and selected. This treatise can be regarded as part of a tradition stressing the universal nature of natural selection, known as 'evolutionary epistemology' or 'universal Darwinism' (Plotkin, 1982, 1994).

In humans innovation is typically regarded as a personality trait or characteristic, found among individuals that are particularly creative or less sensitive to social cues. Innovators have been portrayed as idiosyncratic outsiders, intelligent, well educated, and risk-loving (Rogers, 1995), with innovation regarded by some as qualitatively different from 'ordinary' thinking (Goldenburg *et al.*, 1999). Relative to other animals, humans are frequently characterized as inquisitive, constantly innovating, and having ideas (Fagan, 1993). The Upper-Palaeolithic, with its revolution in stone-tool-making traditions, the emergence of bone and antler tools, and the appearance of art, is often regarded as a historical watershed, perhaps reflecting the evolution of an innovative capability in modern *Homo sapiens*. For instance, Diamond (1991, p. 43) suggests that 'the most important innovation came with our rise to humanity: namely, the capacity for innovation itself'. While the contributors to this volume would argue that non-human animals also innovate, it is not clear whether the same processes, particularly those postulated to explain insight and creativity, underlie innovation in animals.

Behavioural flexibility

Innovation allows modification of an animal's behavioural repertoire, and can be considered a component, and perhaps indicator, of behavioural flexibility or plasticity (Lefebvre *et al.*, 1997; Reader and Laland, 2002). Bateson and Martin (1999) succinctly define plasticity as 'changeability' (p. 182), and Bateson (1983) describes behavioural plasticity as the capacity to modify behaviour. Behavioural plasticity is subsumed within phenotypic plasticity, the capacity of a particular genotype to produce a different phenotype in response to environmental conditions (West-Eberhard, 1989; Schlichting and Pigliucci, 1998). Determining the exact relationship of innovation to phenotypic plasticity may prove important because there is a body of theoretical and empirical work that examines the conditions under which phenotypic plasticity may evolve, and the influence of phenotypic plasticity on evolution itself (e.g. West-Eberhard, 1989; DeWitt *et al.*, 1998; Dukas, 1998, see Chapter 3). Determination of the influence of environmental factors on innovation rates (a 'norm of reaction' approach) might be profitable in this regard. Contemporaneous, plastic, and ephemeral traits such as behaviour will have a particularly strong evolutionary role (West-Eberhard, 1989). As Sewell Wright wrote back in 1932, 'Individual adaptability is ... not only of the greatest significance as a factor of evolution in damping out the effects of selection ... but is itself perhaps the chief object of selection' (in Stearns, 1989).

While apparently conceptually straightforward 'behavioural flexibility' is less easy to quantify. Comparative measures using, for example, 'diet diversity' or 'niche breadth' are difficult to make reliably, particularly where very different species are compared, and so are perhaps only useful when comparing individuals of similar lifestyles. For example, is a filter-feeding baleen whale more flexible than a toothed whale because it consumes many more prey species, or less flexible because it subsists on fewer types of food (largely plankton, compared with the fish, seal, cephalopods, and crustacea included in the odontocete diet)? A similar confusion could arise from a generalist species that is a facultative specialist, utilizing a single food source where this option is most profitable (West-Eberhard, 1989). Field reports of innovative behaviour may prove of value in quantifying behavioural flexibility, since the examination of departures from the behavioural norm may provide a more robust comparative index. However, innovation should not be equated with behavioural flexibility in totality. 'Flexibility' can arise through a variety of routes, and it is possible, even likely, that plasticity in one domain will not correlate with that in another. For clarity it is best to state exactly what aspect of behavioural flexibility is measured, which in this case is innovation rate. These issues and the utility of the innovation concept to measures of behavioural flexibility will be addressed throughout this book, particularly in Chapters 2–4.

The costs and benefits of innovation

Innovation is likely to carry costs, such as an increased predation risk or hazard during exploration, or the investment of time and energy involved in exploiting a new resource. Conversely, innovation potentially offers long-term benefits, in terms of new resources to utilize, more efficient exploitation of the environment, or increased status or mating success. These costs and benefits will vary with the condition and age of the innovator and the foraging ecology and life history of its species (Greenberg and Mettke-Hofman, 2001). For example, group living may decrease the risk of predation or starvation, perhaps promoting innovation, but will also increase the risk of scrounging from innovators, which may discourage innovation if innovators are more susceptible to scrounging than individuals using established techniques (Morand-Ferron and Lefebvre, unpublished data). Whether an animal innovates will thus depend upon the balance between such risks and payoffs. Of course, virtually all learning has costs and benefits, but a consideration of the basic economics may prove particularly insightful in cases of innovation, where the risks may be enhanced.

Let us consider the costs and benefits of foraging innovation as an illustrative example. When animals are highly stressed, such as during periods of food deprivation, the benefits of finding a novel food source and avoiding starvation are likely to be substantial when compared to the costs of innovation, such as increased predation risks. Many animals appear willing to trade an increased risk of predation for access to food in times of food deprivation (e.g. Godin and Smith, 1988). Hence, animals would be expected to be more likely to innovate under such conditions. In contrast, when an animal is not food-deprived, a 'time of plenty' in the parlance of Kummer and Goodall (1985), the benefits of innovation are likely to be diminished. Locating extra food may increase fitness, but this payoff when not food-deprived may be small in comparison with the benefit of avoiding starvation. Hence, in general, the incidence of innovation would be expected to be higher in food-deprived animals than non-food-deprived individuals.

However, in what may be quite limited circumstances, the costs of innovation may be reduced in a 'time of plenty' situation. For example, animals in captivity may be conditioned to assess the risks of predation and hazard as quite low, compared with animals in the wild. Rats are known to decrease their exploratory behaviour when they perceive pre-dation risks to be high (Rosellini and Widman, 1989). Even in natural populations, in a period of food abundance with a number of established food sources available, the energy expended during exploration for novel food patches is diminished and so the fitness costs of exploration may be decreased. In such conditions there may be a selective advantage in certain taxa for the facultative expression of exploratory behaviour. This may help to explain why many innovations are observed in captive, provisioned primate populations (Kummer and Goodall, 1985). Further discussion of the costs and benefits of innovation can be found in Chapter 12.

The ecology and evolution of innovation

In birds, the incidence of behavioural innovation is most closely correlated with the relative size of the neostriatum–hyperstriatum ventrale complex (Timmermans *et al.*, 2000). More recently Reader and Laland (2002) have applied Lefebvre *et al.*'s (1997) methods to primates, using the reported incidence of behavioural innovation, social learning and tool use, to show that brain size and these measures of cognitive capacity are correlated in primates. These results support the idea of convergent evolution in these two taxa, with the primate neocortex and the avian neostriatum–hyperstriatum ventrale complex taking similar integrative roles (Rehkämper and Zilles, 1991). Reader and Laland also found that innovation and social learning frequencies covary across species, consistent with the view that these capacities have evolved together. The ability to invent new behav-iour may have played a pivotal role in primate and avian brain evolution. Individuals capa-ble of inventing new solutions to ecological challenges, or exploiting the discoveries and inventions of others, may have had a selective advantage over less able conspecifics, which generated selection for the brain regions that facilitate such behaviour. This can be regarded as evidence for Allan Wilson's behavioural drive hypothesis. Additional support is rapidly accumulating, particularly through further comparative studies on avian innova-tion (see Chapters 2 and 3). For instance, there is now evidence that innovative birds are more successful in invading new habitats and constructing new niches, and more speciose than less innovative avian taxa (Sol and Lefebvre, 2000; Sol *et al.*, 2002; Nicolakakis *et al.*, 2003; Sol, Chapter 3).

The patterns and causes of innovation

It is unclear whether animal innovation should be regarded as a personality trait (for instance, associated with particularly clever or creative individuals) or as a state-dependant variable (e.g. foraging innovation may be driven by hunger) or perhaps innovation is not a characteristic of individuals at all: perhaps it results from exposure to pertinent ecologi-cal stimuli (for instance, a sudden change in the environment). Little is known about which individuals form new behaviour patterns, what causes them to do so, and what ecological variables influence innovation (Kummer and Goodall, 1985; Lee, 1991).

The cost-benefit, developmental, and psychological theories outlined above lead to a number of predictions regarding individual differences in innovative tendency. It is likely that individual differences will be pervasive and will have significant implications for the spread of innovations and individual survivorship.

Is innovation a personality trait?

'No one supposes that all individuals of the same species are cast in the very same mould'. (Darwin 1859, p. 102)

Innovation in animals is not often associated with clever or creative individuals, although there are exceptions such as Imo, described as 'a monkey genius' by Edward Wilson and John Tyler Bonner and as 'gifted' by Jane Goodall (Wilson, 1975; Bonner, 1980; Kummer and Goodall, 1985). However, to the extent that Imo's achievements can be attributed to her personality, it is not clear whether she should be characterized as unusually 'intelligent' and 'creative', or as being the possessor of some less cerebral trait, such as 'boldness', 'non-conformism', 'perseverance', or 'explorative'. A number of studies have identified bold/shy, or risk prone/risk averse dimensions of variability in animal behaviour (Daly and Wilson, 1983; Wilson et al., 1994), variability that may influence the propensity to innovate. Comparatively few studies have explored whether innovation is a personality trait. Fragaszy and Visalberghi (1990) found no evidence that particular individuals in a captive group of capuchin monkeys possessed a 'characteristic propensity' to innovate. Unfortunately, most of these studies of innovation as a personality trait do not rule out state-dependent variables as confounding sources of variation. One study that does take steps to eliminate this confound did report individual differences in innovative tendency in the guppy (Laland and Reader, 1999; Laland and van Bergen, Chapter 7). Pfeffer et al. (2002) report a correlation between high levels of faecal corticosterone and enhanced problem-solving ability in greylag geese (Anser anser), and argue that this hormonal correlate may underpin an innovative personality. Studies of the effects of environmental enrichment on primate infant development suggest that differences between individuals in problem-solving ability could result from variability in the richness of the developmental environment (Schneider et al., 1991). At this stage it is not clear whether any stable differences between individuals in problem-solving ability reflect variation in mental abilities (e.g. intelligence, creativity), sociality (e.g. a tendency to stay with or leave the group), boldness (e.g. a tendency to approach unfamiliar objects), exploratory behaviour (e.g. a tendency to investigate unfamiliar spaces), activity levels, some other factor, or some combination of these factors.

Is innovation explained by state-dependent factors?

Although compelling evidence is scarce, observations of animal populations suggest that particular classes of individuals may be prone to innovation. There is at least anecdotal evidence that innovators often differ from the remainder of the group in some characteristics, such as sex, age, or social rank.

Sex differences

To account for sex differences in guppy foraging innovation, Laland and Reader (1999) noted that, in guppies, females are the larger sex and have unconstrained growth, whereas males stop growing at sexual maturity. In fish, female fecundity increases with accelerating returns with increasing body length, while a male's ability to obtain matings probably increases linearly or with diminishing returns with body length (Sargent and Gross, 1993; Magurran and Garcia, 2000). Female guppies suffer greater metabolic costs of growth and pregnancy, yet female fecundity is directly related to body size, so there are accelerating returns to foraging success. This means a conservative foraging strategy is less likely to be adaptive in females than in males and suggests that, in fish, females are more likely than males to innovate in a foraging context. This explanation may equally apply to foraging innovations in other species where maternal investment is greater than paternal investment, that is, to most mammals, including primates (Davies, 1991). However, depending on the taxa or context, different functions may describe the relationships between foraging success and reproductive success. The shape of this function will be of critical importance in determining the payoffs of innovation, and in some circumstances a more conservative strategy for females may be favoured by selection (see Laland and van Bergen, Chapter 7).

In some primates males benefit from having a large body size relative to other males, for instance, and this may be an advantage in competition for females. In such circumstances it may pay males, particularly those of low status, to take risks and try out novel solutions that either allow access to the foods they require to fuel the growth they need to be an effective competitor, or constitute alternative strategies to reproductive success. In particular, male primates might be predicted to show more innovation than females in contexts linked directly or indirectly to mating success, such as sexual or courting behaviour (Laland and Reader, 1999). Kummer and Goodall (1985; Chapter 10) provide a number of examples among male chimpanzees. For instance, Satan rose earlier than other chimpanzees to shake branches at females and lead them away to copulate secretly, Mike used empty paraffin cans to augment his displays, and Shadow developed a new and effective courtship display by flipping his upper lip over his nostrils. Reader and Laland's (2001) survey of primate behaviour collated from the published literature found that the reported incidence of innovation was significantly higher in males and lower in females than would be expected by chance given the estimated relative proportions of these groups. This finding conflicts with the general impression in the primate behaviour literature that innovators are more likely to be female than male. It is possible that this impression is misleading. Box (1991, 1997; Chapter 9) notes that females of some marmoset and tamarin species appear more adaptively responsive to environmental change than males, but that alternative explanations to greater female responsiveness, such as females having priority of access to food, may be responsible. Alternatively, the reports may be a biased sample of cases of innovation as a whole, or reflect a disproportionate effort on the part of researchers towards studying males (Reader and Laland, 2001).

Age differences

Reader and Laland (2001) also found that across all primates there were more reports of innovation in adults, and less in non-adults, than expected by chance given the proportions of adults and non-adults in populations of the species analysed. Again this was surprising, and the possibility remains that the data are a biased sample. Several authors have presented anecdotal evidence or suggested that young individuals are more likely to be innovatory (see Russon, Box, Chapters 9 and 13). For example, Kummer and Goodall (1985, p. 209) state 'many innovations appear during childhood when a youngster is cared for and protected by its mother and thus has much time for carefree play and exploration', and Hauser (1988) reported that newly observed behaviours are first recorded in young animals 'in most cases' of invention and transmission in non-human primates. Whiten and Byrne (1988) suggest that social play is important in the discovery of deceptive techniques. The greater than expected extent of adult innovation may result because innovation frequently builds on other skills, and requires a degree of experience, maturity, or competence that is more common in adults than younger primates. Russon and Box (Chapters 9 and 13) both illustrate how anatomical features (such as size, strength, manual dexterity, and dentition), features that vary throughout development and will differ in infants, juveniles and adults, may play a critical role in facilitating innovation.

Status differences
Individuals whose position in a dominance hierarchy restricts their access to mates or resources may be expected to turn to novel methods to gain access to resources. Katzir (1982, 1983) found that mid-to low-ranking jackdaws, *Corvus monedula*, were quicker to enter a novel space and exploit a novel food source than top-ranking birds. High-ranking birds may avoid the risks of innovation by careful monitoring of subordinates to rapidly take advantage of any beneficial discovery. The scrounging behaviour of dominant individuals has also been documented by Caraco *et al.* (1989), and may reduce the benefits of innovation in subordinates (Kummer and Goodall, 1985; Fragaszy and Visalberghi, 1990; Giraldeau, 1997).

Primate studies appear to indicate that innovators are frequently on the outskirts of the social group (Kummer and Goodall, 1985). For instance, peripheral female hamadryas baboons, who tend to be low-ranking, were significantly better at learning novel water-finding tasks than central females (Sigg, 1980), and subordinate macaques made fewer errors learning a buzzer task and mastered a reversal learning task quicker than dominants (Strayer, 1976; Bunnell *et al.*, 1980; Bunnell and Perkins, 1980). In chimpanzees, three times as much innovation is reported on the part of low-ranking individuals than high-ranking individuals (Reader and Laland, 2001). Where access to mates is restricted by social rank, innovation abilities in low-ranking individuals may serve to decrease differences in reproductive success between high- and low-rankers. Such studies support the argument that necessity drives animals to innovate (Laland and Reader, 1999).

Does the environment evoke innovation?

Comparatively few studies have mentioned the ecological context in which innovation occurred. Reader and Laland (2001) found that innovation was prompted by ecological challenges, such as periods of food shortage, dry seasons, or habitat degradation, in 17 of the 36 cases where data were available. Of course, this data may be subject to a reporting bias, with researchers more likely to mention when there was a significant change in the environment than when there was no change. Additionally, there are also many examples of innovations reported in provisioned or captive populations, an apparent conflict with the 'necessity drives innovation' argument that we discussed earlier. Nevertheless, this finding raises the possibility that changes in the environment may precede much primate innovation (Hauser, 1988; Lee, 1991). If it were the case that individuals satisfice with established behaviour unless they are unable to do so, then innovations may be more likely to spread on occasions when the entire population is stressed (as opposed to just the weakest individuals).

The diffusion of novel behaviour patterns

Diffusion dynamics

Mathematical models of cultural transmission have been constructed to predict the pattern of spread of novel traits as a result of social learning processes in humans (Cavalli-Sforza and Feldman, 1981; Boyd and Richerson, 1985), and these, together with optimal foraging models, have been applied to animal traditions (Giraldeau *et al.*, 1994; Laland *et al.*, 1996; Giraldeau and Caraco, 2000). Researchers have speculated as to whether the shape of the diffusion curve may reveal something about the learning processes involved, although this exercise may be more complicated than it appears (Lefebvre, 1995; Laland and Kendal, in press). Most models predict that the diffusion of cultural traits will exhibit a sigmoidal pattern over time, with the trait initially increasing in frequency slowly, then going through a period of rapid spread, and finally tailing-off. The reason this pattern is anticipated is that as the trait spreads the number of demonstrators increases (enhancing the opportunity for social learning in the remaining observing individuals), but the number of individuals left to learn decreases. Early and later on in the process the opportunities for social learning are limited because there are too few demonstrators and then too few observers, respectively; however growth is rapid during the intermediate stages. This prediction has received considerable empirical support from the human literature on the diffusion of innovations (Rogers, 1995), and has been recognized for some time:

> 'A slow advance in the beginning, followed by rapid and uniformly accelerated progress, followed again by progress that continues to slacken until it finally stops: These are the three ages of … invention'. (Gabriel Tarde, 1903, The Laws of Imitation, p. 127)

Everett Rogers classic text *Diffusion of Innovations*, updated over 40 years, carefully documents how technological innovations spread among human populations. Among the important research questions addressed by such diffusion scholars are (1) how the early adopters differ from the later adopters of an innovation, (2) how the perceived attributes of an innovation, such as relative advantage or compatibility affect its rate of adoption, and (3) why the S-shaped diffusion curve 'takes off' at about 10–25 per cent adoption, when interpersonal networks become activated so that a critical mass of adopters begins using an innovation. Rogers plots a normal distribution called the 'innovativeness dimension', as measured by the time at which an individual adopts an innovation. The innovativeness variable is partitioned into five adopter categories, (1) innovators (2.5% of the population, consisting of well-educated, risk-loving individuals), (2) early adopters, (3) early majority, (4) late majority, and (5) laggards. Unfortunately, researchers of animal innovation have paid little attention to this literature.

Why do so few novel behaviour patterns spread?

Kummer and Goodall (1985) observed that while there was no shortage of innovation on the part of individual animals, their inventions rarely spread through the population, despite their apparent utility to the inventor and potential utility to the rest of the group. This seems to contrast with human innovation, where many, though not all, advantageous innovations spread. Earlier Goodall (1973) had noted that most innovative behaviour that she observed in chimpanzees disappeared after a few years of being in fashion. This was regarded as something of a puzzle. For example, chimpanzee Mike's use of kerosene cans to augment his dominance display coincided with a dramatic rise in his social rank, yet while some chimpanzees investigated the cans no other males completely adopted Mike's novel behaviour. Why? Tomasello (1994; Tomasello and Call, 1997) asserts that the social learning process in chimpanzees differs from that in human beings, as non-human apes lack essential cognitive abilities (theory of mind, imitation, teaching). Is the fact that many innovations never progress beyond idiosyncratic behaviours related to this claimed cognitive difference between humans and chimpanzees? The answer remains to be elucidated, but there are alternative explanations as to why innovations may not spread.

First, it has yet to be established what proportion of innovations are of utility to other members of the population, in the sense that they confer greater reinforcement than established behaviour and ultimately are associated with a fitness benefit. If most inventions are not generally useful, or if the utility of innovations is frequency-dependent (e.g. if rareness itself brought advantages), then there is no reason to expect innovations to spread.

Second, as noted previously, in hierarchical societies a disproportionate amount of innovation appears to be carried out by low-status individuals. Such low-rankers often occupy the periphery of the social group, are attended to less than high-ranking individuals, have less influence on group behaviour, or tend to move away when approached by higher-ranking group members and are more vulnerable to scrounging than high-ranking individuals. Additionally, subordinate individuals may be inhibited to perform learned behaviours in the presence of more dominant individuals (Drea and Wallen, 1999). All these characteristics would mitigate against the diffusion of innovations generated by low rankers.

Third, in more egalitarian societies social learning appears commonly to exhibit a positive frequency dependence as individuals conform to the behaviour of the majority. Although few empirical studies have directly addressed these issues, evidence for conformist social learning has been observed in every species for which data exists (see Day *et al.*, 2001 and references therein). Social learning in which the probability of adopting a pattern of behaviour increases with the proportion of demonstrators has been previously demonstrated in guppies, *Poecilia reticulata* (Sugita, 1980; Laland and Williams, 1997; Lachlan *et al.*, 1998), golden shiners, *Notemigonus crysoleucas* (Reebs, 2000), rats, *Rattus norvegicus* (Beck and Galef, 1989), and pigeons, *Columba livia* (Lefebvre and Giraldeau, 1994). These observations are supported by theoretical analyses, which have found that in most circumstances where natural selection favours reliance on social learning, conformity is also favoured (Boyd and Richerson, 1985). One consequence of conformist social learning is that individuals would be expected to be less likely than chance to adopt novel behavioural variants.

Galef (Chapter 6) notes that, at various times through history, social learning has been regarded as a 'conservative force' that acts to inhibit the acquisition of novel behaviour, and a conformist view of social learning is in line with this perspective. However, Galef also points out that at other times social learning has also been viewed as a 'progressive force', which propagates novel variants, a view that is inconsistent with conformity being of importance. Galef is able to muster evidence for both perspectives, which implies either that conformity is not universal, or that it is not always sufficiently powerful to prevent diffusion.

Fourth, the human innovation literature has identified a number of characteristics that affect whether an innovation spreads (Rogers, 1995), some of which are likely to be relevant to animals. These include:

1. Relative advantage—the greater the perceived relative advantage of an innovation, the more rapid will be its rate of adoption.
2. Compatibility—an idea that is incompatible with the values and norms of a social system will not be adopted as rapidly as an innovation that is compatible.
3. Complexity—new ideas that are simple to understand are adopted more rapidly than innovations that require the adopter to develop new skills and understandings.
4. Trialability—new ideas that can be tried out will generally be adopted more quickly than innovations that cannot.
5. Observability—the easier it is for individuals to see the results of an innovation, the more likely they are to adopt it.

Moreover, it is important to remember that early hominid technological culture, at least as represented by the fossil record, did not change rapidly and that conservatism was the rule rather than the exception. Oldowan and Acheulean hand-axe technologies lasted without significant changes for over one million years each, which perhaps suggest that most innovations of pre-*sapiens* hominids vanished without being socially transmitted over generations (Nishida, 1994). This lack of change is still somewhat of a mystery, and suggests that there may be processes that insulate human cultural traits from alteration (Heyes, 1994; Miller, 2000).

Outstanding questions

By way of summary and to point to future research avenues, we list 10 unanswered or only partially answered questions of animal innovation:

1. What are the functions of animal innovation?
2. What are the psychological and neurological mechanisms underlying animal innovation, and are they distinct from other cognitive processes?
3. Who are the innovators?
4. Has the capacity for innovation been directly favoured by natural selection, or is it a by-product of selection for other attributes, such as behavioural flexibility or social cognition?

5. What is the evolutionary history of innovation and which life-history traits are associated with an enhanced reliance on innovation?
6. To what extent do similar processes underlie human and animal innovation?
7. Do animals primarily innovate by learning from accidental happenings, or do they more frequently actively seek novel solutions to problems?
8. How does an individual's capacity for innovation develop and change over its life-time, and how is it influenced by experience? Is a propensity to innovate heritable?
9. Which personality traits (if any) best account for individual differences in animal innovation?
10. What are the principal costs and benefits of innovation and how are they utilized by animals in decision-making?

We hope that this book will provide a stimulus to generating answers to some of these interesting and important questions.

Acknowledgements

We thank Louis Lefebvre, Celia Heyes, Daniel Sol, Johan Bolhuis, Rachel Day, Carolyn Hall, Julie Morand-Ferron, Tamsin Rothery, David Ryan, Gray Stirling, and Yvonne Vadboncoeur for helpful comments and discussion. KNL and SMR were funded by The Royal Society.

References

Archer, J. and **Birke, L. I. A.** (ed.) (1983). *Exploration in animals and humans.* Wokingham, UK: Van Nostrand Reinhold.

Barnett, S. A. (1970). Search and explorations. In *Contemporary readings in behaviour* (ed. J. R. Young and E. M. Strock), pp. 33–45. New York: McGraw-Hill, Inc.

Bateson, P. P. G. (1983). Genes, environment and the development of behaviour. In *Animal behaviour, genes, development and learning* (ed. T. R. Halliday and P. J. B. Slater), Vol. 3, pp. 52–81. Oxford: Blackwell.

Bateson, P. P. G. and **Martin, P.** (1999). *Design for a life: How behaviour develops.* London: Jonathan Cape.

Beck, B. B. (1980). *Animal tool behavior: The use and manufacture of tools by animals.* New York: Garland.

Beck, M. and **Galef, B. G. Jr.** (1989). Social influences on the selection of a protein-sufficient diet by Norway rats (*Rattus norvegicus*). *Journal of Comparative Psychology,* **103,** 132–9.

Berlyne, D. E. (1950). Novelty and curiosity as determinants of exploratory behaviour. *British Journal of Psychology,* **41,** 68–86.

Berlyne, D. E. (1960). *Conflict, arousal and curiosity.* London: McGraw-Hill.

Bierens de Haan, J. A. (1933). Der Stieglitz als Schopfer. *Journal für Ornithologie,* **1,** 22.

Birke, L. I. A. and **Archer, J.** (1983). Some issues and problems in the study of animal exploration. In *Exploration in animals and humans* (ed. J. Archer and L. I. A. Birke), pp. 1–21. Wokingham, UK: Van Nostrand Reinhold.

Roakes, R. (1984). *From Darwin to behaviourism: Psychology and the minds of animals.* Cambridge: Cambridge University Press.

Boinski, S. (1988). Use of a club by a wild white-faced capuchin (*Cebus capucinus*) to attack a venomous snake (*Bothrops asper*). *American Journal of Primatology,* **14,** 177–9.

Bonner, J. T. (1980). *The evolution of culture in animals.* Princeton, NJ: Princeton University Press.

Bouchard, J. (2002). *Is social learning correlated with innovation in birds? An inter- and an intraspecific test.* M.Sc. Thesis, McGill University, Montréal.

Box, H. O. (1991). Training for life after release: Simian primates as examples. *Symposia of the Zoological Society of London,* **62**, 111–23.

Box, H. O. (1997). Foraging strategies among male and female marmosets and tamarins (Callitrichidae): New perspectives in an underexplored area. *Folia Primatologica,* **68**, 296–306.

Box, H. O. and **Gibson, K. R.** (ed.) (1999). *Mammalian social learning: Comparative and ecological perspectives.* Cambridge: Cambridge University Press.

Boyd, R. and **Richerson, P. J.** (1985). *Culture and the evolutionary process.* Chicago, IL: University of Chicago.

Bunnell, B. N., Gore, W. T., and **Perkins, M. N.** (1980). Performance correlates of social behaviour and organization: Social rank and reversal learning in crab-eating macaques (*M. fascicularis*). *Primates,* **21**, 376–88.

Bunnell, B. N. and **Perkins, M. N.** (1980). Performance correlates of social behaviour and organization: Social rank and complex problem solving in crab-eating macaques (*M. fascicularis*). *Primates,* **21**, 515–23.

Cambefort, J. P. (1981). A comparative study of culturally transmitted patterns of feeding habits in the chacma baboon *Papio ursinus* and the vervet monkey *Cercopithecus aethiops. Folia Primatologica,* **36**, 243–63.

Caraco, T., Barkan, C., Beacham, J. L., Brisbin, L., Lima, S., Mohan, A. *et al.* (1989). Dominance and social foraging: A laboratory study. *Animal Behaviour,* **38**, 41–58.

Cavalli-Sforza, L. and **Feldman, M. W.** (1981). *Cultural transmission and evolution: A quantitative approach.* Princeton, NJ: Princeton University Press.

Coleman, K. and **Wilson, D. S.** (1998). Shyness and boldness in pumpkinseed sunfish: Individual differences are context-specific. *Animal Behaviour,* **56**, 927–36.

Cousteau, J. Y. (1958). *The silent world.* Harmondsworth, UK: Penguin.

Daly, M. and **Wilson, M.** (1983). *Sex, evolution and behavior,* 2nd edn. Belmont, CA: Wadsworth.

Darwin, C. (1859). *The origin of species (1968 edn.).* London: Penguin.

Darwin, C. (1897). *The descent of man and selection in relation to sex (rev. edn.).* New York: Appleton (originally published 1871).

Davey, G. (1981). *Animal learning and conditioning.* London: Macmillan.

Davies, N. B. (1991). Mating systems. In *Behavioural ecology: An evolutionary approach,* 3rd edn. (ed. J. R. Krebs and N. B. Davies), pp. 263–94. Oxford: Blackwell Scientific.

Day, R. L., Coe, R. L., Kendal, J. R., and **Laland, K. N.** (2003). Neophilia, innovation and social learning: A study of intergeneric differences in callitrichid monkeys. *Animal Behaviour,* **65**, 559–71.

Day, R. L., MacDonald, T., Brown, C., Laland, K. N., and **Reader, S. M.** (2001). Interactions between shoal size and conformity in guppy social foraging. *Animal Behaviour,* **62**, 917–25.

DeWitt, T. J., Sih, A., and **Wilson, D. S.** (1998). Costs and limits of phenotypic plasticity. *Trends in Ecology and Evolution,* **13**, 77–81.

Diamond, J. (1991). *The rise and fall of the third chimpanzee.* London: Vintage.

Dingemanse, N. J., Both, C., Drent, P. J., Van Oers, K., and **Van Noordwijk, A. J.** (2002). Repeatability and heritability of exploratory behaviour in great tits from the wild. *Animal Behaviour,* **64**, 929–38.

Drea, C. M. and **Wallen, K.** (1999). Low-status monkeys 'play dumb' when learning in mixed social groups. *Proceedings of the National Academy of Sciences, USA,* **96**, 12965–9.

Drent, P., Van Oers, K., and **Van Noordwijk, A. J.** (2003). Realised heritability of personalities in the great tit (*Parus major*). *Proceedings of the Royal Society of London Series B*, **270**, 45–51

Drent, P. J. (1997). Realised heritability of a behavioural syndrome in the Great Tit. *Progress Report of the Netherlands Institute of Ecology*, 44–7.

Dukas, R. (1998). Evolutionary ecology of learning. In *Cognitive ecology: The evolutionary ecology of information processing and decision making* (ed. R. Dukas), pp. 129–74. Chicago and London: University of Chicago Press.

Eaton, G. G. (1976). The social order of Japanese macaques. *Scientific American*, **234**, 96–106.

Eysenck, H. J. (1995). *Genius: The natural history of creativity*. Cambridge: Cambridge University Press.

Fagan, B. M. (1993). *World prehistory: A brief introduction*, 2nd edn. New York: Harper & Collins.

Fisher, J. and Hinde, R. A. (1949). The opening of milk bottles by birds. *British Birds*, **42**, 347–57.

Fragaszy, D. M. and Perry, S. (ed.) (in press). *The biology of traditions: Models and evidence*. Cambridge: Cambridge University Press.

Fragaszy, D. M. and Visalberghi, E. (1990). Social processes affecting the appearance of innovative behaviours in capuchin monkeys. *Folia Primatologica*, **54**, 155–65.

Galef, B. G. Jr. (1992). The question of animal culture. *Human Nature*, **3**, 157–78.

Galef, B. G. Jr. and Giraldeau, L. (2001). Social influences on foraging in vertebrates: Causal mechanisms and adaptive functions. *Animal Behaviour*, **61**, 3–15.

Giraldeau, L. A. (1997). The ecology of information use. In *Behavioural Ecology* (ed. J. R. Krebs and N. B. Davies), 4th edn, pp. 42–68. Blackwell, Oxford.

Giraldeau, L.-A. and Caraco, T. (2000). *Social foraging theory*. Princeton, NJ: Princeton University Press.

Giraldeau, L.-A., Caraco, T., and Valone, T. J. (1994). Social foraging: Individual learning and cultural transmission of innovations. *Behavioural Ecology*, **5**, 35–43.

Glickman, S. E. and Sroges, R. W. (1966). Curiosity in zoo animals. *Behaviour*, **26**, 151–8.

Godin, J.-G. J. and Smith, S. A. (1988). A fitness cost to foraging in the guppy. *Nature*, **333**, 69–71.

Goldenburg, J., Mazursky, D., and Solomon, S. (1999). Creative sparks. *Science*, **285**, 1495–6.

Goodall, J. (1964). Tool-using and aimed throwing in a community of free-living chimpanzees. *Nature*, **201**, 1264–6.

Goodall, J. (1973). Cultural elements in a chimpanzee community. In *Symp. IVth Int. Congr. Primat. Vol. 1: Precultural primate behaviour* (ed. E. Menzel), pp. 144–84. Basel: Karger.

Goodall, J. (1986). *The chimpanzees of gombe: Patterns of behaviour*. Cambridge, MA: Belknap Press.

Gould, J. L. and Gould, C. G. (1994). *The animal mind*. New York: Freeman.

Greenberg, R. and Mettke-Hofman, C. (2001). Ecological aspects of neophobia and exploration in birds. *Current Ornithology*, **16**, 119–78.

Harlow, H. (1953). Motivation as a factor in new responses. In *Current theory and research in motivation: A symposium.* (ed. J. S. Brown, H. F. Harlow, L. J. Postman, V. Nowlis, O. H. Mowrer, and T. M. Newcomb), pp. 24–49. Lincoln, NE: University of Nebraska Press.

Harlow, H. F., Harlow, M. K., and Meyer, D. R. (1950). Learning motivated by a manipulation drive. *Journal of Experimental Psychology*, **40**, 228–34.

Hauser, M. D. (1988). Invention and social transmission: New data from wild vervet monkeys. In *Machiavellian intelligence: Social expertise and the evolution of intellect in monkeys, apes and humans* (ed. R. W. Byrne and A. Whiten), pp. 327–43. Oxford: Oxford University Press.

Heinrich, B. (1995). Neophilia and exploration in juvenile common ravens, *Corvus corax*. *Animal Behaviour*, **50**, 695–704.

Heinrich, B. (2000). Testing insight in ravens. In *The evolution of cognition* (ed. C. Heyes and L. Huber), pp. 289–305. Cambridge, MA: MIT Press.

Heyes, C. M. (1994). Imitation and culture: Longevity, fecundity and fidelity in social transmission. In *Behavioral aspects of feeding* (ed. P. Valsecchi, M. Mainardi, P. Valsecchi, and D. Mainardi), pp. 271–87. London: Harwood Academic.

Heyes, C. M. and Galef, B. G. Jr. (1996). *Social learning in animals: The roots of culture.* London: Academic Press.

Hinde, R. A. and Fisher, J. (1951). Further observations on the opening of milk bottles by birds. *British Birds*, **44**, 393–6.

Hosey, G. R., Jacques, M., and Pitts, A. (1997). Drinking from tails: Social learning of a novel behaviour in a group of ring-tailed lemurs (*Lemur catta*). *Primates*, **38**, 415–22.

Janik, V. M. and Slater, P. J. B. (2000). The different roles of social learning in vocal communication. *Animal Behaviour*, **60**, 1–11.

Katzir, G. (1982). Relationships between social structure and response to novelty in captive jackdaws, *Corvus monedula*. I. Response to novel space. *Behaviour*, **81**, 231–64.

Katzir, G. (1983). Relationships between social structure and response to novelty in captive jackdaws, *Corvus monedula*. II. Response to novel palatable food. *Behaviour*, **87**, 183–208.

Kawai, M. (1965). Newly-acquired pre-cultural behavior of the natural troop of Japanese monkeys on Koshima Islet. *Primates*, **6**, 1–30.

Köhler, W. (1925). *The mentality of apes.* London: Kegan Paul.

Kummer, H. and Goodall, J. (1985). Conditions of innovative behaviour in primates. *Philosophical Transactions of the Royal Society of London Series B*, **308**, 203–14.

Lachlan, R. F., Crooks, L., and Laland, K. N. (1998). Who follows whom? Shoaling preferences and social learning of foraging information in guppies. *Animal Behaviour*, **56**, 181–90.

Laland, K. N. and Kendal, J. R. (in press). What the models say about social learning. In *The biology of traditions: Models and evidence* (ed. D. M. Fragaszy and S. Perry). Cambridge: Cambridge University Press.

Laland, K. N. and Plotkin, H. C. (1990). Social learning and social transmission of digging for buried food in Norway rats. *Animal Learning & Behavior*, **18**, 246–51.

Laland, K. N. and Reader, S. M. (1999). Foraging innovation in the guppy. *Animal Behaviour*, **57**, 331–40.

Laland, K. N. and Williams, K. (1997). Shoaling generates social learning of foraging information in guppies. *Animal Behaviour*, **53**, 1161–69.

Laland, K. N., Richerson, P. J., and Boyd, R. (1996). Developing a theory of animal social learning. In *Social learning in animals: The roots of culture* (ed. C. M. Heyes and B. G. Jr., Galef), pp. 129–54. London: Academic Press.

Lee, P. (1991). Adaptations to environmental change: An evolutionary perspective. In *Primate responses to environmental change* (ed. H. O. Box), pp. 39–56. London: Chapman & Hall.

Lefebvre, L. (1995). Culturally-transmitted feeding behaviour in primates: Evidence for accelerating learning rates. *Primates*, **36**, 227–39.

Lefebvre, L. (2000). Feeding innovations and their cultural transmission in bird populations. In *The evolution of cognition* (ed. C. Heyes and L. Huber), pp. 311–28. Cambridge, MA: MIT Press.

Lefebvre, L. and Giraldeau, L.-A. (1994). Cultural transmission in pigeons is affected by the number of tutors and bystanders present. *Animal Behaviour*, **47**, 331–7.

Lefebvre, L., Reader, S. M., and Webster, S. J. (2001). Novel food use by Gray Kingbirds and Red-necked pigeons in Barbados. *Bulletin of the British Ornithologists' Club*, **121**, 247–9.

Lefebvre, L., Whittle, P., Lascaris, E., and **Finkelstein, A.** (1997). Feeding innovations and forebrain size in birds. *Animal Behaviour*, **53**, 549–60.

Lloyd Morgan, C. (1912). *Instinct and experience.* London: Methuen.

Magurran, A. E. and **Garcia, M.** (2000). Sex differences in behaviour as an indirect consequence of mating system. *Journal of Fish Biology*, **57**, 839–57.

Marchetti, C. and **Drent, P. J.** (2000). Individual differences in the use of social information in foraging by captive great tits. *Animal Behaviour*, **60**, 131–40.

McDougall, W. (1936). *An outline of psychology*, 7th edn. London: Methuen.

McFarland, D. (ed.) (1981). *The Oxford companion to animal behaviour.* Oxford: Oxford University Press.

McGrew, W. C. (1994). Tools compared: The material of culture. In *Chimpanzee cultures* (ed. R. W. Wrangham, W. C. McGrew, B. M. de Waal, and P. Heltne), pp. 25–39. Cambridge, MA: Harvard University Press.

Mettke-Hofmann, C. (2000). Changes in exploration from courtship to the breeding state in red-rumped parrots (*Psephotus haematonotus*). *Behavioural Processes*, **49**, 139–48.

Miller, G. (2000). *The mating mind: How sexual choice shaped the evolution of human nature.* London: William Heinemann.

Nicol, C. J. and **Guildford, T.** (1991). Exploratory activity as a measure of motivation in deprived hens. *Animal Behaviour*, **41**, 333–41.

Nicolakakis, N., Sol, D., and **Lefebvre, L.** (2003). Behavioural flexibility predicts species richness in birds, but not extinction risk. *Animal Behaviour*, **65**, 445–52.

Nishida, T. (1994). Review of recent findings on Mahale chimpanzees. In *Chimpanzee Cultures* (ed. R. W. Wrangham, W. C. McGrew, F. B. M. de Waal and P. G. Heltne), pp. 373–96. Cambridge, MA: Harvard University Press.

Paquette, D. (1992). Discovering and learning tool-use for fishing honey by captive chimpanzees. *Human Evolution*, **7**, 17–30.

Pfeffer, K., Fritz, J., and **Kotrschal, K.** (2002). Hormonal correlates of being an innovative greylag goose, *Anser anser. Animal Behaviour*, **63**, 687–95.

Pitocchelli, J. (1985). Apparent insight learning by some common grackles breeding in Central Park, New York. *Kingbird*, **35**, 32–4.

Plotkin, H. C. (ed.) (1982). *Learning, development, and culture: Essays in evolutionary epistomology.* Chicester: Wiley.

Plotkin, H. C. (1994). *Darwin machines and the nature of knowledge.* London: Penguin.

Reader, S. M. and **Laland, K. N.** (2001). Primate innovation: Sex, age and social rank differences. *International Journal of Primatology*, **22**, 787–805.

Reader, S. M. and **Laland, K. N.** (2002). Social intelligence, innovation and enhanced brain size in primates. *Proceedings of the National Academy of Sciences, USA*, **99**, 4436–41.

Reader, S. M., Nover, D., and **Lefebvre, L.** (2002). Locale-specific sugar packet opening by Lesser Antillean bullfinches in Barbados. *Journal of Field Ornithology*, **73**, 82–5.

Reebs, S. G. (2000). Can a minority of informed leaders determine the foraging movements of a fish shoal? *Animal Behaviour*, **59**, 403–9.

Rehkämper, G. and **Zilles, K.** (1991). Parallel evolution in mammalian and avian brains: Comparative cytoarchitectonic and cytochemical analysis. *Cell and Tissue Research*, **263**, 3–28.

Renner, M. J. (1987). Experience-dependent changes in exploratory behavior in the adult rat (*Rattus norvegicus*): Overall activity level and interactions with objects. *Journal of Comparative Psychology*, **101**, 94–100.

Renner, M. J. (1988). Learning during exploration: The role of behavioral topography during exploration in determining subsequent adaptive behavior. *International Journal of Comparative Psychology*, **2**, 43–56.

Rogers, E. M. (1995). *Diffusion of innovations*, 4th edn. New York: Free Press.

Rosellini, R. A. and Widman, D. R. (1989). Prior exposure to stress reduces the diversity of exploratory behaviour of novel objects in the rat (*Rattus norvegicus*). *Journal of Comparative Psychology*, **103**, 339–46.

Rowe, N. (1996). *The pictorial guide to the living primates*. New York: Pogonias Press.

Russell, P. A. (1983). Psychological studies of exploration in animals: A reappraisal. In *Exploration in animals and humans* (ed. J. Archer and L. I. A. Birke), pp. 22–54. Wokingham, UK: Van Nostrand Reinhold.

Sargent, R. C. and Gross, M. R. (1993). Williams' principle: An explanation of parental care in teleost fishes. In *Behaviour of teleost fishes* (ed. by T. J. Pitcher), 2nd edn, pp. 333–61. London: Chapman and Hall.

Schiller, P. H. (1952). Innate constituents of complex responses in primates. *Psychological Review*, **59**, 177–91.

Schlichting, C. D. and Pigliucci, M. (1998). *Phenotypic evolution: A reaction norm perspective*. Sunderland, MA: Sinauer.

Schneider, M. L., Moore, C. F., Suomi, S. J., and Champoux, M. (1991). Laboratory assessment of temperament and environmental enrichment in rhesus monkey infants (*Macaca mulatta*). *American Journal of Primatology*, **25**, 137–55.

Seferta, A., Guay, P.-J., Marzinotto, E., and Lefebvre, L. (2001). Learning differences between feral pigeons and zenaida doves: The role of neophobia and human proximity. *Ethology*, **107**, 281–93.

Shettleworth, S. J. (2001). Animal cognition and animal behaviour. *Animal Behaviour*, **61**, 277–86.

Sigg, H. (1980). Differentiation of female positions in hamadryas one-male units. *Zietschrift für Tierpsychologie*, **53**, 265–302.

Sol, D. and Lefebvre, L. (2000). Behavioural flexibility predicts invasion success in birds introduced to New Zealand. *Oikos*, **90**, 599–605.

Sol, D., Lefebvre, L., and Timmermans, S. (2002). Behavioural flexibility and invasion success in birds. *Animal Behaviour*, **63**, 495–502.

Stearns, S. C. (1989). The evolutionary significance of phenotypic plasticity. *Bioscience*, **39**, 436–45.

Sternberg, R. J. (ed.) (1999). *Handbook of creativity*. Cambridge: Cambridge University Press.

Strayer, F. F. (1976). Learning and imitation as a function of social status in Macaque monkeys (*Macaca nemestrina*). *Animal Behaviour*, **24**, 835–48.

Sugita, Y. (1980). Imitative choice behavior in guppies. *Japanese Psychological Research*, **22**, 7–12.

Suzuki, A. (1965). An ecological study of wild Japanese monkeys in snowy areas focused on their food habits. *Primates*, **6**, 1965.

Tarde, G. (1903). *The laws of imitation (E.C. Parsons, translator)* New York: Holt (*reprinted 1969, Chicago: University of Chicago Press*).

Taylor, P. M. (1972). Hovering behavior by house finches. *The Condor*, **74**, 219–21.

Thompson, C. F., Ray, G. F., and Preston, R. L. (1996). Nectar robbing in Blue Tits *Parus caeruleus*: Failure of a novel feeding trait to spread. *Ibis*, **138**, 552–3.

Thorpe, W. H. (1943). A type of insight learning in birds. *British Birds*, **37**, 29–31.

Thorpe, W. H. (1956). *Learning and instinct in animals*. London: Methuen.

Timmermans, S., Lefebvre, L., Boire, D., and Basu, P. (2000). Relative size of the hyperstriatum ventrale is the best predictor of feeding innovation rate in birds. *Brain, Behavior and Evolution*, **56**, 196–203.

Tinbergen, N. (1963). On aims and methods of ethology. *Zietschrift für Tierpsychologie*, **20**, 410–33.

Tomasello, M. (1994). The question of chimpanzee culture. In *Chimpanzee cultures* (ed. R. W. Wrangham, W. C. McGrew, F. B. M. de Waal, and P. G. Heltne), pp. 301–17. Cambridge, MA: Harvard University Press.

Tomasello, M. and Call, J. (1997). *Primate cognition.* New York: Oxford University Press.

Verbeek, M. E. M., Boon, A., and Drent, P. J. (1996). Exploration, aggressive behavior and dominance in pair-wise confrontations of juvenile male great tits. *Behaviour*, **133**, 945–63.

Verbeek, M. E. M., DeGoede, P., Drent, P. J., and Wiepkema, P. R. (1999). Individual behavioural characteristics and dominance in aviary groups of great tits. *Behaviour*, **136**, 23–48.

Verbeek, M. E. M., Drent, P. J., and Wiepkema, P. R. (1994). Consistent individual differences in early exploratory behavior of male great tits. *Animal Behaviour*, **48**, 1113–21.

Visalberghi, E. and Fragaszy, D. M. (1990). Food-washing behaviour in tufted capuchin monkeys, *Cebus apella*, and crabeating macaques, *Macaca fascicularis. Animal Behaviour*, **40**, 829–36.

Webster, S. J. and Lefebvre, L. (2001). Problem solving and neophobia in a Columbiforme—Passeriforme assemblage in Barbados. *Animal Behaviour*, **62**, 23–32.

West-Eberhard, M. J. (1989). Phenotypic plasticity and the origins of diversity. *Annual Review of Ecology and Systematics*, **20**, 249–78.

Whiten, A. and Byrne, R. (1988). Taking (Machiavellian) intelligence apart: Editorial. In *Machiavellian intelligence: Social expertise and the evolution of intellect in monkeys, apes and humans* (ed. R. W. Byrne and A. Whiten), pp. 50–65. Oxford: Oxford University Press.

Whiten, A. and Ham, R. (1992). On the nature and evolution of imitation in the animal kingdom: Reappraisal of a century of research. *Advances in the Study of Behavior*, **21**, 239–83.

Wilson, A. C. (1985). The molecular basis of evolution. *Scientific American*, **253**, 148–57.

Wilson, D. S. (1998). Adaptive individual differences within a single population. *Philosophical Transactions Royal Society of London Series B*, **353**, 199–205.

Wilson, E. O. (1975). *Sociobiology.* Cambridge, MA: Harvard University Press.

Wilson, D. S., Clark, A. B., Coleman, K., and Dearstyne, T. (1994). Shyness and boldness in humans and other animals. *Trends in Ecology and Evolution*, **9**, 442–6.

Wyles, J. S., Kunkel, J. G., and Wilson, A. C. (1983). Birds, behaviour, and anatomical evolution. *Proceedings of the National Academy of the USA*, **80**, 4394–7.

Zentall, T. R. and Galef, B. G. Jr., (ed.) (1988). *Social learning: Psychological and biological perspectives.* Hillsdale, NY: Erlbaum.

COMPARATIVE AND EVOLUTIONARY ANALYSES OF INNOVATION

POSITIVE AND NEGATIVE CORRELATES OF FEEDING INNOVATIONS IN BIRDS: EVIDENCE FOR LIMITED MODULARITY

LOUIS LEFEBVRE AND JOHAN J. BOLHUIS

Introduction

When ornithologists witness a bird using a new or unusual feeding behaviour, they routinely report this in the short notes section of avian journals. All other things being equal, taxonomic groups that eat more food types (generalists), take advantage of more feeding possibilities (opportunists) and use more complex or novel food handling techniques should appear more often in these notes. Corrected for confounding variables, the number of reports per taxon (termed 'innovation rate') is a useful quantitative estimate of feeding flexibility differences between animals. In this chapter, we review current work on feeding innovations in birds, focusing on correlations between innovation rate and a set of cognitive, neurobiological, ecological, and evolutionary variables. We first provide a short history of the use of anecdotes in the study of feeding flexibility and then present an overview of recent findings. Finally, we discuss the implications of inter-taxon correlations between cognitive and neural traits. We argue that these correlations provide clues to the organization of animal cognition and that the current evidence supports Fodor's (1983) view that non-modular problem-solving systems co-occur with specialized input modules in the animal brain.

Anecdotes and correlates of feeding flexibility: a short history

In 1949, Fisher and Hinde published their classic paper on the opening of milk bottles by birds. The paper initiated interest in the cultural transmission of feeding innovations, an interest stimulated in the following two decades by field work on Japanese macaques (Itani, 1965; Kawai, 1965). Fisher and Hinde's (1949) paper was published in *British Birds*, but it was only one of many examples of novel feeding behaviours reported in that journal.

As early as 1943, W. H. Thorpe, the founder of the Madingley field station at Cambridge University (where Hinde also worked), was interested in field studies of learning (Thorpe, 1943, 1944). In a 1956 article on novel feeding methods by wild birds, Thorpe encouraged both amateur and professional ornithologists to note 'examples of the production of original or unusual actions by birds, however small the change' (p. 389), to evaluate the influence of learning on the then-dominant area of interest in European ethology, 'instinct'.

Reports of unusual behaviours by birds had long been a feature of ornithological journals, particularly in English-speaking countries. Illustrative use of anecdotes on clever animals had been popularized by Romanes (1882), but heavily criticized in the following years (Thorndike, 1911). The problem with the illustrative method is that cases favourable to a hypothesis (e.g. 'dogs are intelligent, fish are not') are more likely to be retained than cases that are unfavourable and that anecdotes are often over-interpreted. A more cautious, quantitative use of anecdotes was advocated a few years ago by Whiten and Byrne (1988; see, however, the open peer comments that follow their article), who collected all possible cases of tactical deception in primates and looked at the taxonomic distribution of their frequencies. This procedure still relies on anecdotes, but it quantifies all the evidence available without picking and choosing, relying instead on the comparative method to correlate the variable with traits predicted to evolve with it, such as brain size (Byrne, 1993).

Thorpe's focus on avian learning in the field was framed in more ecological terms by Peter Klopfer, who did postdoctoral work at Madingley in the 1950s. In a series of papers published between 1960 and 1967, Klopfer laid the groundwork for the ecological study of feeding flexibility in birds. Some of Klopfer's articles were coauthored with Robert H. MacArthur, one of the founders of optimality theory (MacArthur and Pianka, 1966). Despite his association with Thorpe and Hinde, Klopfer did not focus on novel feeding techniques in the field, but on theoretical predictions and experimental tests on captive birds. In his series of papers, Klopfer made ten important predictions about the relationship between behaviour, ecology, evolution, and neurobiology across species. Klopfer proposed that more variable environments (e.g. temperate ones) should favour more flexible species than should stable environments (e.g. tropical; Klopfer and MacArthur, 1960). Klopfer also suggested that greater flexibility should be associated with opportunism and wider ecological niches (Klopfer and MacArthur, 1960), a higher population density per species (Klopfer, 1962), and a greater variance in the morphological traits used for feeding (Klopfer, 1962). Ecosystems that favour more flexible species should also be characterized by a less diverse avifauna (Klopfer and MacArthur, 1960). Klopfer further proposed that behavioural flexibility should be greater in taxa with larger and more complex brains (Klopfer, 1962), as well as in species that colonize low-competition areas like islands (Klopfer, 1967). Finally, Klopfer made three predictions concerning social learning: that opportunistic species should be better at social learning than conservative ones (Klopfer, 1961) and that solitary species and birds that feed in mixed species flocks should have less of their behaviour established through social learning (Klopfer, 1961). Klopfer's ten predictions are summarized in Figure 2.1.

Klopfer tested several of these hypotheses with experiments on learning in greenfinches and great tits (Klopfer, 1961), as well as habitat selection in chipping sparrows (Klopfer, 1963) and bananaquits (Klopfer, 1967). He also used data from the literature for some of

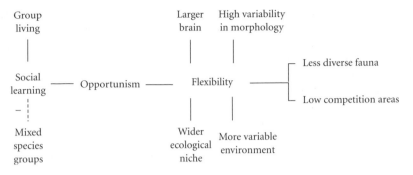

Figure 2.1 Relationships between behavioural flexibility and other variables predicted by Klopfer. Full lines indicate a positive association; the stippled line with the minus sign indicates a negative association.

the ecological predictions (Klopfer and MacArthur, 1960). The results of these tests were often unclear, however, and 'one major problem' that Klopfer (1965) identified was 'the actual measurement of behavioural stereotypy' (p. 376; Klopfer often used the opposite term to flexibility, stereotypy). He operationalized flexibility as either learning probability (Klopfer, 1961), change in habitat preference under hand-rearing conditions (Klopfer, 1963), or number of habitats exploited in the wild (Klopfer, 1967). All these measures have disadvantages, however, and may possibly be the cause of the ambiguous results that Klopfer obtained. Comparative tests in captivity are slow and work-intensive. Gathering data on only a few species can take years and an animal's response in captivity is not always clearly related to its behaviour in the wild. When one tests two or three species, a simple reversal in the predicted performance of one species on one variable can have enormous negative consequences for the hypothesis.

Many of these problems can be solved by adopting Thorpe's (1956) focus on novel feeding techniques used by birds in the wild. When examples of these new techniques, as well as new food types exploited, are systematically collected, they can provide a statistically unbiased sample of taxonomic differences in opportunistic generalism and flexible feeding. Using the comparative correlational method, the taxonomic distribution of cases can then be used to test a series of predictions in neurobiology, cognition, ecology and evolution, many of them similar to those made by Klopfer. Since 1997, Lefebvre and colleagues have collected over 2200 reports of this type by searching through the short notes section of 67 ornithological journals in six parts of the world: western Europe, North America, Australia, New Zealand, the Indian subcontinent, and southern Africa. These areas include temperate and tropical climates, dry and humid conditions, insular and continental environments, urbanized and pristine habitats, as well as the northern and southern hemispheres. Taken together, the reports provide a worldwide sample of taxonomic variation in innovative feeding. Behaviours described in the reports range from simple observations of a new food type incorporated in the diet of a species to the adoption of complex and spectacular feeding techniques (see examples in the tables published in Lefebvre et al., 1997, 1998). A normally frugivorous, anthropophobic red-necked pigeon feeding on spilled maize at a warehouse in Barbados (Lefebvre *et al.*, 2001*a*), a normally insectivorous yellowhead in New Zealand seen for the first time eating bush lily fruits (Child, 1978), are some of the more mundane examples of novel food types. A southern skua taking milk from a lactating seal alongside the pups (Johnston, 1973) is a more spectacular case. Examples of novel feeding techniques would be a herring gull using its normal shell-dropping technique to kill rabbits (Young, 1987), house sparrows flying in front of an electric eye to open the door of a cafeteria (Breitwisch and Breitwisch, 1991), and a giant petrel (Cox, 1978) killing its prey by drowning.

For the moment, innovation data on birds are restricted to new feeding behaviours for three reasons: sample size, reliability, and presumed association with cognitive traits. Since the method is based on frequency distributions, large samples are required lest some taxonomic groups end up with too few cases; a minimum expected frequency of five cases per category is the norm in studies of this type. Feeding is the second most frequently reported behaviour in the short notes, before nesting and after range expansion and sightings of vagrants. Clear predictions exist in the literature on the relationship between cognition and feeding, but not vagrancy or nesting. For example, opportunistic generalists are predicted to learn more readily, both individually (Rozin and Kalat, 1971; Roper, 1983; Sasvàri, 1985*a*) and socially (Klopfer, 1961; Sasvàri, 1985*b*; Laland and Plotkin, 1990) than are conservative specialists. Ranging and nesting have less obvious links with cognition: nesting is presumed to be under genetic or hormonal control (Hansell, 1984), while range expansion is often a simple product of demographic pressure and vagrancy in birds an artifact of wind direction. In line with these assumptions, a test on nesting reports failed to show any significant association with neural trait size, while feeding reports showed the predicted positive correlation (Nicolakakis and Lefebvre, 2000).

Feeding innovation frequencies can theoretically be obtained on all avian species in a region that has sufficient journal coverage. In North America, for example, the current database includes 619 reports on 256 species. Because the number of species involved is so large, an outlier will have no more effect on overall trends than a single data point has in a large sample. Tests are thus more likely to be robust than they are in comparative experiments with two or three species. The major disadvantage of the measure is its anecdotal nature, which may lead to several biases. For example, ornithologists may be more likely to notice and report a new feeding technique in a taxon that is heavily studied or includes many species. One would also expect more reports in a taxon that has a large population size or is expected to show 'intelligent' behaviour. Most biases can be factored into multivariate analyses, however, and their possible effects removed before a biologically important prediction is tested (Lefebvre *et al.*, 1997, 1998, 2001*b*; Nicolakakis and Lefebvre, 2000; Nicolakakis *et al.*, 2003). To date, the effects of research effort (number of full-length papers published on a particular taxon, easily obtained with a reference tool like *The Zoological Record*), species number (a parvorder like Passerida, with over 3500 species worldwide, is likely to yield more reports than a parvorder like Odontophorida, with only six species), avian population size, degree of appeal to ornithologists (frequency of photographs in birding magazines), and likeliness to notice and report (a questionnaire filled out at a national ornithological meeting) have all been assessed. The effects of journal source (journals with many reports, e.g. *British Birds* and *The Wilson Bulletin*, vs others), historical period (1930 to 1960 vs 1960 to the present), reader reliability, and phylogeny have also been measured. Finally, the 'trawling' (a term used by a referee commenting on one of the innovation papers) nature of the data collection technique may reduce sampling bias through its focus on a low impact section of the literature. Full-length papers in high-impact factor journals might be biased towards spectacular cases in large-brained species, for instance the report in *Nature* on leaf and hook tool-making in New Caledonian crows (Hunt, 1996) that confirmed the short note published a quarter of a century earlier by Orenstein (1972). A more mundane observation on partridges eating leeks is more likely to appear in the short notes of *Scottish Birds* than it is in *Nature* or *Science*. For all these reasons, innovation rate, after correction for confounding variables like species number and research effort, can be assumed to yield an unbiased estimate of feeding flexibility and opportunistic generalism in the field.

Correlates of feeding innovations: a summary of recent results

Neurobiology

In birds, innovation rate has been shown to correlate with four sets of variables: neurobiological, ecological, evolutionary, and cognitive. These relationships are summarized in Figure 2.2. Several elements in the figure are similar to those predicted by Klopfer. Some of the relationships have also been found in primates (see Reader and MacDonald, Chapter 4). In particular, the relationship between relative size of the isocortex and innovation rate that

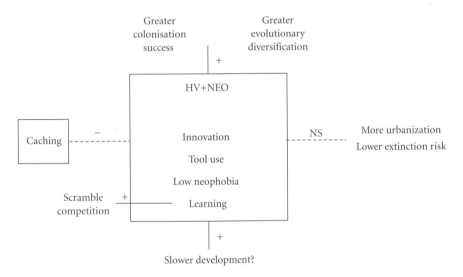

Figure 2.2 Relationships between feeding innovations and cognitive, neural, ecological, developmental, and evolutionary variables in birds. Full lines indicate a positive association; the stippled line with the minus sign indicates a negative association; the stippled line with NS indicates a non-significant relationship. HV + NEO: hyperstriatum ventrale and neostriatum.

Reader and Laland (2002) have found in primates has a striking parallel in birds. The avian equivalents of the mammalian isocortex, the hyperstriatum ventrale and neostriatum (Karten, 1991; Rehkämper and Zilles, 1991; Dubbeldam, 1998), are the structures whose size is most strongly correlated with innovation rate (Timmermans et al., 2000). Detailed data on telencephalic areas are available for only 17 parvorders and 32 species (Boire, 1989; Rehkämper et al., 1991). However, because the neostriatum/hyper-striatum ventrale complex (Neo–HV) is so large (60–65% of telencephalic volume, Boire, 1989) and because taxa with bigger brains are also those that have evolved a large Neo–HV (Rehkämper et al., 1991), less detailed measures of neural substrate can also be used. Relative size of the cerebral hemispheres is available for 140 species (Portmann, 1947) and predicts 99 per cent of the variance in relative size of Neo–HV. Relative size of the whole brain predicts 96 per cent of the variance in Neo–HV and is available for 767 species in 36 orders, infraorders and parvorders (Mlikovsky, 1989a,b,c, 1990). Many of the data taken by Mlikovsky are endocranial volume measurements, but Iwaniuk and Nelson (2002) have shown that this technique yields reliable estimates of brain size. All three estimates (brain, forebrain, Neo–HV) of neural substrate size are positively correlated with innovation rate (Lefebvre et al., 1997; Timmermans et al., 2000; Nicolakakis et al., 2003), with partial cor-relations in the 0.50–0.70 range. About a third of the variance in innovation rate can thus be predicted by neural substrate size. Common ancestry accounts for about 12 per cent of this relationship, based on the average difference between regressions done on phyletically uncorrected parvorders and those conducted on independent contrasts (Lefebvre et al., 1998, 2001b, 2002; Timmermans et al., 2000).

The relationship between innovations and brain size is not an artifact of juvenile development mode, as Bennett and Harvey (1985) found for other ecological variables. Despite the known relationship between brain size and development mode in birds (on average, precocial birds have smaller brains as adults than do altricial ones; Portmann, 1946), inclusion of precocial vs altricial development modes in multiple regressions does not remove the correlation between innovations and neural substrate size (Lefebvre et al., 1998, 2001; Nicolakakis and Lefebvre, 2000). Beyond the precocial/altricial dichotomy, however, continuous variation in development time is associated with both brain size and innovation rate. Goslings, for example, depend on their parents for a very long time even though they are precocial; pigeons and doves develop very quickly even though they are altricial. Continuous variation in development time, specifically in residual time to fledging and independence from the parent (after removal of allometric body size effects), is positively associated with innovation rate (Ryan and Lefebvre, unpublished data). This suggests that longer development time (and the possible higher juvenile mortality associated with it) might be one of the costs of innovative feeding.

Ecology and evolution

In ecology and evolution, innovation rate in the zone of origin is positively associated with the colonization success of introduced species (Sol, Chapter 3; Sol and Lefebvre, 2000; Sol et al., 2002). If innovative birds are better able to adapt to the new conditions of the zone in which they are introduced, one might also predict that they would more readily settle in urbanized habitats (Timmermans, 1999) and be less vulnerable to extinction (Nicolakakis *et al.*, 2003). However, neither of these predictions has been supported for the moment. The number of endangered species per parvorder is instead predicted by the number of species there are in that taxon in the first place (Nicolakakis *et al.*, 2003), while urbanization is associated with colonization success (Sol *et al.*, 2002).

The ability to invade and colonize new habitats is one of the traits thought to favour evolutionary diversification (Sol, Chapter 3). Two decades ago, Allan Wilson (Wyles *et al.*, 1983; Wilson, 1985) proposed that innovation, coupled with social learning, could make large-brained species encounter a wider array and a higher rate of potential selective pressures. For example, a mutation in the enzyme system facilitating digestion of lactose in birds would disappear with its bearer if it occurred in a grouse, but not in a tit. Because the tit and its descendants open milk bottles, but grouse do not, the innovation increases the probability that mutant individuals will encounter situations where the genetic change increases fitness. Following this idea, which Wilson called 'behavioural drive', the number of species per taxon, which depends in part on the rate of evolutionary diversification, might be positively associated with innovation rate. Nicolakakis *et al.* (2003) have found that this is indeed the case at the parvorder level in birds. The relationship is still significant when the largest avian orders are removed: Passeriformes, where the existence of imitated song in oscines may play an isolating role favouring speciation (Slater and Lachlan, Chapter 5), and Ciconiiformes, where molecular research (Sibley and Ahlquist, 1990) has shown surprising similarity between taxa previously classified as separate orders. Sol (Chapter 3) has also found that innovative passerine species tend to have more subspecies than non-innovative ones. Using an ordinal measure of the number of food types eaten per family, Owens *et al.* (1999) have obtained similar results to those of Nicolakakis *et al.* (2003); one would normally predict a positive association between diet breadth and innovation rate, although this prediction has not yet been tested. The positive relationship between species per taxon and behavioural flexibility appears to contradict Klopfer's prediction on this point (Figure 2.1), but the two ideas actually refer to different frames of reference. Klopfer compares more productive tropical ecosytems to less productive temperate ones and predicts more species per unit space in tropical ecosystems. Wyles *et al.* (1983), Owens *et al.* (1999), and Nicolakakis *et al.* (2003) predict that, averaged over different ecosystems, taxa with more flexible behaviour should contain more species. Both ideas are supported by the data in birds.

Cognition

Inter-taxon correlations between cognitive abilities may yield clues to the way animal cognition is organized. If correlations between abilities are negative, this might suggest a trade-off between them. If correlations are positive, this might mean that the abilities are part of the same system. Of the six cognitive measures that have been tested against innovation rate in birds and primates, five yield positive correlations in one or in both groups: tool use, social learning, individual learning, neophobia, and problem solving. In birds, the number of tool use reports (particularly that of proto-tools, i.e. parts of the environment used as feeding implements, but not detached from the substrate and directly manipulated with the beak or foot) shows a similar taxonomic distribution to the one shown by innovation rate (Figure 2.3(a); Lefebvre *et al.*, 2002). The relationship still holds when relative brain size is entered in multivariate models. Within the telencephalon, tool use in birds is best predicted by size of the neostriatum, followed by that of the hyperstriatum ventrale (Lefebvre *et al.*, 2002). A very similar result has been found by Reader and Laland (2002) on primates. In this order, tool use report frequency is positively correlated with both innovation rate and the size of the mammalian equivalent of the avian Neo–HV complex, the isocortex.

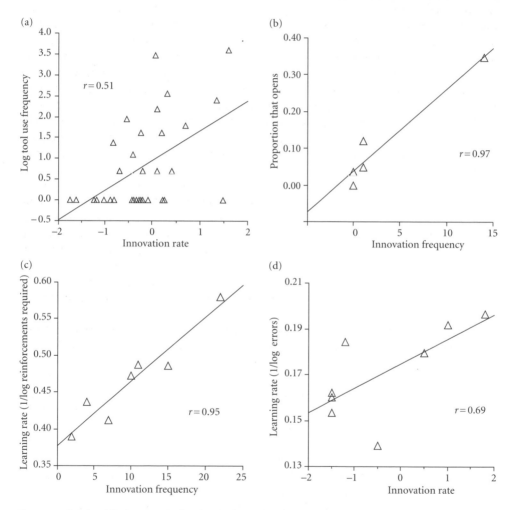

Figure 2.3 Relationships between feeding innovation and (a) tool use reports (data from Lefebvre *et al.*, 2000); (b) proportion of subjects that solve an innovative feeding problem (data from Webster and Lefebvre, 2001); (c) learning speed (data from Sasvàri, 1985a); (d) reversal learning performance (data from Gossette, 1968) in different avian taxa.

The two largest comparative data sets on learning in birds, the seven-species sample of Sasvàri (1985a) and the eight-species sample of Gossette (1968), both reveal high positive correlations between innovation rate and learning speed (Figure 2.3(c) and (d); Timmermans *et al.*, 2000). A similar positive correlation between learning and innovation is found in primates (Reader and MacDonald, Chapter 4). Learning differences between the seven avian species studied by Sasvàri correlate with urbanized opportunism and with a second measure of evolutionary diversification, the number of subspecies per species (Sol, Chapter 3). Primates show another positive correlation, between the taxonomic distribution of innovation and social learning reports (Reader and Laland, 2002). In birds, the data are more difficult to interpret and, for the moment, the relatively small data set available does not allow clear conclusions: different statistical techniques yield different answers and the vast majority of presumed social learning cases in the wild are concentrated in the suborder Passeri, causing outlier problems for both normal regressions and independent contrasts (Bouchard, 2002). Indirect evidence for positive correlations is provided by Lefebvre and Giraldeau's (1996) reanalysis of Sasvàri's (1985a,b) data. In the five species tested by Sasvàri, as well as the two Columbiformes tested by Lefebvre *et al.* (1996), social and individual learning covary. The only study that suggests a trade-off between social and individual learning is the one by Templeton *et al.* (1999) on jays.

Low neophobia (Greenberg, 1983, 1990b; Greenberg and Mettke-Hoffman, 2001) is positively associated with social (Whittle, 1996; Bouchard, 2002) and individual learning speed (Daly *et al.*, 1982; Seferta *et al.*, 2001), as well as generalism (Greenberg, 1984, 1989, 1990*a*; Webster and Lefebvre, 2000). It is also associated with inter-taxon differences in an innovative problem-solving test (Webster and Lefebvre, 2001). This test, given to five Barbadian species in the field and in captivity, revealed a striking positive correlation between innovation frequencies in the wild and probability of success in the experiment (Figure 2.3(b)), as well as a positive correlation between low neophobia and various measures of problem solving.

A final set of predictions that are similar to the ones made by Klopfer have been tested with an experimental program on feral pigeons, Zenaida doves and Carib grackles. The experiments confirm that social learning, individual learning and neophobia are all positively correlated (Lefebvre *et al.*, 1996; Seferta *et al.*, 2001; Bouchard, 2002). Klopfer's (1961) predictions that intraspecific gregariousness favours social learning, but that mixed species flocking does not, are not well supported in the literature (Reader and Lefebvre, 2001), except for the study of Templeton *et al.* (1999). Instead, Klopfer's two predictions can be fused into one: scramble competition, whether in the context of homospecific groups or mixed species associations, appears to favour both social (Dolman *et al.*, 1996; Lefebvre *et al.*, 1996, 1999; Carlier and Lefebvre, 1997) and individual learning (Carlier and Lefebvre, 1996; Seferta *et al.*, 2001), while aggressive interference competition does not (Dolman *et al.*, 1996; Lefebvre *et al.*, 1996). Overall, the data also support Klopfer's idea that opportunistic generalism is positively correlated with social (Sasvàri, 1985b; Lefebvre and Giraldeau, 1996; Reader and Lefebvre, 2001) as well as individual learning (Sasvàri, 1985a; Lefebvre and Palameta, 1988).

Correlations between innovation rate and other cognitive systems: evidence for limited modularity?

The dominant pattern in the previous section is obviously the large number of positive correlations between cognitive measures. Inter-taxon correlations may be a useful tool for testing the idea that animals' minds and brains are modular. That is, when there is a positive correlation between certain cognitive abilities (or between cognitive abilities and brain parameters), it may be that they are part of the same cognitive module. In contrast, a negative correlation between cognitive traits may imply that they belong to different modules (cf. Reader and Laland, 2002; Seyfarth and Cheney, 2002). In recent discussions it is not always clear what is meant by 'modularity' (e.g. Pinker, 1997; Shettleworth, 2000; Bolhuis and Macphail, 2001; Flombaum *et al.*, 2002), thus it is prudent to discuss this concept in some detail. The concept of modularity first came to prominence through the philosopher Jerry Fodor's monograph *The Modularity of Mind* (Fodor, 1983). Fodor (1983) provides a general description of what modular systems are, when he says that 'Roughly, modular cognitive systems are domain-specific, innately specified, hardwired, autonomous and not assembled', and '...modular systems are domain-specific computational mechanisms...' (p. 37). Later in the book he adds informational encapsulation to the list, and it becomes clear that this is in fact the defining feature of modules. More recently, Fodor (2000) made this explicit when he said that 'Modules are informationally encapsulated by definition' (p. 63). Informational encapsulation means basically that individual modules cannot obtain information from other modules, or from other cognitive systems. Domain specificity refers to the limited amount of input that can be processed by a module. Fodor (1983, p. 103) summarizes this as 'Roughly, domain specificity has to do with the range of questions for which a device provides answers (the range of inputs for which it computes analyses); whereas encapsulation has to do with the range of information that the device consults in deciding what answers to provide'.

It is important to realize that Fodor (1983, 2000) explicitly limits modularity to what he terms 'input systems', which roughly correspond with perception and language processing. They go beyond the traditional modalities such as sight and smell, and could, for example, include systems concerned with conspecific face or voice recognition. Fodor contrasts these modular input systems with what he calls 'central systems', which are distinctly non-modular. These central systems are concerned with cognitive processes such as 'thought' and 'problem solving'. In Fodor's account, the fact that central systems are non-modular means that there is not very much that the cognitive scientist can say about them. In his own words '...the limits of modularity are also likely to be the limits of what we are going to be able to understand about the mind...'. A problem with Fodor (1983) is that he is sometimes vague in his use of terminology, even going so far as explicitly saying that he is 'not in the business of defining... terms' (p. 37).

Some recent work in evolutionary psychology has made extensive use of the concepts and assumptions of modularity. Many of these uses go far beyond Fodor's focus on 'input systems'. For instance, Shettleworth (1998) says that '...the animal mind contains a variety of adaptively specialized cognitive modules', the latter being 'distinguishable cognitive mechanisms' (p. 566). Like Fodor, Pinker (1997) is also somewhat vague about defining modules, which he describes as 'metaphors' and sees as synonyms of 'intuitive theories' or 'ways of knowing' (p. 315). Some evolutionary psychologists (e.g. Cosmides and Tooby, 1994; Pinker, 1997; Shettleworth, 1998, 2000) suggest that most, if not all, of the mind consists of modules, a view that was dubbed 'massive modularity' by Fodor (2000). For instance, Shettleworth (2000), after briefly discussing Fodor's (1983) account, says that 'More recent discussions of modularity in developmental and evolutionary psychology, however, see cognition as modular right through from input to decision processes..., and this view is taken here'. (p. 54). This view is shared by Pinker (1994, 1997) and Cosmides and Tooby (1994), who provide long lists of putative modular systems, including 'cheater detection' and 'intuitive biology'.

Fodor (2000) himself has criticized Cosmides and Tooby's version of massive modularity. Interestingly, Pinker (1997) suggests that informational encapsulation is not a defining characteristic of modules. Pinker states that '...mental modules need not be tightly sealed off from one another, communicating only through a few narrow pipelines (that is a specialized sense of 'module' that many cognitive scientists have debated, following a definition by Jerry Fodor)' (p. 31). Similarly, Shettleworth (2000) suggests that modules cannot be 'completely encapsulated', as that might lead to 'conflicting decisions from different modules' (p. 55). Finally, Flombaum et al. (2002) have suggested that domain-specificity might also not be a definitional feature of modularity. In summary, evolutionary psychologists differ in their use of 'modularity' in the way they define the characteristics of modules (encapsulation, domain specificity) and in the way they regard the whole mind to be modular, in contrast to Fodor's limited view of modularity as applying to input systems only.

In neuroecology (functional and evolutionary principles applied to questions in neuro-biology), the cognitive systems that are usually seen as modular and localized neurally are spatial memory and song learning. Other neuroecological studies are often based on much more general assumptions about neural substrates. In many of these cases, hypotheses are tested at the level of the whole brain, for example, the work of Madden (2001) on bower building, Iwaniuk et al. (2001) on play, Gittleman (1994) on parental care, Byrne (1993) on tactical deception, Clutton-Brock and Harvey (1980), Harvey et al. (1980), Eisenberg and Wilson (1978) and Gittleman (1986) on diet. In other cases, neural substrates are more restricted, but still involve large, broadly defined parts of the brain like the isocortex and striatum for primate social intelligence (Dunbar, 1998), innovation, tool use and social learning (Reader and Laland, 2002) or the neostriatum and hyperstriatum ventrale for innovation and tool use in birds (Timmermans et al., 2000; Lefebvre et al., 2002).

In cases where modularity is invoked, three assumptions are usually made: cognitive systems are based on localized neural substrates, they are domain-specific and are predicted to vary between taxa as a result of natural, artificial, or sexual selection in different ecological contexts. The first feature, modular neural substrates, has been clearly addressed by Fodor (1983), who talks of 'hardwired' modular systems that are 'associated with specific, localized, and elaborately structured neural systems' (p. 37). Later he states that 'there is a characteristic [fixed] neural architecture associated with each of what I have been calling the input systems' (p. 98). Researchers applying modularity ideas to neuroecology also assume that modules are 'hardwired', and that cognitive modularity is accompanied by neural modularity. For variation between taxa, five related predictions are usually made in neuroecology: across phyletically independent groups or species, there should be a positive correlation between variation in a natural history trait and (1) variation in a cognitive ability presumed to be selected in this ecological context, as well as (2) variation in a key neural substrate of the ability. This would necessarily imply a (3) positive correlation between substrate and ability. There should also be a (4) negative or zero correlation between variation in the neural substrate and variation in the rest of the brain or telencephalon and (5) a negative or zero correlation between performance on the selected cognitive ability and other abilities not thought to be involved in that ecological context.

In investigations of food-storing birds it is, implicitly or explicitly, assumed that the hippocampus is a key neural substrate for information stored in spatial memory (Krebs *et al.*, 1989; Sherry *et al.*, 1989; Healy and Krebs, 1992; Smulders and DeVoogd, 2000). Likewise, in the study of bird song learning, it is usually assumed that some of the so-called 'song control nuclei' (particularly those in the rostral pathway) are a key substrate of auditory memories (Nottebohm, 1981; DeVoogd, 1994). For both song and food-storing, there has been some recent debate on the exact localization of the neural substrates (song: Nottebohm, 2000; Solis *et al.*, 2000; Bolhuis *et al.*, 2000a; storing: Bolhuis and Macphail, 2001; Macphail and Bolhuis, 2001).

Several behaviours have been examined in neuroecological studies of spatial memory, for example, artificial selection for homing in Rehkämper *et al.* (1988), sexual selection for range size in rodents (Jacobs *et al.*, 1990), gender-specific natural selection for host selection in brood parasitic cowbirds (Sherry *et al.*, 1993; Reborada *et al.*, 1996). Food-storing is best suited for detailed study, however, because it provides the largest data base for comparative work, with 19 avian species studied up to date by Basil *et al.* (1996), Healy and Krebs (1992, 1996) and Hampton *et al.* (1995). Overall, the data on food-storing birds provide fairly good evidence for predictions 2 and 4 outlined above. Basil *et al.* (1996), Healy and Krebs (1992, 1996), and Hampton *et al.* (1995) all report a positive correlation between relative size of the hippocampus and degree of food-storing in Corvidae and Paridae, as well as a zero or negative correlation between size of the hippocampus and that of the rest of the telencephalon.

Evidence for predictions 1 and 3 is more controversial. Differences in spatial memory performance between storing and non-storing Paridae are often not consistent (see Biegler *et al.*, 2001 and Macphail and Bolhuis, 2001, for a detailed discussion). In Corvidae, there does not appear to be a one-to-one relationship between degree of food-storing and performance in spatial memory tests. As pointed out by Macphail and Bolhuis (2001) in their table 2, Clark's nutcracker, the New World Corvid that relies most heavily on food-storing, comes out behind the Pinyon jay (who relies on stores to a lesser extent) in four of the seven tests available up to now. In a similar vein, scrub jays come out ahead of Mexican jays (who rely on storing more than do scrub jays) in two of the three tests where the species are featured. When average rank of the four storing Corvids and the non-storer control is calculated over the seven tests available, three clusters seem to emerge: non-storing jackdaws and pigeons do worst on the memory tests, high-storing nutcrackers and pinyon jays do best and the two intermediate storers fall in between.

Comparative data on song learning in Passeriformes supports prediction 3, but not prediction 4. Relative size of the telencephalic nucleus HVC (High Vocal Centre) is positively associated with the number of songs in the species' repertoire (DeVoogd *et al.*, 1993) and with individual differences in song imitation (Ward *et al.*, 1998). It has been suggested that these correlations are due to a limitation of learning capacity by the neural substrate, rather than variation in the neural parameter being the result of learning (Ward *et al.*, 1998). Larger repertoires are sexually selected through female choice and increased extra-pair paternity (Hasselquist *et al.*, 1996). Both repertoire and HVC size are heritable (Airey *et al.*, 2000). Contrary to prediction (4), the relative size of HVC appears to be positively correlated with that of the whole telencephalon (DeVoogd, personal communication; Smulders, 2002). Predictions 1, 2 and 5 have not been tested.

If the idea of adaptive specialization in 'central systems' is open to controversy, the debate may be simpler with regard to 'input systems'. Food-storing birds are obviously better at food storing than non-storers, and songbirds are better at singing complex songs than non-songbirds. In the case of food-storing birds, it may be that spatial navigation or spatial stimulus processing is such a module. Likewise, songbirds could have a module for auditory perception, perhaps not unlike Fodor's (1983) suggestion of 'computational systems...that detect the melodic or rhythmic structure of acoustic arrays' (p. 47). At the same time, there are often obvious neural differences between food storers and non-storers and between songbirds and non-songbirds, suggesting a kind of parallel neural modularity akin to Fodor's 'hardwired' modules.

Predictions 4 and 5 (negative or zero correlations between modules) are relevant to the issue of dómain specificity. Assumptions on this point seem to vary from one study to another in neuroecology, as evidenced by the diversity of controls and predictions in the studies. For example, Templeton *et al.* (1999) clearly state that the predicted difference in social learning between the social jay they are studying and the non-social one should not also apply to a non-social task. An earlier study by Bednekoff and Balda (1996) on the same hypothesis did not have such a clear statement of assumptions nor such a well-controlled design. Controls like the one used by Templeton *et al.* (1999) are relatively rare in comparative tests of ecological predictions on learning (Lefebvre and Giraldeau, 1996).

In the field of spatial memory, different studies seem to use different assumptions concerning domain specificity. For example, the pigeon is used as a control in Olson's (1991) comparative experiments on memory for hidden food in storing Corvids. Pigeons do not store food, but they do use landmark memory in homing and there is evidence for artificial selection on relative hippocampus size in pigeon strains bred for racing ability (Rehkämper *et al.*, 1988). Astié *et al.* (1998) predict a gender-specific enhancement of spatial memory in a hidden food task in shiny cowbirds. This prediction is based on Reborada *et al.*'s (1996) finding that shiny cowbird females have a larger relative hippocampus size than do males; in this brood parasitic species, females, but not males, search out suitable nests in which to dump their eggs. Males and females of the non-parasitic bay-winged cowbird or of the parasitic screaming cowbird (in which both males and females search for suitable nests) do not show gender differences in hippocampal size. Astié *et al.*'s prediction would seem to imply that spatial memory selected in a nesting context could be used in a feeding context, while Olson (1991) implies that spatial memory selected in a homing context could not. If spatial memory is indeed specific to a very restricted context, then it is reassuring to note that, contrary to their prediction, Astié *et al.* (1998) found no evidence for gender-specific memory enhancement in the feeding task they gave their cowbirds.

Predictions 4 and 5 clearly imply that if different cognitive systems are subject to different natural, sexual, or artificial selection histories, then inter-taxon correlations between them should not all be positive. If they are, then this either means that the systems are all in the same module and have been selected as one unit, or that the modules they belong to have been selected together. If the cognitive systems are traded-off against one another because memory and brain space are limited (Sherry and Schacter, 1987), negative correlations should be expected between at least some cognitive systems measured in the same taxa. If the systems simply have independent selection histories and are not necessarily incompatible, then some zero correlations between systems would be expected.

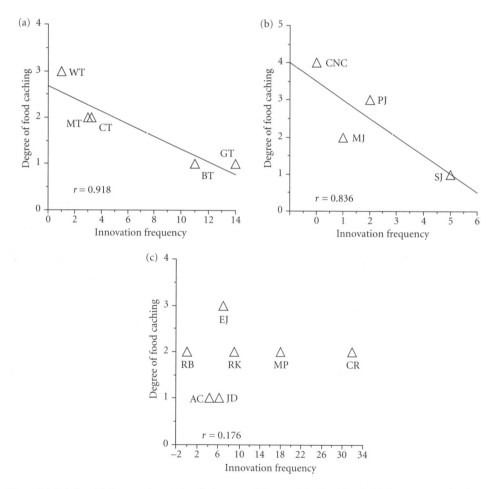

Figure 2.4 Relationship between innovation frequency and degree of food caching in (a) European Paridae (WT, willow tit; MT, marsh tit; GT, great tit; BT, blue tit; CT, coal tit). (b) North American Corvidae (CNC, Clark's nutcracker; PJ, pinyon jay; MJ, Mexican jay; SJ, scrub jay). (c) Eurasian Corvidae (EJ, Eurasian jay; RB, red-billed blue magpie; RK, rook; MP, magpie; CR, carrion crow; AC, Alpine chough; JD, jackdaw).

The difficulty here is to obtain enough data for different cognitive systems in enough taxa. This is where the innovation literature can be interesting, because it provides data on a large number of species and upper-level taxa. Sufficient innovation data ($n = 116$ reports) are available on Corvidae and Paridae (17 species) from the three largest studies on food-storing, those of Basil *et al.* (1996) and Healy and Krebs (1992, 1996). Contrary to the positive correlations mentioned earlier between innovation and tool use, problem-solving, learning and low neophobia (Figure 2.3), the correlations between innovation frequency and degree of food-storing are strongly negative in the case of Paridae (Figure 2.4(a)) and North American corvids (Figure 2.4(b)) and zero in the case of European corvids (Figure 2.4(c)). In Paridae, non-storing blue tits and great tits show 10 and 11 innovation reports respectively in the European data base (Figure 2.4(a)), a level consistent with their faster learning speed in Sasvàri's (1985) experiments. Among the North American corvids, the scrub jay, who relies the least on storing, shows the highest number of innovations (Figure 2.4(b)). In Europe, crows and magpies show the highest rates of innovation ($n = 26$ and 15, respectively), but do not rely heavily on stored food. Low innovation frequencies are shown both by non-storing (jackdaw, Alpine chough) and high storing (Eurasian jay) European corvids (Figure 2.4(c)).

Avian innovations and neurocognitive modularity

The finding that innovation rate is positively correlated with four measures of learning and cognition, while at the same time showing negative or zero correlations with food storing, supports the suggestion that storing birds have an adaptively specialized spatial module (Shettleworth, 1998, 2000) and that innovation, tool use and the learning tasks studied by Gossette (repeated reversals) and Sasvàri (a single change in the place where a food reward is delivered) could all be processed by the central systems envisioned by Fodor (1983). The zero correlation suggests independent evolution, while the negative correlation is consistent with the idea that there has been a trade-off between the demands of storing and innovation. In the case of negative correlations, this may mean that selection for one system has occurred at the expense of the other, or that animals that have fewer stores to rely on need more opportunistic feeding. More simply, it could be that the time spent in storing and retrieving is time that must necessarily be taken from time spent searching for new feeding opportunities. The correlations obviously describe an average relationship over several species and do not preclude the use of innovative feeding by storers in special circumstances. For example, Clayton and Jollife (1996) have seen marsh tits use adhesive paper to pick up powdered food and store it, folding the paper to use it as a wrapper, a clear case of innovative tool use.

The negative and zero correlations are consistent with the view that food storing is modular, but they do not tell us which aspect of food storing (dead reckoning, landmark use (Shettleworth, 2000), navigation, working memory, long term memory, stimulus perception) constitutes the 'spatial module', nor whether the module is central or restricted to the input level, *sensu* Fodor. Innovations can be described as new ways to solve problems, and, as we have seen, Fodor (1983) considers 'problem solving' as an operation performed by central systems. As Bolhuis and Macphail (2001) have argued, the modularity that has been claimed for food-storing birds and songbirds may not be a modularity of memory systems, but of input systems. Often (but not always: Bolhuis and Macphail, 2001), a positive relationship is found between the volume of the hippocampus and the degree of food storing between avian species. It may well be that the positive correlations between food storing and hippocampal volume are not due to variation in spatial learning and memory, but to variation in some kind of input system such as spatial navigation (Bolhuis and Macphail, 2001). We are left with an interesting state of affairs where innovation rate and learning (which might be called operations of central systems) show positive correlations with the volume of large sections of the brain or whole brain size. On the other hand, performance involving adaptively specialized 'input systems' such as those involved in bird song and food storing, correlates with the volume of rather restricted brain regions. This is broadly consistent with a Fodorian view of 'hardwired' modularity restricted to input systems, with central systems being non-modular and having an 'equipotential' (Fodor, 1983) representation in the brain.

Fodor's (1983, 2000) somewhat gloomy view of the scientific accessibility of central systems does not bode well for the cognitive neuroscience of learning and memory. Nevertheless, there is plenty of evidence to suggest that considerable progress has been made in this field. We suggest that if the activity of central systems such as learning and memory does not correlate with the volume of restricted brain regions, this does not necessarily imply that the information storage involved is not in some way localized neurally. Thus, there appears to be considerable localization of function in learning and memory, for instance in filial imprinting (Horn, 1985, 2000), eye blink conditioning (King and Thompson, 2000), olfactory learning (Brennan and Keverne, 2000) and song learning (Bolhuis et al., 2000a, 2001). In the case of imprinting (McCabe and Horn, 1988, 1994; Sheu et al., 1993) and song learning (Bolhuis et al., 2000a, 2001), there are significant positive correlations between measures of the strength of learning and neuronal activation and plasticity in restricted brain regions. It is likely that the behavioural measures used in these studies do indeed reflect the strength of learning, and not some kind of input variable (Bolhuis et al., 2000b), although it is perhaps impossible to rule this out completely, a limitation inherent to any neurobiological approach to the study of learning and memory. It may well be, then, that 'equipotentiality' with a distributed neural representation of central systems for learning and memory can nevertheless involve localized neural representation of information stored in memory.

Summary

Behavioural flexibility has been predicted to correlate with several neurobiological, cognitive, ecological and evolutionary traits. Four decades ago, Peter Klopfer provided the first comprehensive framework for thinking about these correlations. A recently developed operational measure of feeding flexibility, innovation rate, allows several tests of the positive and negative relationships anticipated by Klopfer. In birds, innovation rate is positively associated with colonization success and species richness. It is also positively correlated with relative size of the whole brain, of the forebrain and of the neostriatum/hyperstriatum ventrale complex, as well as tool use report frequency, individual learning, and problem-solving. In contrast, innovation rate shows negative or zero correlations with degree of food-storing in 17 species of Paridae and Corvidae. This pattern of positive correlations between different problem-solving measures and negative correlations with a specialized ability are compatible with Fodor's views on limited modularity in animal cognition.

References

Airey, D. C., Castillo-Juarez, H., Casella, G., Pollak, E. J., and **DeVoogd. T. J.** (2000). Variation in the volume of zebra finch song control nuclei is heritable: Developmental and evolutionary implications. *Proceedings of the Royal Society of London, Series B*, **267**, 2099–104.

Astié, A. A., Kacelnik, A., and **Reboreda, J. C.** (1998). Sexual differences in memory in shiny cowbirds. *Animal Cognition*, **1**, 77–82.

Basil, J. A., Kamil, A. C., Balda, R. P., and **Fite, K. V.** (1996). Differences in hippocampal volume among food storing corvids. *Brain, Behavior and Evolution*, **47**, 156–64.

Bednekoff, P. A. and **Balda, R. J.** (1996). Observational spatial memory in Clark's nutcracker and Mexican jays. *Animal Behaviour*, **52**, 833–9.

Bennett, P. M., and **Harvey, P. H.** (1985). Relative brain size and ecology in birds. *Journal of Zoology London (A)*, **207**, 151–69.

Biegler, R., McGregor, A., Krebs, J. R., and **Healy, S. D.** (2001). A larger hippocampus is associated with longer-lasting spatial memory. *Proceedings of the National Academy of Sciences of the USA*, **98**, 6941–4.

Boire, D. (1989). Comparaison quantitative de l'encéphale, de ses grandes subdivisions et de relais visuels, trijumaux et acoustiques chez 28 espèces d'oiseaux. Ph.D. Thesis, Université de Montréal.

Bolhuis, J. J. and **Honey, R. C.** (1998). Imprinting, learning, and development: From behaviour to brain and back. *Trends in Neurosciences*, **21**, 306–11.

Bolhuis, J. J. and **Macphail, E. M.** (2001). A critique of the neuroecology of learning and memory. *Trends in Cognitive Sciences*, **5**, 426–33.

Bolhuis, J. J. and **Macphail, E. M.** (2002). Everything in neuroecology makes sense in the light of evolution. *Trends in Cognitive Sciences*, **6**, 7–8.

Bolhuis, J. J., Zijlstra, G. G. O., Den Boer-Visser, A. M., and **Van der Zee, E. A.** (2000a). Localized neuronal activation in the zebra finch brain is related to the strength of song learning. *Proceedings of the National Academy of Sciences of the USA*, **97**, 2282–5.

Bolhuis, J. J., Cook, S., and **Horn, G.** (2000b). Getting better all the time: Improving preference scores reflect increases in the strength of filial imprinting. *Animal Behaviour*, **59**, 1153–9.

Bolhuis, J. J., Hetebrij, E., Den Boer-Visser, A. M., De Groot, J. H., and **Zijlstra, G. G. O.** (2001). Localized immediate early gene expression related to the strength of song learning in socially reared zebra finches. *European Journal of Neuroscience*, **13**, 2165–70.

Bouchard, J. (2002). Is social learning correlated with innovation in birds? An inter- and an intraspecific test. M.Sc. thesis, McGill University.

Breitwisch, R. and **Breitwisch, M.** (1991). House sparrows open an automatic door. *Wilson Bulletin*, **103**, 725.

Brennan, P. A. and **Keverne, B. E.** (2000). Neural mechanisms of olfactory recognition memory. In *Brain, perception, memory. Advances in cognitive neuroscience* (ed. J. J. Bolhuis), pp. 93–112. Oxford: Oxford University Press.

Byrne, R. W. (1993). Do larger brains mean greater intelligence? *Behavioral and Brain Sciences*, **16**, 696–7.

Carlier, P. and Lefebvre, L. (1996). Differences in individual learning between group-foraging and territorial zenaida doves. *Behaviour*, **133**, 15–16.

Carlier, P. and Lefebvre, L. (1997). Ecological differences in social learning between adjacent, mixing populations of zenaida doves. *Ethology*, **103**, 772–84.

Child, P. (1978). Yellowhead not entirely insectivorous. *Notornis*, **25**, 252–3.

Clayton, N. S. and Jollife, A. (1996). Marsh tits *Parus palustris* use tools to store food. *Ibis*, **138**, 554.

Clutton-Brock, T. and Harvey, P. H. (1980). Primates, brains and ecology. *Journal of Zoology London*, **190**, 309–23.

Cosmides, L. and Tooby, J. (1994). Origins of domain specificity: The evolution of functional organization. In *Mapping the mind* (ed. L. Hirschfeld and S. Gelman), pp. 85–116. Cambridge: Cambridge University Press.

Cox, J. B. (1978). Albatross killed by giant petrel. *Emu*, **78**, 94–5.

Daly, M., Rauschenberger, J., and Behrends, P. (1982). Food aversion learning in kangaroo rats: A specialist–generalist comparison. *Animal Learning and Behavior*, **10**, 314–20.

DeVoogd, T. J. (1994). The neural basis for the acquisition and production of bird song. In *Causal mechanisms of behavioural development* (ed. J. A. Hogan and J. J. Bolhuis), pp. 49–81. Cambridge: Cambridge University Press.

DeVoogd, T. J., Krebs, J. R., Healy, S. D., and Purvis, A. (1993). Relations between song repertoire size and the volume of brain nuclei related to song: Comparative evolutionary analysis amongst oscine birds. *Proceedings of the Royal Society of London, Series B*, **254**, 75–82.

Dolman, C., Templeton, J., and Lefebvre, L. (1996). Mode of foraging competition is related to tutor preference in Zenaida aurita. *Journal of Comparative Psychology*, **110**, 45–54.

Dubbeldam, J. L. (1998). Birds. In *The central nervous system of vertebrates* (ed. R. Nieuwenhuys, H. J. TenDonkelaar, and C. Nicholson), pp. 1525–620. Berlin: Springer.

Dunbar, R. I. M. (1998). The social brain hypothesis. *Evolutionary Anthropology*, **6**, 178–90.

Eisenberg, J. F. and Wilson, D. E. (1978). Relative brain size and feeding strategies in the Chiroptera. *Evolution*, **32**, 740–51.

Fisher, J. and Hinde, R. A. (1949). The opening of milk bottles by birds. *British Birds*, **42**, 347–57.

Flombaum, J. I., Santos, L. R., and Hauser, M. D. (2002). Neuroecology and psychological modularity. *Trends in Cognitive Sciences*, **6**, 106–8.

Fodor, J. A. (1983). *The modularity of mind: An essay on faculty psychology*. Cambridge, MA: MIT Press.

Fodor, J. A. (2000). *The mind doesn't work that way. The scope and limits of computational psychology*. Cambridge, MA: MIT Press.

Gittleman, J. L. (1986). Carnivore brain size, behavioral ecology and phylogeny. *Journal of Zoology* (London), **190**, 309–23.

Gittleman, J. L. (1994). Female brain size and parental care in carnivores. *Proceedings of the National Academy of Science of the USA*, **91**, 5495–7.

Gossette, R. L. (1968). Examination of retention decrement explanation of comparative successive discrimination reversal learning by birds and mammals. *Perceptual and Motor Skills*, **27**, 1147–52.

Greenberg, R. (1983). The role of neophobia in foraging specialization of some migrant warblers. *American Naturalist*, **122**, 444–53.

Greenberg, R. (1984). Differences in feeding neophobia in the tropical migrant wood warblers *Dendroica castanea* and *D. pensylvanica*. *Journal of Comparative Psychology*, **98**, 131–6.

Greenberg, R. (1989). Neophobia, aversion to open space, and ecological plasticity in song and swamp sparrows. *Canadian Journal of Zoology*, **67**, 1194–9.

Greenberg, R. (1990a). Feeding neophobia and ecological plasticity: A test of the hypothesis with captive sparrows. *Animal Behaviour*, **39**, 375–9.

Greenberg, R. (1990b). Ecological plasticity, neophobia, and resource use in birds. *Studies in Avian Biology*, **13**, 431–7.

Greenberg, R. and Mettke-Hoffman, C. (2001). Ecological aspects of neophobia and exploration in birds. *Current Ornithology*, **16**, 119–78.

Hampton, R. R., Sherry, D. F., Shettleworth, S. J., Khurgel, M., and Ivy, G. (1995). Hippocampal volume and food-storing behavior are related in birds. *Brain, Behavior and Evolution*, **45**, 54–61.

Hansell, M. H. (1984). *Animal architecture and building behaviour*. London: Longman.

Hasselquist, D., Bensch, S., and von Schantz, T. (1996). Correlation between male song repertoire, extra-pair paternity and offspring survival in the great reed warbler. *Nature*, **381**, 229–32.

Harvey, P. H., Clutton-Brock, T. H., and Mace, G. M. (1980). Brain size and ecology in small mammals and primates. *Proceedings of the National Academy of Sciences of the USA*, **77**, 4387–9.

Healy, S. D. and Krebs, J. R. (1992). Food storing and the hippocampus in corvids: Amount and volume are correlated. *Proceedings of the Royal Society of London, Series B*, **248**, 241–5.

Healy, S. D. and Krebs, J. R.. (1996). Food storing and the hippocampus in Paridae. *Brain, Behavior and Evolution*, **47**, 195–9.

Horn, G. (1985). *Memory, imprinting, and the brain*. Oxford: Clarendon Press.

Horn, G. (2000). In memory. In *Brain, perception, memory. Advances in cognitive neuroscience* (ed. J. J. Bolhuis), pp. 329–63. Oxford: Oxford University Press.

Hunt, G. R. 1996. Manufacture and use of hook-tools by New Caledonian crows. *Nature*, **379**, 249–51.

Itani, J. (1965). On the acquisition and propagation of a new food habit in the troop of Japanese monkeys at Takasakiyama. In *Japanese monkeys: A collection of translations* (ed. K. Imanishi and S. Altmann), pp. 52–65. Edmonton: University of Alberta Press.

Iwaniuk, A. N. and Nelson, J. E. (2002). Can endocranial volume be used as an estimate of brain size in birds? *Canadian Journal of Zoology*, **80**, 16–23.

Iwaniuk, A. N., Nelson, J. E., and Pellis, S. M. (2001). Do big-brained animals play more? Comparative analyses of play and relative brain size in mammals. *Journal of Comparative Psychology*, **115**, 29–41.

Jacobs, L. F., Gaulin, S. J. C., Sherry, D. F., and Hoffman, G. E. (1990). Evolution of spatial cognition: Sex specific patterns of spatial behavior predict hippocampal size. *Proceedings of the National Academy of Sciences of the USA*, **87**, 6349–52.

Johnston, G. C. (1973). Predation by southern skuas on rabbits at Macquarie island. *Emu*, **73**, 25–6.

Karten, H. J. (1991). Homology and evolutionary origins of the 'neocortex'. *Brain, Behavior and Evolution*, **38**, 264–72.

Kawai, M. (1965). Newly acquired pre-cultural behavior of the natural troop of Japanese monkeys on Koshima islet. *Primates*, **6**, 1–30.

King, D. A. T. and Thompson, R. F. (2000). Skill learning. The role of the cerebellum. In *Brain, perception, memory. Advances in cognitive neuroscience* (ed. J. J. Bolhuis), pp. 215–31. Oxford: Oxford University Press.

Klopfer, P. H. (1961). Observational learning in birds: The establishment of behavioral modes. *Behaviour*, **17**, 71–80.

Klopfer, P. H. (1962). *Behavioral aspects of ecology*. Englewood Cliffs, NJ: Prentice Hall.

Klopfer, P. H. (1963). Behavioral aspects of habitat selection: The role of early experience. *Wilson Bulletin*, **75**, 15–22.

Klopfer, P. H. (1965). Behavioral aspects of habitat selection: A preliminary report on stereotypy in foliage preferences of birds. *Wilson Bulletin*, **77**, 376–81.

Klopfer, P. H. (1967). Behavioral stereotypy in birds. *Wilson Bulletin*, **79**, 290–300.

Klopfer, P. H. and MacArthur, R. H. (1960). Niche size and faunal diversity. *American Naturalist*, **94**, 293–300.

Krebs, J. R., Sherry, D. F., Healy, S. D., Perry, V. H., and Vaccarino, A. L. (1989). Hippocampal specialization of food-storing birds. *Proceedings of the National Academy of Sciences of the USA*, **86**, 1388–92.

Laland, K. N. and Plotkin, H. C. (1990). Social learning and social transmission of foraging information in Norway rats (*Rattus norvegicus*). *Animal Learning and Behavior*, **18**, 246–51.

Lefebvre, L. (2000). Feeding innovations and their cultural transmission in bird populations. In *The evolution of cognition* (ed. C. M. Heyes and L. Huber), pp. 311–28. Cambridge, MA: MIT Press.

Lefebvre, L. and Giraldeau, L. A. (1996). Is social learning an adaptive specialisation? In *Social learning in animals: The roots of culture* (ed. C. M. Heyes and B. G. Galef), pp. 107–28. New York, NY: Academic Press.

Lefebvre, L. and Palameta, B. (1988). Mechanisms, ecology, and population diffusion of socially-learned, food-finding behavior in feral pigeons. In *Social learning: Psychological and biological perspectives* (ed. T. R. Zentall and B. G. Galef), pp. 141–64. Hillsdale, NJ: Erlbaum.

Lefebvre, L, Palameta, B. and Hatch, K. K. (1996). Is group-living associated with social learning? A comparative test of a gregarious and a territorial columbid. *Behaviour*, **133**, 241–61.

Lefebvre, L., Whittle, P., Lascaris, E., and Finkelstein, A. (1997). Feeding innovations and forebrain size in birds. *Animal Behaviour*, **53**, 549–60.

Lefebvre, L., Gaxiola, A., Dawson, S., Timmermans, S., Rozsa, L., and Kabai, P. (1998). Feeding innovations and forebrain size in Australasian birds. *Behaviour*, **135**, 1077–97.

Lefebvre, L., Templeton, J., Koelle, M. and Brown, K. (1999). Carib grackles imitate conspecific and zenaida dove tutors. *Behaviour*, **134**, 1003–17.

Lefebvre, L., Reader, S. M., and Webster, S. J. (2001a). Novel food use by grey kingbirds and red-necked pigeons in Barbados. *Bulletin of the British Ornithologists' Club*, **121**, 247–8.

Lefebvre, L., Juretic, N., Timmermans, S., and Nicolakakis, N. (2001b). Is the link between innovation rate and forebrain size caused by confounding variables? A test on North American and Australian birds. *Animal Cognition*, **4**, 91–7.

Lefebvre, L., Nicolakakis, N., and Boire, D. (2002). Tools and brains in birds. *Behaviour*, **139**, 939–73.

MacArthur, R. H. and Pianka, E. R. (1966). On optimal use of a patchy environment. *American Naturalist*, **100**, 603–9.

Macphail, E. M. and Bolhuis, J. J. (2001). The evolution of intelligence: Adaptive specialisations versus general process. *Biological Reviews*, **76**, 341–64.

McCabe, B. J. and Horn, G. (1988). Learning and memory: Regional changes in N-methyl-D-aspartate receptors in the chick brain after imprinting. *Proceedings of the National Academy of Sciences of the USA*, **85**, 2849–53.

McCabe, B. J. and Horn, G. (1994). Learning-related changes in Fos-like immunoreactivity in the chick forebrain after imprinting. *Proceedings of the National Academy of Sciences of the USA*, **91**, 11417–21.

Madden, J. (2001). Sex, bowers and brains. *Proceedings of the Royal Society of London, Series B*, **268**, 833–8.

Mlikovsky, J. (1989a). Brain size in birds: 1. Tinamiformes through ciconiiformes. *Vestnik Ceskoslovenske Spolenosti Zoologicke*, **53**, 33–47.

Mlikovsky, J. (1989b). Brain size in birds: 2. Falconiformes through gaviiformes. *Vestnik Ceskoslovenske Spolenosti Zoologicke*, **53**, 200–13.

Mlikovsky, J. (1989c). Brain size in birds: 3. Columbiformes through piciformes. *Vestnik Ceskoslovenske Spolenosti Zoologicke*, **53**, 252–64.

Mlikovsky, J. (1990). Brain size in birds: 4. Passeriformes. *Acta Societatis Zoologicae Bohemoslovacae*, **54**, 27–37.

Nicolakakis, N., and **Lefebvre, L.** (2000). Forebrain size and innovation rate in European birds: Feeding, nesting and confounding variables. *Behaviour*, **137**, 1415–27.

Nicolakakis, N., Sol, D., and **Lefebvre, L.** (2003). Innovation rate predicts species richness in birds, but not extinction risk. *Animal Behaviour*, **65**, 445–52.

Nottebohm, F. (1981). A brain for all seasons: Cyclical anatomical changes in song control nuclei of the canary brain. *Science*, **214**, 1368–70.

Nottebohm, F. (2000). The anatomy and timing of vocal learning in birds. In *The design of animal communication* (ed. M. D. Hauser and M. Konishi), pp. 63–110. Cambridge, MA: MIT Press.

Olson, D. J. (1991). Species differences in spatial memory among Clark's nutcrackers, scrub jays and pigeons. *Journal of Experimental Psychology, Animal Behavior Processes* **17**, 363–76.

Orenstein, R. I. (1972). Tool use by the New Caledonian crow, *Corvus moneloduloides*. *Auk*, **89**, 674–6.

Owens, I. P. F., Bennett, P. M., and **Harvey, P. H.** (1999). Species richness among birds: Body size, life history, sexual selection or ecology. *Proceedings of the Royal Society of London, Series B*, **266**, 933–9.

Pinker, S. (1994). *The language instinct*. London: Penguin Books.

Pinker, S. (1997). *How the mind works*. New York, NY: W.W. Norton.

Portmann, A. (1946). Etude sur la cérébralisation des oiseaux I. *Alauda*, **14**, 2–20.

Portmann, A. (1947). Etude sur la cérébralisation chez les oiseaux II. *Alauda*, **15**, 1–15.

Reader, S. M. and **Laland, K. N.** (2002). Social intelligence, innovation and enhanced brain size in primates. *Proceedings of the National Academy of Sciences of the USA*, **99**, 4436–41.

Reader, S. M. and **Lefebvre, L.** (2001). Social learning and sociality. *Behavioral and Brain Sciences*, **24**, 353–4.

Reboreda, J. C., Clayton, N. S., and **Kacelnik, A.** (1996). Species and sex differences in hippocampus size in parasitic and non-parasitic cowbirds. *Neuroreport*, **7**, 505–8.

Rehkämper, G., Frahm, H. D., and **Zilles, K.** (1991). Quantitative development of brain structures in birds (Galliformes and Passeriformes) compared to that in mammals (Insectivores and Primates). *Brain, Behavior and Evolution*, **37**, 125–43.

Rehkämper, G., Haase, E., and **Frahm, H. D.** (1988). Allometric comparison of brain weight and brain structure volumes in different breeds of the domestic pigeon, *Columba livia* (fantails, homing pigeons, strassers). *Brain, Behavior and Evolution*, **31**, 141–9.

Rehkämper, G. and **Zilles, K.** (1991). Parallel evolution in mammalian and avian brains: Comparative cytoachitectonic and cytochemical analysis. *Cell and Tissue Research*, **263**, 3–28.

Romanes, G. J. (1882). *Animal intelligence*, 2nd edn. London: Kegan, Paul, Trench.

Roper, T. J. (1983). Learning as a biological phenomenon. In *Animal behaviour. Vol 3: Genes, development and learning* (ed. T. R. Halliday and P. J. B. Slater), pp. 178–212. Oxford: Blackwell.

Rozin, P. and **Kalat, J. W.** (1971). Specific hungers and poison avoidance as adaptive specializations of learning. *Psychological Review*, 78, 459–86.

Sasvàri, L. (1985a). Keypeck conditioning with reinforcement in two different locations in thrush, tit, and sparrow species. *Behavioral Processes*, **11**, 245–52.

Sasvàri, L. (1985b). Different observational learning capacity in juvenile and adult individuals of congeneric bird species. *Zeitschrift für Tierpsychologie*, **69**, 293–304.

Seferta, A., Guay, P. J., Marzinotto, E., and Lefebvre, L. (2001). Learning differences between feral pigeons and zenaida doves: The role of neophobia and human proximity. *Ethology*, **107**, 281–93.

Seyfarth, R. M. and Cheney, D. L. (2002). What are big brains for? *Proceedings of the National Academy of Sciences of the USA*, **99**, 4141–2.

Sherry, D. F. and Schacter, D. L. (1987). The evolution of multiple memory systems. *Psychological Review*, **94**, 439–54.

Sherry, D. F., Vaccarino, A. L., Buckenham, K., and Herz, R. S. (1989). The hippocampal complex of food-storing birds. *Brain, Behavior and Evolution*, **34**, 308–17.

Sherry, D. F., Forbes, M. R. L., Kurgel, M., and Ivy, G. O. (1993). Females have a larger hippocampus than males in the brood parasitic brown-headed cowbird. *Proceedings of the National Academy of Sciences of the USA*, **90**, 7839–43.

Shettleworth, S. J. (1998). *Cognition, evolution, and behaviour*. Oxford: Oxford University Press.

Shettleworth, S. J. (2000). Modularity and the evolution of cognition. In *The evolution of cognition*. (ed. C. Heyes and L. Huber), pp. 43–60. Cambridge, MA: MIT press.

Sheu, F.-S., McCabe, B. J., Horn, G., and Routtenberg, A. (1993). Learning selectively increases protein kinase C substrate phosphorylation in specific regions of the chick brain. *Proceedings of the National Academy of Sciences of the USA*, **90**, 2705–9.

Sibley, G. C. and Ahlquist, J. E. (1990). *Phylogeny and classification of birds: A Study in molecular evolution*. New Haven, CT: Yale University Press.

Smulders, T. V. (2002). Natural breeding conditions and artificial increases in testosterone have opposite effects on the brains of adult male songbirds: A meta-analysis. *Hormones and Behavior*, **41**, 156–69.

Smulders, T. V. and DeVoogd, T. J. (2000). The avian hippocampal formation and memory for hoarded food: Spatial learning out in the real world. In *Brain, perception, memory. advances in cognitive neuroscience* (ed. J. J. Bolhuis), pp. 127–48. Oxford: Oxford University Press.

Sol, D. and Lefebvre, L. (2000). Forebrain size and foraging innovations predict invasion success in birds introduced to New Zealand. *Oikos*, **90**, 599–605.

Sol, D., Lefebvre, L., and Timmermans, S. (2002). Behavioural flexibility and invasion success in birds. *Animal Behaviour*, **63**, 495–502.

Solis, M. M., Brainard, M. S., Hessler, N. A., and Doupe, A. J. (2000). Song selectivity and sensorimotor signals in vocal learning and production. *Proceedings of the National Academy of Sciences of the USA*, **97**, 11836–42.

Templeton, J. J., Kamil, A. C., and Balda, R. P. (1999). Sociality and social learning in two species of corvids: The pinyon jay (*Gymnorhinus cyanocephalus*) and the Clark's nutcracker (*Nucifraga columbiana*). *Journal of Comparative Psychology*, **113**, 450–5.

Thorndike, E. L. (1911). *Animal intelligence*. New York, NY: MacMillan.

Thorpe, W. H.. (1943). A type of insight learning in birds. *British Birds*, **37**, 29–31.

Thorpe, W. H. (1944). Further notes on a type of insight learning in birds. *British Birds*, **38**, 46–9.

Thorpe, W. H. (1956). Records of the development of original and unusual feeding methods by wild passerine birds. *British Birds*, **49**, 389–95.

Timmermans, S. (1999). Opportunism and the neostriatal/ hyperstriatum ventrale complex in birds. M.Sc. Thesis, McGill University.

Timmermans, S., Lefebvre, L., Boire, D., and Basu, P. (2000). Relative size of the hyperstriatum ventrale is the best predictor of innovation rate in birds. *Brain, Behavior and Evolution*, **56**, 196–203.

Ward, B. C., Nordeen, E. J., and **Nordeen, K. W.** (1998). Individual variation in neuron number predicts differences in the propensity for avian vocal imitation. *Proceedings of the National Academy of Sciences of the USA*, **95**,1277–82.

Webster, S. J. and **Lefebvre, L.** (2000). Neophobia in the Lesser-Antillean Bullfinch, a foraging generalist, and the Bananaquit, a nectar specialist. *Wilson Bulletin*, **112**, 424–7.

Webster, S. J. and **Lefebvre, L.** (2001). Problem solving and neophobia in a Passeriforme-Columbiforme assemblage in Barbados. *Animal Behaviour*, **62**, 23–32.

Whiten, A. and **Byrne, R. W.** (1988). Tactical deception in primates. *Behavioral and Brain Sciences*, **11**, 233–73.

Whittle, P. J. (1996). The relationship between scramble competition and social learning: A novel approach to testing adaptive specialisation theory. M.Sc. Thesis, McGill University.

Wilson, A. C. (1985). The molecular basis of evolution. *Scientific American*, **253**, 148–57.

Wyles, J. S., Kunkel, J. G., and **Wilson, A. C.** (1983). Birds, behavior and anatomical evolution. *Proceedings of the National Academy of Sciences of the USA*, **80**, 4394–97.

Young, H. G. (1987). Herring gull preying on rabbits. *British Birds*, **80**, 630.

CHAPTER 3

BEHAVIOURAL INNOVATION: A NEGLECTED ISSUE IN THE ECOLOGICAL AND EVOLUTIONARY LITERATURE?

DANIEL SOL

Introduction

Living in a changing world, animals often have to face novel environmental conditions. Organisms are generally well adapted to common conditions but poorly adapted to novel ones, as traits that are adaptive with respect to one situation are unlikely to be so with respect to others (Hoffmann and Hercus, 2000). However, animals have, to different extents, the ability to respond to changes in the environment by inventing new behaviours or adjusting established behaviours to the novel conditions (Klopfer, 1962; Plotkin and Odling-Smee, 1979; Morse, 1980; Johnston, 1982; Lefebvre, 2000; Reader and Laland, Chapter 1). They may for instance develop novel anti-predatory responses when faced with new predators (e.g. Berger *et al.*, 2001), shift to new resources in cases of food shortage (e.g. Estes *et al.*, 1998) or adjust their breeding behaviour to the prevailing ecological conditions (e.g. Brooker *et al.*, 1998). Innovative behaviours[1] can be applied in many different contexts (Balda *et al.*, 1996) and may thus be an important part of the adaptive arsenal with which animals cope with the many different demands of a changing world.

The central theme of this chapter is that innovative behaviours are of considerable importance to understanding the ecology and, in turn, the evolution of animals. This has long been appreciated by some authors (Klopfer, 1962; Mayr, 1965; Fagen, 1982; Wyles *et al.*, 1983), but has been the subject of little empirical verification. Yet, the potential implications of variation in this form of behavioural flexibility lie at the heart of a key problem in ecology and evolution: Why are there so many kinds of organisms? From an

---

[1] Throughout this chapter I will use the following terminology: (1) innovative behaviour, a behaviour pattern that has been invented or modified to cope with an ecological demand on the animal; (2) innovative species, species with high propensities to respond to environmental challenges by inventing new behaviours or adjusting previous behaviour to the new conditions; and (3) behavioural flexibility, the general ability of an animal to invent new behaviour patterns or modify its behaviour in an adaptive way. These are operational definitions for purposes of clarity; more general definitions can be found in Reader and Laland (Chapter 1).

ecological perspective, innovatory propensity may influence biodiversity patterns through its influence on some of the processes that determine the gain and loss of species within the community. From an evolutionary standpoint, it may affect the rate of evolutionary divergence by altering the selective pressures to which individuals are exposed. The growing need to understand the consequences of innovative behaviours and novel improvements in methods to measure innovation have stimulated a number of recent studies. Here, I will use these new advances to illustrate the importance of behavioural innovation in some of the ecological and evolutionary processes that govern biodiversity. I will begin by examining the adaptive value of behaving in a flexible way and discuss new developments in the methods of quantifying innovatory capacity. Then, I will present a series of comparative studies that seek to understand the role of this capacity in three different biological processes: the invasion of new environments, the risk of extinction, and the rate of evolutionary diversification. Finally, I will advance some testable predictions in order to encourage future research.

The adaptive value of innovative behaviours

The important role of behavioural flexibility as an adaptive response to environmental changes is suggested by a number of case studies (e.g. Brooker *et al.*, 1998; Estes *et al.*, 1998; Dukas and Bernays, 2000; Berger *et al.*, 2001). In the North American Yellowstone region, for example, moose *Alces alces* have been unfamiliar with wolves *Canis lupus* for at least 50 years. Although moose are highly vulnerable to this re-colonising predator, within a single generation they have developed behavioural adjustments to reduce predation (Berger *et al.*, 2001). For instance, moose mothers who have lost juveniles to re-colonising wolves develop hypersensitivity to wolf howls. Such rapid behavioural adjustments may have prevented the extinction of the population.

Innovative behaviours may also provide benefits by allowing animals to exploit resources in completely novel ways. One well-known example is the practice of bait fishing observed in some bird species (Lefebvre *et al.*, 2002). In the green heron *Butorides striatus*, for example, some individuals drop small objects onto the surface of the water. The objective is to lure hungry fish that hope that the object is a prey item to the surface, where the fish may be easily snatched by the heron.

Animals with a high propensity for innovative behaviours have a great potential for modifying their ecological niche and entering new adaptive zones. An example is provided by the detailed studies of Terkel and collaborators of social learning in black rats *Rattus rattus* (Terkel, 1996). In Israel, most natural oak and pine forests were cut down around the end of the nineteenth century, and have subsequently been replaced with Jerusalem pines. Pinewoods provide a valuable food source, pine seeds, but many animals present in the region cannot take advantage of this resource because of their inability to open pine cones. However, black rats have learned to open the cones, which coupled with cultural transmission, has allowed them to extend their range of habitats to include Jerusalem pine forests.

The ecological significance of innovative behaviours may on occasion go beyond the ecology of a single species and extend to the ecosystem level. For example, the sea otter population in western Alaska Aleutian island has crashed by 90 per cent since 1990.

Figure 3.1 In Alaska, killer whales have recently shifted to feeding on sea otters as a result of a decline in their traditional prey, seals and sea lions. This behavioural innovation is having a devastating effect on the whole ecosystem. (Photograph courtesy of Cristina Sanchez.)

The reason appears to be that killer whales *Orcinus orca* (Figure 3.1), never before known to eat sea otters, have shifted to this novel prey as a result of a decline in their traditional prey: seals and sea lions (Estes *et al.*, 1998). This has had devastating effects on the ecosystem because, by eating sea urchins, otters contribute to the preservation of the kelp forests that sustain the local diversity of species. These changes are likely to affect the evolutionary dynamics of the entire system.

The studies described previously show that important progress can be made in our understanding of the consequences of behavioural flexibility by a detailed examination of the relationship between ecology, evolution, and innovative behaviours in single species. These studies indicate that innovations may have important ecological and evolutionary consequences through their influences on the way animals use resources and interact with other species (Figure 3.2). However, determining general principles regarding the relationships between ecology, evolution, and innovative behaviours that apply across taxa requires a comparative approach, based upon information drawn from a variety of species that differ in their propensity for innovative behaviours. This task is by no means easy but, as I will show, it is possible.

Do animals differ in their innovative propensities?

Understanding general ecological and evolutionary consequences of innovative behaviours requires one to demonstrate, first, that such flexibility consistently varies among lineages and, second, that this variation is associated with variation in ecological and evolutionary factors. Attempting to solve the first aspect has proved elusive, due essentially to the

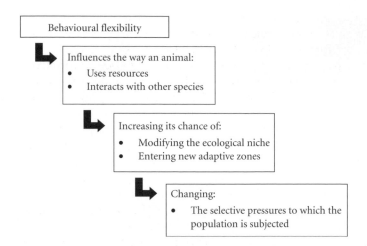

Figure 3.2 Behavioural flexibility may shape the ecology and evolution of animals through its influence on the way they exploit the resources and interact with other species.

difficulties of quantifying differences in innovatory capacity among taxa. Yet recent progress suggests that innovative propensities may be measured (Lefebvre and Bolhuis, Reader and MacDonald, Chapters 2 and 4). The most direct quantitative measure of innovation available is the rate of reported behavioural innovation (Lefebvre *et al.*, 1997). This measure is based on exhaustive reviews of the short note and other sections of journals, noting those behaviours that are considered new or unusual in the repertoire of species (Lefebvre and Bolhuis, Reader and MacDonald, Chapters 2 and 4; see also Lefebvre *et al.*, 1997, 1998; Timmermans *et al.*, 2000; Nicolakakis and Lefebvre, 2001). The rationale of the measure is that, all other things being equal, a species that eats a wider range of food types, that more readily exploits new feeding possibilities and that uses more sophisticated handling techniques should yield a higher frequency of innovations. The use of innovation rate to operationalize behavioural flexibility is not exempt from problems. The frequency of behavioural innovation is based on opportunistic observations and may thus be prone to potential biases. For example, the frequency of innovations may be influenced by differences in the research effort devoted to different species. Yet one may minimize these problems by considering the potential confounding effects in the analysis. Reporting bias may, for instance, be controlled by including in the analysis the number of published papers for each species. Other possible biases, like phylogenetic inertia, population size, or editorial bias in journals, appear to have a negligible effect on the measure (Nicolakakis and Lefebvre, 2000; Lefebvre and Bolhuis, Chapter 2). The use of behavioural innovations to quantify behavioural flexibility was first developed for foraging behaviours in birds by Lefebvre *et al.* (1997), but the idea has recently been extended to primates by Reader and Laland (2002; see also Reader and MacDonald, Chapter 4; Reader in press).

In addition to innovation rate, the size of the brain can also be used as an indirect measure of behavioural flexibility, after controlling for the allometric effect of body size (see Reader and MacDonald, Chapter 4). In vertebrates, learning and cognitive capacities are thought to be related to the size of the neural structures, on the grounds that enhanced neural volumes allow the increased information processing required for increased cognitive capacities (Jerison, 1973; Johnston, 1982; Madden, 2001). In birds, for example, there is good evidence suggesting that the volume of the song nuclei is correlated with birdsong repertoire size (e.g. DeVoogd *et al.*, 1993), hippocampus size with spatial memory (e.g. Krebs *et al.*, 1989), and total brain size with bower complexity (Madden, 2001). These studies focus on specific behaviours, but recent evidence suggests that there is also a link between the size of neural structures and general behavioural flexibility. In birds, variation in relative brain size, in particular the *hyperstriatum ventrale* and *neostriatum*, is correlated with the rate of foraging innovation and learning ability (Lefebvre and Bolhuis, Chapter 2; Lefebvre *et al.*, 1997, 1998; Timmermans *et al.*, 2000; Nicolakakis and Lefebvre, 2001). Similarly, Reader and Laland (2002) have reported that the relative size of the 'executive' brain (neocortex, striatum) is positively associated with innovation frequency, social learning and tool use in primates.

These recent developments have uncovered important differences between species and higher taxonomic levels in their degree of behavioural flexibility (e.g. Lefebvre, 1997, 1998; Reader and Laland, 2002). For example, groups like crows (Corvidae) or parrots (Psittacidae) have big brains and tend be highly innovative, while others like pheasants (Phasianidae) or doves (Columbidae) have smaller brains and are generally less innovative. The existence of important differences in the degree of behavioural flexibility suggests that lineages respond differentially to changes in the environment. Now, the question is: does this variation in behavioural flexibility correspond to variation in the ecological and evolutionary processes that govern biodiversity?

In the following sections, I will describe a series of comparative studies in birds using foraging innovations and brain size to test several ecological and evolutionary hypotheses. The adoption of two different ways to measure the ability for behavioural change has two major advantages in these studies. First, identifying consistent patterns with two different measures increases our confidence that we are observing a robust pattern. Second, foraging innovations and brain size appear to measure different aspects of flexible behaviour. Variation in innovation frequency is mostly found at low taxonomic levels (Figure 3.3), suggesting that this variable largely reflects the current degree of foraging flexibility of a species. Conversely, variation in brain size is mostly concentrated at higher taxonomic levels (Figure 3.3) and thus appears to describe differences in cognitive abilities that evolved in the early diversification of avian lineages. Therefore, the use of the two measures can facilitate a broader understanding of the ecological and evolutionary significance of behavioural flexibility.

Ecological implications of behavioural innovation

Invasion of novel environments

Biological invasions are central to the understanding of species diversity. This is not only because the ecological diversity in a given region depends on the number of species that have been able to invade that region, but also because the invasion of novel environments

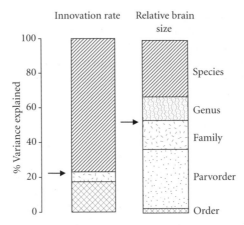

Figure 3.3 Distribution of variation in innovation rate and relative brain size throughout the taxonomy in Palearctic passerines. The arrows separate the lowest taxonomic levels (species and genus) from the highest taxonomic levels (families, orders, and parvorders). Taxonomic levels are shown in varying shades, according to the key given on the right. The percentage of variance accounted for by each taxonomic level has been estimated with a nested ANOVA (Harvey and Pagel, 1991).

is known to play a major role in extinction and evolutionary diversification processes (see below). Ecologists recognize that species differ enormously in their ability to establish in novel regions (Mayr, 1965; Ehrlich, 1989), but until recently the reasons behind such differences remained poorly understood.

A species that invades a new environment will generally face many different and novel environmental challenges. Differences in invasion success will thus depend to a large extent on whether the invader can rapidly cope with these new challenges (Morse, 1980). Invasions generally involve a small number of individuals, and while the population remains low it is highly vulnerable to stochastic extinction. Rapid behavioural adjustments to adopt new food sources or avoid predation may therefore be essential to prevent the extinction of the population. Not surprisingly, behavioural flexibility has long been argued to be a characteristic of successful invaders (Mayr, 1965; Morse, 1980; Greenberg, 1990). For example, in his diagnostic of avian invaders Mayr (1965) argued that a successful invader should show ecological flexibility, a tendency to discover unoccupied habitats, and an ability to shift habitat preferences. Likewise, Morse (1980) hypothesized that bird species vary in the degree to which they are able to respond to new resources through learned changes and suggested that these differences may determine how readily birds colonize islands.

It is easy to imagine a number of ways through which the ability to invent or modify behaviours may be beneficial during invasions. For instance, the adoption of new resources may increase the probability of surviving temporary food shortages, as well as reduce the chance that other species are already exploiting the available resources. Rapid behavioural adjustments to avoid predation or nest parasitism might also be important to increase the average individual fitness during the early stages of invasion, and thus prevent the extinction of the population. Finally, because innovative species are likely to be more exploratory (Greenberg and Mettke-Hofmann, 2001), they may also have higher chances of discovering new habitats or new resources that may be important to survive and reproduce in the novel environment. Nevertheless, despite sound theoretical arguments the hypothesis that behavioural flexibility may influence invasions has remained empirically untested.

How can we test the hypothesis that behaviourally flexible species are better invaders than less flexible ones? An ideal testing method would consist of introducing, under identical conditions, species with varying degrees of flexibility in places where they did not occur before. Unfortunately, the ecological problems often associated with exotic species, as well as methodological issues (Pimm, 1991), often preclude this type of experiment. As an alternative, we can rely on past, human-driven introductions. Birds offer particularly good opportunities to study the factors influencing invasions, as a long-standing tradition in ornithological reporting provides an impressive quantitative data set of introduced species and the biotas they have or have not been able to invade (Long, 1981). In New Zealand, for example, over 140 species of birds were introduced at the end of the nineteenth century by acclimatization societies, 34 of which were able to successfully establish themselves on the islands (Long, 1981). Sol and Lefebvre (2000) used the New Zealand introduction data to test the hypothesized link between behavioural flexibility and invasion success. If innovative behaviours represent an adaptive way to deal with new challenging conditions, one would expect innovative species to be more successful at establishing themselves in New Zealand than less flexible species. As predicted, successful species tend to have relatively larger brains and to be more innovative in their area of origin than unsuccessful species (Figure 3.4(a)). This trend is not biased by the phylogeny or by other possible correlates of invasion success, such as number of individuals released, migratory habits, mode of development, or plumage dichromatism.

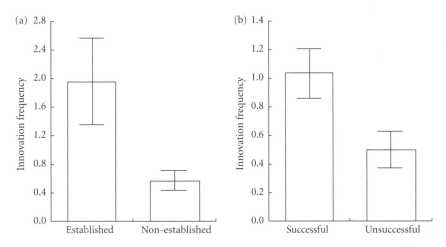

Figure 3.4 Avian innovation frequency in species' area of origin compared with their introduction success in New Zealand (a) and to different regions of the world (b). [Based on data from Sol and Lefebvre (2000) and Sol *et al.* (2002).]

Because the previous study is based on introductions carried out in a single geographic zone, the results allow little generalization. A species that succeeds at establishing itself in a particular region is not necessarily a successful invader in all habitats, as some traits may simply pre-adapt it to live in that particular region. Blackburn and Duncan (2001), for example, have recently shown that species are more likely to succeed at invading regions when the new environmental conditions are similar to those the species encounter in its native range. In light of this possibility, Sol *et al.* (2002) conducted a second test examining introductions of species across the globe. The results of this study confirm and generalize the previous findings: successful invaders tend to be more innovative and larger-brained than unsuccessful invaders (Figure 3.4(b)). For example, species like the House Sparrow *Passer domesticus* are innovative, relatively large-brained, and tend to be successful invaders, whereas others like the Common Quail *Coturnix coturnix* are less innovative, relatively small-brained, and tend to be unsuccessful invaders.

Extinction risk

Contemporary species represent just over 0.1 per cent of the species that have existed on earth. All other species are now extinct and many contemporary species will also die out in the coming years, many as a result of human perturbations. Extinction is thus a major determinant of species diversity.

While it is widely recognized that not all species are equally vulnerable to extinction risk (Bennett and Owens, 1997, 2002; McKinney, 1997), the causes of such differences are just beginning to be understood (see Bennett and Owens, 2002). Ecological theory makes a number of predictions regarding the characteristics of species that correlate with their risk of extinction. This predicts, for example, that the smaller the geographical range, the more likely it is that the species will go extinct (Rosenzweig, 1995). The role of behaviour in extinctions has not generally been considered, yet what evidence is available indicates that in some species behaviour can play a critical role (Arcese *et al.*, 1997; Sutherland, 1998; Reed, 1999). Because extinctions are often caused by changes in the environment, one could argue that risk should be lower for innovative species than for less flexible ones. For example, high flexibility may increase the ability of a species to shift to other resources or even move to more favourable localities when faced to changes in their niche. The predicted larger distribution of flexible species (see below) could also, if confirmed, make them less vulnerable to extinctions than less flexible species (Rosenzweig, 1995; McKinney, 1997). The link between behavioural innovation and extinction risk is supported in a number of case studies. The story of the Mauritius Kestrel *Falco punctatus*, one of the most endangered birds in the world, is particularly illustrative. The decline of the species began with the introduction of long-tailed macaques *Macaca fascicularis* in Mauritius (Arcese *et al.*, 1997). Kestrels nested in tree-holes and, after Macaques were introduced, they experienced high rates of nest failure. The situation became particularly dramatic in 1972, when the population was reduced to only two pairs. Fortunately, the picture changed substantially in 1974; that year, one pair nested in a cliff and successfully raised a nestling. Since then, many other individuals have nested and raised their offspring on the cliffs. The behavioural innovation of cliff-nesting has contributed to a partial recovery of the population and could save the species from extinction.

More general, although indirect, support for a link between behavioural innovation and extinction risk comes from the finding that innovation ability influences introduction success. Invaders may be viewed as populations faced with a novel situation to which the species may be unable to adapt, resulting in extinction. Hence, variation in introduction success between species may also be revealing differences in extinction risk.

The relationship between behavioural innovation and extinction risk has only recently been directly addressed in comparative studies, and to date there is little evidence in support of this link: highly flexible lineages, measured in terms of relative brain size and foraging innovation rate, do not tend to contain fewer endangered or extinct species than less flexible lineages (Nicolakakis *et al.*, 2003). This result may be interpreted as showing that innovative behaviours do not influence extinction risk or, alternatively, that other factors have a stronger effect than innovative behaviours on extinction risk. This latter possibility is plausible because we know that the causes of extinction, not considered in Nicolakakis *et al.* (2003), are multiple and vary geographically (Manne *et al.*, 1999). A recent analysis by Owens and Bennett (2000), for instance, has revealed that habitat generalism is associated with extinction risk incurred by habitat loss, but not with that incurred through persecution and introduced predators. Consequently, to clarify the exact relationship between behavioural flexibility and extinction risk, we need additional tests that take into account the causes of extinction. Indeed, behaviourally flexible species might be expected to be more tolerant to habitat loss than less flexible species, but not necessarily more tolerant to human persecution (see also Reader and MacDonald, Chapter 4).

Evolutionary implications of behavioural flexibility

Lineages vary considerably in species richness and this pattern is not simply a consequence of random branching patterns (e.g. Owens *et al.*, 1999). Recent studies on a wide array of taxa suggest that the evolutionary processes associated with shifts in ecology or invasion of new habitats can cause extremely rapid evolutionary divergence (Smith and Skúlason, 1996; Orr and Smith, 1998; Price, 1999). For example, species that invade new regions may diverge rapidly from ancestors through a combination of divergent natural selection, genetic drift, divergence under uniform selection, and geographic isolation interrupting gene flow with ancestors. Because the ability to produce ecological shifts or to invade new habitats is predicted to be higher for innovative species, it can be argued that behaviourally flexible species should evolve at a faster rate than less flexible species (see Mayr, 1963; Wyles *et al.*, 1983; Fitzpatrick, 1988; Greenberg, 1990). Wyles *et al.* (1983) noted, for instance, that the rapid adaptive radiation found in birds and mammals compared to other vertebrate groups, coincides with their relatively large brain sizes and their more developed social learning skills. However, this argument cannot be used as a satisfactory proof in support of a link between behavioural innovation and evolutionary rate, as these vertebrate classes differ in many other aspects in addition to their behavioural flexibility. Moreover, some theoreticians predict the contrary pattern, that is, a reduction instead of an increase in the rate of evolution in species with high ability for behavioural change (Lynch, 1990; Robinson and Dukas, 1999). For example, the strength of natural selection could be reduced if individuals tend to respond to new ecological challenges by means of behavioural modifications rather than by heritable variation (Robinson and Dukas, 1999).

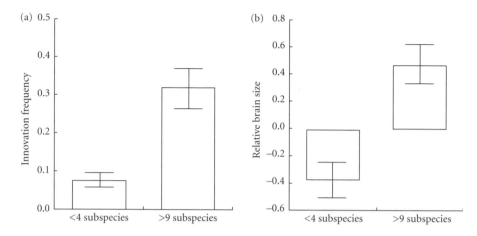

Figure 3.5 Innovation rate (controlling for differences among species in research effort) and relative size of the brain in Palearctic passerines that differ in the degree of subspecies diversification (GLM with Poisson error, $P < 0.004$ in both cases). Sol and Lefebvre, unpublished data.

Recent analyses of birds provide the first evidence in support of the hypothesis that innovative behaviours enhance the rate of evolutionary diversification. In Palearctic passerines, preliminary results indicate that species with a higher frequency of foraging innovations and bigger brains have undergone a more extensive subspecies diversification than less innovative, small-brained species (Sol and Lefebvre, unpublished data; Figure 3.5). Similar results were obtained in an analysis of Nearctic species (Sol and Lefebvre, unpublished data). Subspecies are recognisable geographical units within a species (Roselaar, 1995) and are thought to represent early stages of morphological differentiation that may on occasions lead to speciation (Møller and Cuervo, 1998). Thus, a species containing many subspecies is assumed to have undergone a greater evolutionary divergence than another with few subspecies.

If the propensity to innovate enhances the rate of evolutionary diversification, why are most species not highly flexible in behaviour? In fact, they are. Species-rich parvorders have been found to be more innovative (after adjusting for differences in research effort) and have relatively larger brains than species-poor parvorders (Nicolakakis et al., 2003; Figure 3.6). Assuming that the number of species reflects speciation processes rather than variation in extinction rates, which seems valid in this case (Nicolakakis et al., 2003), these results suggest that behavioural flexibility can enhance evolutionary rate. Even when there appear to be a trend for increased number of species in innovative lineages, there are still many lineages that thrive despite containing species with relatively unsophisticated cognitive abilities. Although a high ability for behavioural innovation may be beneficial under certain ecological contexts, such as changing environmental conditions, it may be costly under other conditions (Lefebvre, Reader and Sol, unpublished data). Costs of reliance on innovative behaviours may include less efficiency in exploiting specific resources, increased exposure to interspecific interactions (e.g. predators and parasites),

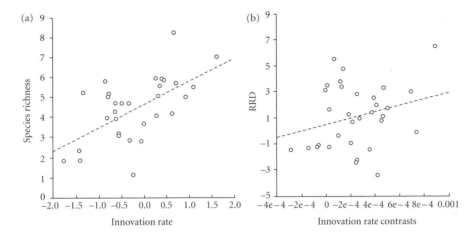

Figure 3.6 (a) Relationship between innovation rate (adjusting for research effort), and species richness across avian orders and parvorders (non-contrast data, based on Nicolakakis *et al.* 2003), reprinted with permission from Elsevier. (b) Relationship between innovation rate (adjusting for research effort) and species richness across avian orders and parvorders, after controlling for phylogenetic effects. In contrast to Nicolakakis *et al.* (2003), independent contrasts were re-calculated with a program designed to study correlates of species richness (MacroCAIC, Agapow and Isaac, 2002). Relative rate difference (RRD) is the index of diversity difference between sister taxa. Using MacroCAIC, the regression of innovation rate against species richness through the origin is significant ($P = 0.007$), replicating the results of Nicolakakis *et al.*'s (2003) independent contrast analysis.

and high risk of eating poisoned food. In addition to costs, there are probably some constraints associated with the cognitive capacities necessary to innovate (Lefebvre, Reader and Sol, unpublished data). For example, large brains are energetically expensive to produce and maintain and require lengthy development periods, which could constrain the evolution of flexibility in animals that rely on energetically poor food sources or live in regions with high predatory pressure. The cost and constraints of innovative behaviour are an important area for future research.

Future prospects

In the previous sections, I have discussed the few comparative studies that have to date explored the link between innovative behaviours, ecology and evolution. Below, I provide a number of hypotheses and predictions for possible future investigation (see Table 3.1). Some of these hypotheses will require refinements and, in some cases, even reformulation. However, I advance these ideas with the hope that this will stimulate further thinking and new research. The topics include niche expansion, geographic range size, migratory behaviours, population regulation, and species coexistence.

Niche expansion

Fundamental to understanding biodiversity is the need to determine why there is not an ultimate organism adapted to exploit all ecological niches (Timms and Read, 1999).

Table 3.1 Predictions concerning the importance of innovative behaviour for the ecology and evolution of animals, provided that all other effects are held constant (see also Figure 3.7). Comparative studies supporting and not supporting the predictions are presented. Note that many predictions have yet to be tested

Predictions	Supporting evidence	Non-supporting evidence
Link 1: Niche breadth		
Innovative species are predicted to have wider diets and foraging techniques than less innovative species	Timmermans (1999)	—
They should also be able to occur in wider variety of habitats than less innovative species	—	—
Link 2: Invasion of novel environments		
Innovative species should be more successful invaders than less innovative species	Sol and Lefebvre (2000), Sol *et al.* (2002)	—
Link 3: Geographical range size		
Innovative species should attain larger geographic ranges than less innovative species	—	Reader and MacDonald (Chapter 4)
Species that undergo geographical range expansions should be more innovative than those that do not show similar expansions	—	—
Link 4: Demographic factors		
Innovative species should be less subjected to - density dependent mortality and/or birth rate than less innovative species	—	—
Populations of innovative species should be less variable than those of less innovative species	—	—
Link 5: Migratory habits		
Migratory species should be less innovative than resident species	—	—
Migratory habits should have mostly evolved in lineages that are poor innovators	—	—
Link 6: Species coexistence		
Innovative species are predicted to have a higher modulating effect on resources available for other species than less innovative species	—	—
Those species should also experience more interspecific interactions than less innovative ones	—	—
Innovative species should be less efficient in using specific resources than more specialized species	—	—
Link 7: Extinction rate		
Innovative species should be less vulnerable to extinction than less innovative species	—	Nicolakakis *et al.* (2003)
Innovative species should be less vulnerable to extinction due to changes in the environment, but not necessarily for other causes	—	—

(*Continued*)

Table 3.1 (*Continued*)

Predictions	Supporting evidence	Non-supporting evidence
Link 8: Evolutionary rate		
Innovative species should show a higher morphological and physiological divergence than less innovative species	Sol and Lefebvre (unpublished data)	—
Innovative lineages should speciate at a faster rate than less innovative lineages	Nicolakakis *et al.* (2003)	—

Some authors, notably Klopfer (1962; Klopfer and MacArthur, 1960; Sheppard *et al.*, 1968) and Greenberg (1983, 1990), have suggested that the ability of species to incorporate additional habitats or resources depend partially upon their degree of behavioural flexibility. The 'Neophobia Threshold Hypothesis' suggests, for example, that individuals or species that avoid approaching and exploring novel objects are less likely to incorporate new foods, as specialists, than those that are more flexible, which would tend to behave as generalists (Greenberg, 1990, Chapter 8; but see Marples and Kelly, 1999). Behavioural innovation is, in the final analysis, the way animals incorporate new resources in their habitual repertoire, and hence it may be reasoned that species with a high propensity for innovative behaviours should show wider ecological niches than those that are less innovative. There is indeed some comparative data that support the hypothesis. For example, in the North American warblers studied by Greenberg (1979), species that use a higher repertoire of foraging techniques in winter are also those for which more innovative behaviours have been reported in the ornithological literature (Timmermans, 1999). However, broader comparative evidence is required. It is worth noting that one would not expect a perfect correspondence between innovative behaviours and niche breadth; this is because the degree of specialization may depend upon other factors than innovative ability, such as the degree of morphological and physiological specialization, the availability of resources, and the current level of interspecific competition (Morse, 1980; Greenberg, 1983). In Darwin finches, for example, the ability to shift diet during a severe dry period is mostly constrained by bill morphology, which permits individuals to efficiently exploit some types of seeds but not others (Grant and Grant, 1989). Other investigators (e.g. Werner and Sherry, 1987) have shown that individuals may behave as specialists even in species that have a broader feeding niche. More research is needed to clarify the exact relationship between the degree of behavioural flexibility and the specialist–generalist continuum, an aspect that is central to understanding many other ecological and evolutionary issues.

Geographic range size

Why do some species have small geographic ranges whereas others occur almost everywhere? From an ecological point of view, the geographic distribution of a species is the result of a combination of many physical and biotic variables that are required for the sur-vival and reproduction of its individuals. It is easy to imagine that a flexible species that can cope with a wide array of conditions and acquire sufficient resources so as to reach high population densities, should also be able to expand its range and occur in many more places than a less flexible species. The finding that innovative species are successful invaders clearly points in this direction. One can therefore argue that, all other things being equal, behaviourally flexible species should have larger geographical ranges than less flexible species. This prediction has recently been tested in primates (Reader and MacDonald, Chapter 4), and the results are not consistent with the hypothesis. More comparative tests are required.

Migratory movements

Migrations are extensive movements that animals undergo as a response to periodic changes in the environment. Migratory behaviours are pervasive phenomena in animals and affect a number of important ecological processes, such as the invasion of new envir-onments (Veltman et al., 1996), the risk of extinction (Berthold, 1993), and the rate of evo-lutionary diversification (Belliure et al., 2000). However, the adaptive reasons that lead some species to undergo extensive migratory movements are incompletely understood (Bennett and Owens, 2002).

In birds, migrations appear to be a response to seasonally changing conditions in species where individuals survive better by moving out for the winter than by staying in their breeding areas year round (Lack, 1954). The most common reason that species cannot stay in the breeding areas in winter is shortage in food sources (e.g. Jansson et al., 1981; Newton and Dale, 1996). It is reasonable to expect that an animal that specializes in a reduced variety of food types, that rarely exploits new feeding possibilities and that shows little tendency to develop new foraging techniques should be less able to cope with contrasting seasonal changes in environmental conditions than a more flexible species. A prediction is then that migratory habits should have evolved in lineages that lack sufficient foraging plasticity to cope with the demands of seasonal changes in environmental conditions. Extensive mobility would thus be seen as a way to reduce variation in a key environmental variable (food availability) that cannot be afforded by behavioural adjustments.

Demographic factors

The propensity to innovate might also have implications for demographic factors. Specifically, it could lead to a reduction of the density-dependence of mortality and/or birth rate of the species' own population (see Dall and Cuthill, 1997). The idea is that an innova-tive species is less likely to experience a reduction in both the influence of a particular prey type on its own population dynamics relative to the appropriate specialists. MacArthur (1955), for example, predicted that populations of specialists should be more variable than populations of generalists, because they are more susceptible to population crashes brought on by the vagaries of weather acting upon their single food resource. Density-dependence in demographic parameters is thought to be important in determining population stability, which in turn may affect the community as a whole (Pimm, 1991). The relationship between behavioural flexibility and population stability has received some credence from theoretical studies (Wilson and Yoshimura, 1994), but has never been empirically tested.

Species coexistence

Diversity at an ecosystem scale depends on the ability of the different species to coexist in the same community, despite the effect of competition, predation, or parasitism. Ecologists have tried to identify the factors that make possible the coexistence of generalists and specialists (Rosenzweig, 1995). Some authors have suggested that the potential for coexistence is critically dependent on the flexible habitat choice of species (e.g. Wilson and Yoshimura, 1994). Generalists tend to perform worse in a given habitat than specialists, but are less vulnerable to changes in the habitat because of their increased ability to use alternative habitats. When resources in one habitat become scarce, an innovatory species may be more likely to shift to another habitat than a less flexible species. However, by shifting to a different habitat, generalists favour coexistence because they reduce the level of competition experienced by the specialist.

Concluding remarks

Attempting to study the ecological and evolutionary consequences of flexible behaviours has proved difficult, reflected by the fact that some early hypotheses have remained untested for at least 30 years. Recent times have, however, seen significant developments in measuring innovatory capacity. Coupled with numerous advances in phylogenetic-based analyses (e.g. Harvey and Pagel, 1991; Blackburn and Duncan, 2001), these developments have allowed tests of some of the predictions in a comparative framework. Innovatory propensity has been shown to have important ecological and evolutionary consequences (Figure 3.7, Table 3.1). For instance, a high flexibility in behaviour appears to increase the ability of species to invade new environments (Sol and Lefebvre, 2000; Sol et al., 2002). Evidence also exists supporting the idea that innovatory propensity may, by virtue of the high 'versatility' that it confers to exploit new ecological contexts, act as a 'pacemaker' of evolution (Nicolakakis et al., 2003; Sol and Lefebvre, unpublished data). In addition, theory predicts that innovatory capacities will have an influential role in several other fields, influences that are yet to receive empirical attention. These include extinction risk, niche breadth, geographical distribution, migratory habits, population dynamics, and species coexistence.

The potential of behavioural innovation as a mechanism to influence ecological processes in some taxa has broad implications for our understanding of species diversity, both at ecological and evolutionary levels, and should therefore be incorporated in a more general framework. Such a framework would be more consistent with the increasingly recognized view that animals are not passive agents of selection but, by virtue of their behaviour, actively determine in part their own environment and the external agents of selection that impinge upon them (Lewontin, 1978; Odling-Smee et al., 1996; Futuyma, 1998; Laland et al., 2000). Obviously, the research agenda that seeks to discover general principals in the relationship between behavioural innovation, ecology and evolution is at an early stage and presently we have more questions than answers. Yet, the perspectives for

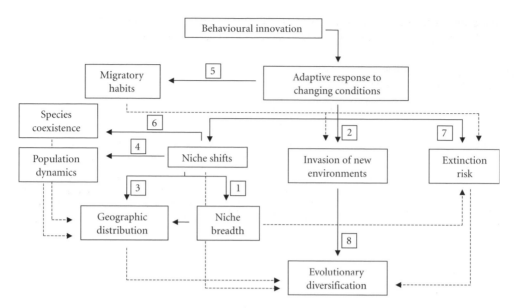

Figure 3.7 Hypothesized ecological and evolutionary consequences of variation among animals in innovatory capacity. Numbers correspond to the predictions depicted in Table 3.1. Direct and indirect links are represented with solid and discontinuous lines, respectively.

future research are promising, and it is unlikely we have to wait another 30 years before the ecological and evolutionary consequences of variation in innovatory propensity are better understood.

Summary

Innovative behaviours are an important means by which animals cope with changes in environmental conditions. They may significantly shape their ecology and evolution by influencing the way animals use the resources and interact with other species. While the ecological and evolutionary significance of this form of behavioural flexibility has long been appreciated by ecologists and evolutionary biologists, attempting to unravel general rules with a comparative approach has confronted the problem of quantifying differences in innovatory propensity among taxa. New comparative evidence based on recent developments in methods to quantify innovatory propensities indicate that variation in such behavioural flexibility does indeed affect a number of ecological and evolutionary processes, such as the invasion of new environments or the rate at which lineages evolve. Theory also predicts an influential effect of innovatory propensity on other factors for which evidence is inconclusive or not yet assembled, including extinction risk, niche breadth, geographic distribution, migratory habits, population regulation, or species coexistence. Behavioural innovation should be regarded as a mechanism that influences the ecological and evolutionary processes that govern diversity patterns, and it should therefore be recognized as playing a more prominent role in ecological and evolutionary theories.

Acknowledgements

This work has benefited from commentaries from or discussion with L. Lefebvre, S. M. Reader, J. D. Ryan, J. Morand-Ferron, D. Réale, N. Nicolakakis, K. N. Laland, R. Jovani, J.-D. Rodriguez-Teijeiro, Ll. Brotons, the members of the McGill and Université du Québec à Montréal Behaviour & Ecology discussion groups, and two anonymous referees. I thank K. N. Laland for organizing the interesting symposium of animal innovation in the I. E. C. of Tübingen that prompted this chapter. The work has been supported by a Québec Ministry of Education Postdoctoral fellowship, and NSERC (Canada) and FCAR (Québec) grants to L. Lefebvre.

References

Agapow, P.-M. and **Isaac, N. J. B.** (2002). MacroCAIC: Revealing correlates of species richness by comparative analysis. *Diversity and Distributions*, **8**, 41–3.

Arcese, P., Keller, L. F., and **Cary, J. R.** (1997). Why hire a behaviorist into a conservation or management team? In *Behavioral approaches to conservation in the wild* (ed. J. R. Clemmons and R. Buchholz), pp. 48–71. Cambridge: Cambridge University Press.

Balda, R. P., Kamil, A. C., and **Bednekoff, P. A.** (1996). Predicting cognitive capacity from natural history: Examples from four species of corvids. In *Current ornithology* (ed. V. Nolan, Jr. and E. D. Ketterson), pp. 33–66. New York: Plenum Press.

Belliure, J., Sorci, G., Møller, A. P., and **Clobert, J.** (2000). Dispersal distance predict subspecies richness in birds. *Journal of Evolutionary Biology*, **13**, 480–7.

Bennett, P. M. and **Owens, I. P. F.** (1997). Variation in extinction risk among birds: Chance or evolutionary predisposition? *Proceedings of the Royal Society of London, Series B*, **264**, 401–8.

Bennett, P. M. and **Owens, I. P. F.** (2002). *Evolutionary ecology of birds*. Oxford: Oxford University Press.

Berger, J., Swenson, J. E., and **Persson, I.-L.** (2001). Recolonizing carnivores and naive prey: Conservation lessons from Pleistocene extinctions. *Science*, **291**, 1036–9.

Berthold, P. (1993). *Bird migration: A general survey*. Oxford: Oxford University Press.

Blackburn, T. M. and **Duncan, R. P.** (2001). Determinants of establishment success in introduced birds. *Nature*, **414**, 195.

Brooker, M. L., Davies, N. B., and **Noble, D. G.** (1998). Rapid decline of host defences in response to reduced cuckoo parasitism: Behavioural flexibility of reed warblers in a changing world. *Proceedings of the Royal Society of London, Series B*, **265**, 1277–82.

Dall, S. R. X. and **Cuthill, I. C.** (1997). The information costs of generalism. *Oikos*, **80**, 197–202.

DeVoogd, T. J., Krebs, J. R., Healy, S. D., and **Purvis, A.** (1993). Relations between song repertoire size and the volume of brain nuclei related to song: Comparative evolutionary analyses amongst oscine birds. *Proceedings of the Royal Society of London, Series B*, **254**, 75–82.

Dukas, R. and **Bernays, E. A.** (2000). Learning improves growth rate in grasshoppers. *Proceedings of the National Academy of Sciences, USA*, **97**, 2637–40.

Ehrlich, P. H. (1989). Attributes of invaders and the invading processes: Vertebrates. In *Biological invasions: A global perspective* (ed. J. A. Drake, H. A. Mooney, F. di Castri, R. H. Groves, F. G. Kruger, M. Rejmánek, and M. H. Williamson), pp. 315–28. Chichester: Wiley.

Estes, J. A., Tinker, M. T., Williams, T. M., and Doak, D. F. (1998). Killer whale predation on sea otters linking oceanic and nearshore ecosystems. *Science*, **282**, 473–6.

Fagen, R. (1982). Evolutionary issues in development of behavioral flexibility. In *Perspectives in ethology* (ed. P. P. G. Bateson and P. H. Klopfer), pp. 365–83. New York: Plenum Press.

Fitzpatrick, J. W. (1988). Why so many passerine birds? A response to Raikow. *Systematic Zoology*, **37**, 76.

Futuyma, D. J. (1998). *Evolutionary Biology*, 3rd edn. Sunderland, MA: Sinauer Associates.

Grant, B. R. and Grant, P. R. (1989). Natural selection in a population of Darwin's finches. *American Naturalist*, **133**, 377–93.

Greenberg, R. S. (1979). Body size, breeding habitat and winter exploitation systems in *Dendroica*. *The Auk*, **96**, 756–66.

Greenberg, R. S. (1983). The role of neophobia in determining the degree of foraging specialization in some migrant warblers. *American Naturalist*, **122**, 444–53.

Greenberg, R. S. (1990). Ecological plasticity, neophobia, and resource use in birds. *Studies in Avian Biology*, **13**, 431–7.

Greenberg, R. S. and Mettke-Hofmann, C. (2001). Ecological aspects of neophobia and neophilia in birds. In *Current ornithology* (ed. V. Nolan, Jr.), pp. 119–78.

Harvey, P. H. and Pagel, M. D. (1991). *The comparative method in evolutionary biology*. Oxford: Oxford University Press.

Hoffmann, A. A. and Hercus, M. J. (2000). Environmental stress as an evolutionary force. *BioScience*, **50**, 217–26.

Jansson, C., Ekman, J., and von Brömssen, A. (1981). Winter mortality and food supply in tits *Parus* spp. *Oikos*, **37**, 313–22.

Jerison, H. J. (1973). *Evolution of the brain and intelligence*. New York: Academic Press.

Johnston, T. D. (1982). Selective costs and benefits in the evolution of learning. *Advances in the Study of Behavior*, **12**, 65–106.

Klopfer, P. H. (1962). *Behavioral aspects of ecology*. London: Prentice-Hall.

Klopfer, P. H. and MacArthur, R. H. (1960). Niche size and faunal diversity. *American Naturalist*, **94**, 300.

Krebs, J. R., Sherry, D. F., Healy, S. D., Perry, V. H., and Vaccarino, A. L. (1989). Hippocampal specialisation of food-storing birds. *Proceedings of the Royal Society of London, Series B*, **86**, 1388–92.

Lack, D. (1954). *The natural regulation of animal numbers*. Oxford: SPB Academic Publishing.

Laland, K. N., Odling-Smee, J., and Feldman, M. W. (2000). Niche construction, biological evolution, and cultural change. *Behavioural and Brain Sciences*, **23**, 131–75.

Lefebvre, L. (2000). Feeding innovations and their cultural transmission in bird populations. In *The evolution of cognition* (ed. C. Heyes and L. Huber), pp. 311–28. Cambridge, MA: MIT Press.

Lefebvre, L., Whittle, P., Lascaris, E., and Finkelstein, A. (1997). Feeding innovations and forebrain size in birds. *Animal Behaviour*, **53**, 549–60.

Lefebvre, L., Gaxiola, A., Dawson, S., Rosza, L., and Kabai, P. (1998). Feeding innovations and forebrain size in Australasian birds. *Behaviour*, **135**, 1077–97.

Lefebvre, L. Nicolakakis, N., and Boire, D. (2002). Tools and brains in birds. *Behaviour*, **139**, 939–73.

Lewontin, R. C. (1978). Adaptation. *Scientific American*, **239**, 169.

Long, J. L. (1981). *Introduced birds of the world.* London: David & Charles.

Lynch, M. (1990). The rate of morphological evolution in mammals from the standpoint of the neutral expectation. *American Naturalist,* **136,** 727–41.

MacArthur, R. H. (1955). Fluctuations of animal populations, and a measure of community stability. *Ecology,* **36,** 533–6.

Madden, J. (2001). Sex, bowers and brains. *Proceedings of the Royal Society of London, Series B,* **268,** 833–8.

Manne, L. L., Brooks, T. M., and Pimm, S. L. (1999). Relative risk of extinction of passerine birds on continents and islands. *Nature,* **399,** 258–61.

Marples, N. M. and Kelly, D. J. (1999). Neophobia and dietary conservatism: Two distinct processes? *Evolutionary Ecology,* **13,** 641–53.

Mayr, E. (1963). *Animal species and evolution.* Cambridge, MA: Belknap Press.

Mayr, E. (1965). The nature of colonising birds. In *The genetics of colonizing species* (ed. H. G. Baker and G. L. Stebbins), pp. 29–43. New York: Academic Press.

McKinney, M. L. (1997). Extinction vulnerability and selectivity: Combining ecological and paleontological views. *Annual Review in Ecology and Systematics,* **28,** 495–516.

Møller, A. P. and Cuervo, J. J. (1998). Speciation and feather ornamentation in birds. *Evolution,* **52,** 859–69.

Morse, D. H. (1980). *Behavioral mechanisms in ecology.* Cambridge, MA: Harvard University Press.

Newton, I. and Dale, L. C. (1996). Relationship between migration and latitude among west European birds. *Journal of Animal Ecology,* **65,** 137–46.

Nicolakakis, N. and Lefebvre, L. (2001). Forebrain size and innovation rate in European birds: Feeding, nesting and confounding variables. *Behaviour,* **137,** 1415–29.

Nicolakakis, N., Sol, D., and Lefebvre, L. (2003). Behavioural flexibility predicts species richness but not extinction risk in birds. *Animal Behaviour,* **65,** 445–52.

Odling-Smee, F. J., Laland, K. N., and Feldman, M. W. (1996). Niche construction. *American Naturalist,* **147,** 641–8.

Orr, M. R. and Smith, T. B. (1998). Ecology and speciation. *Trends in Ecology and Evolution,* **13,** 502–6.

Owens, I. P. F. and Bennett, P. M. (2000). Ecological basis of extinction risk in birds: Habitat loss versus human persecution and introduced predators. *Proceedings of the National Academy of Sciences, USA,* **97,** 12144–8.

Owens, I. P. F., Bennett, P. M., and Harvey, P. H. (1999). Species richness among birds: Body size, life history, sexual selection or ecology. *Proceedings of the Royal Society of London, Series B,* **266,** 933–9.

Pimm, S. L. (1991). *The balance of nature?* Chicago, IL: The University of Chicago Press.

Plotkin, H. C. and Odling-Smee, F. J. (1979). Learning, change and evolution: An enquiry into the telenomy of learning. *Advances in the Study of Behavior,* **10,** 1–41.

Price, T. (1999). Sexual selection and natural selection in bird speciation. In *Evolution of biological diversity* (ed. A. E. Magurran and R. M. May), pp. 93–112. Oxford: Oxford University Press.

Reader, S. M. (in press). Relative brain size and the distribution of innovation and social learning across the nonhuman primates. In *Towards a biology of traditions: Models and evidence* (ed. D. Fragaszy and S. Perry) Cambridge: Cambridge University Press.

Reader, S. M. and Laland, K. N. (2002). Social intelligence, innovation and enhanced brain size in primate. *Proceedings of the National Academy of Sciences, USA,* **99,** 4436–41.

Reed, J. M. (1999). The role of behavior in recent avian extinctions and endangerments. *Conservation Biology,* **13,** 232–41.

Robinson, B. W. and **Dukas, R.** (1999). The influence of phenotypic modifications on evolution: The Baldwin effect and modern perspectives. *Oikos*, **85**, 582–9.

Roselaar, C. S. (1995). *Taxonomy, morphology, and distribution of the songbirds of Turkey: An atlas of biodiversity of turkish passerine birds.* Mountfield: Pica press.

Rosenzweig, M. L. (1995). *Species diversity in space and time.* Cambridge: Cambridge University Press.

Sheppard, D. H., Klopfer, P. H., and **Oelke, H.** (1968). Habitat selection: Differences in stereotype between insular and continental birds. *Wilson Bulletin*, **80**, 452–7.

Smith, T. B. and **Skúlason, S.** (1996). Evolutionary significance of resource polymorphism in fishes, amphibians, and birds. *Annual Review in Ecology and Systematics*, **27**, 111–33.

Sol, D. and **Lefebvre, L.** (2000). Behavioural flexibility predicts invasion success in birds introduced to New Zealand. *Oikos*, **90**, 599–605.

Sol, D., Timmermans, S., and **Lefebvre, L.** (2002). Behavioural flexibility and invasion success in birds. *Animal Behaviour*, **63**, 495–502.

Sutherland, W. J. (1998). The importance of behavioural studies in conservation biology. *Animal Behaviour*, **56**, 801–9.

Terkel, J. (1996). Cultural transmission of feeding behavior in the black rat (*Rattus rattus*). In *Social learning in animals: The roots of culture* (ed. C. M. Heyes and B. G. Galef), pp. 17–47. San Diego, CA: Academic Press.

Timmermans, S. (1999). *Opportunism and neostriatal/hyperstriatum complex in birds.* M.Sc. Thesis, McGill University, Montréal.

Timmermans, S., Lefebvre, L., Boire, D., and **Basu, P.** (2000). Relative size of the hyperstriatum ventrale is the best predictor of feeding innovation rate in birds. *Brain, Behavior and Evolution*, **56**, 196–203.

Timms, R. and **Read, A. F.** (1999). What makes a specialist special? *Trends in Ecology and Evolution*, **14**, 333–4.

Veltman, C. J., Nee, S., and **Crawley, M. J.** (1996). Correlates of introduction success in exotic New Zealand birds. *American Naturalist*, **147**, 542–57.

Werner, T. K. and **Sherry, T. W.** (1987). Behavioral feeding specialization in *Pinaroloxias inornata*, the 'Darwin's finch' of Cocos Island, Costa Rica. *Proceedings of the National Academy of Sciences, USA*, **84**, 5506–10.

Wilson, D. S. and **Yoshimura, J.** (1994). On the coexistence of specialists and generalists. *American Naturalist*, **144**, 692–707.

Wyles, J. S., Kunkel, J. G., and **Wilson, A. C.** (1983). Birds, behavior and anatomical evolution. *Proceedings of the National Academy of Sciences, USA*, **80**, 4394–7.

CHAPTER 4

ENVIRONMENTAL VARIABILITY AND PRIMATE BEHAVIOURAL FLEXIBILITY

SIMON M. READER AND KATHARINE MACDONALD

Introduction

Behavioural innovation may allow animals to cope with environmental change, to exploit a wide range of habitat types, and thus to increase their geographical range. Innovation may also be a good indicator of general behavioural flexibility, which has long been predicted to be an important determinant of these ecological variables. Responses to environmental change are thought to be particularly important in hominid evolution (Richerson and Boyd, 2000), and behavioural flexibility is assumed to be concomitant with a large brain. But what is the evidence for such relationships?

In this chapter, we examine critically the notions that behavioural flexibility can be a useful comparative concept, that innovation frequency is an appropriate measure of behavioural flexibility, and that the reported frequency of novel behaviour is a valid indicator of the 'innovativeness' of a species or population. We argue that innovation can be used to gauge species differences in behavioural flexibility, and demonstrate that innovation frequency correlates with relative brain size in primates. In the past, hypotheses regarding the ecological causes and consequences of enhanced behavioural flexibility have tended to use brain size as a proxy measure. We discuss the utility of brain size measures for these purposes, and note how innovation frequency may provide a more direct measure of behavioural flexibility. As an illustrative example, we show how innovation and brain size measures allow us to make specific tests of the hypotheses that behavioural flexibility, species range size and environmental variability are linked in primates. En route, we will consider links between innovation and brain evolution, problems and solutions for comparative methods, and discuss what further data would be helpful for testing these ideas. Our overarching aim is to examine the evolutionary causes and consequences of innovative capacities and enhanced brain size in primates.

Innovation and brain size

Innovation frequency correlates with relative brain size in both birds and primates (Lefebvre *et al.*, 1997; Reader and Laland, 2002). More specifically, brain areas involved in higher order and multimodal integration (neocortex and striatum in primates, hyperstriatum ventrale and neostriatum in birds) are enlarged in taxa with high reported frequencies of novel behaviour patterns, compared to groups where few innovations are reported (Timmermans *et al.*, 2000; Reader and Laland, 2002; Reader, in press). These analyses operate by collating data from published literature and counting the number of reports of innovation for each group of interest (species in primates, parvorders in birds). In non-human primates over 500 reports of innovation have been collated, and in birds over 2200 (Reader and Laland, 2002; Lefebvre and Bolhuis, Chapter 2). To attempt to avoid biases during data collection, 'keywords' in the articles such as 'novel', 'never seen before' or 'invention' were utilized to classify behaviour patterns as innovations, so that the decision of whether a particular report qualified as an instance of innovation was effectively made by the author of the article. Inter-observer reliabilities using this method are high (Reader and Laland, 2001; Reader, in press). The frequencies are corrected for possible biases, such as differences in the research effort devoted to different primate species, or the number of species in an avian parvorder.

The relationship between the innovation measure and a measure of relative brain size can then be examined using comparative techniques such as independent contrasts (Felsenstein, 1985; Harvey and Pagel, 1991). Such techniques account for the fact that species may show similar characteristics simply because they are closely related rather than because they have evolved independently under similar selection pressures. Thus to treat species as independent data points may overestimate the degrees of freedom and potentially give false positive results. Purvis and Webster (1999) give a clear account of how and why independent contrasts should be used, focusing on the computer program *Comparative Analysis by Independent Contrasts* (CAIC) (Purvis and Rambaut, 1995) that was utilized here for comparative analysis.

We now examine the advantages of the innovation measure over other possible comparative cognitive measures, and discuss potential areas of concern. We focus on primates, but point the interested reader to the publications of Lefebvre and coworkers, who developed these measures in birds (see Lefebvre and Bolhuis, Sol, Chapters 2 and 3). Comparative estimates of learning and cognition are problematic. Approaches such as experimental tests of learning ability can be criticized on the grounds that they are not fair to different species, have little adaptive significance, and do not provide data on a large number of species (Byrne, 1992; Lefebvre and Giraldeau, 1996; Gibson, 1999; Deaner *et al.*, 2000). An alternative approach is to collate data on the incidence of traits associated with complex cognition, such as behavioural innovation, on the assumption that incidence is a reflection of that species' intellectual capability. This approach circumvents the aforementioned problems with experimental studies, and provides a tractable method of collecting data on tens or even hundreds of species. Innovation frequency measures opportunistic departures from the species norm (Lefebvre *et al.*, 1997), and so provides a quantitative, ecologically relevant measure of cognition. So far so good. But can innovation frequency be validated by comparison with other measures of learning or cognition? Figure 4.1 suggests that it can. Riddell and Corl (1977) compiled data on performance in a number of learning tasks for a variety of mammalian species, including several primates. We compiled the data across tasks by assuming that if A outperformed B in task 1, and B outperformed

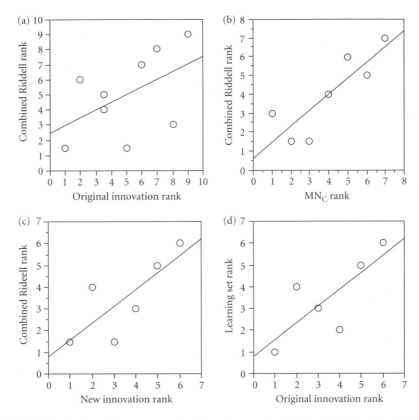

Figure 4.1 Validating the innovation measure using laboratory learning data from Riddell and Corl (1977). They compiled published performance data in six different learning tasks. The largest single data set ('learning sets') includes data on six species of primate (excluding humans), and by comparing performance across the various tasks it was possible to rank nine primate species ('combined Riddell rank'); in ascending rank the species are *Callithrix jacchus* and *Saimiri sciureus* (ranked equally), *Papio hamadryas, Galago senegalensis, Cebus albifrons, Ateles geoffroyi, Macaca mulatta, Cebus capucinus, Pan troglodytes*). We examined the relationships between these learning measures and our innovation measures in order to examine the utility of the innovation measure. As there were clear *a priori* predictions made, all *p* values quoted are one-tailed. Statistics are corrected for ties. (a) Combined Riddell rank vs the innovation measure utilized in Reader and Laland (2002). (Spearman rank correlation: $r_s = 0.51$, $N = 9$, $p = 0.074$.) (b) Combined Riddell rank vs the neocortex ratio MN_C (neocortex/[total brain–neocortex–cerebellum]; see text). Riddell and Corl concluded that brain indices correlate with learning ability, and this conclusion is supported by the reanalysis presented here. (Spearman rank correlation: $r_s = 0.85$, $N = 7$, $p = 0.019$.) (c) Combined Riddell rank vs the innovation measure utilized in this chapter. (Spearman rank correlation: $r_s = 0.78$, $N = 6$, $p = 0.040$.) (d) The learning set data vs the innovation measure utilized in Reader and Laland (2002). (Spearman rank correlation: $r_s = 0.77$, $N = 6$, $p = 0.042$.)

C in task 2, A would also outperform C. A matrix was used to score the various possible interactions and thus rank nine species of primate ('combined Riddell rank', CRR, Figure 4.1). Despite the small number of species involved and the accompanying restrictions on statistical power, the trends are clear. CRR correlates with both the innovation measure utilized by Reader and Laland (2002) and a modified innovation measure described below. Ranked performance in 'learning sets', the learning task that provides the largest single data set, also correlates with Reader and Laland's (2002) innovation measure. CRR correlates with a measure of relative neocortex size, the ratio of neocortex volume to that of the rest of the brain, excluding the cerebellum (a brain measure that is discussed subsequently). The fact that innovation frequencies, corrected for research effort, correlate with laboratory measures of learning, increases our confidence that we have a reliable cognitive index. Additionally, primate innovation frequency correlates with the reported frequency of both social learning and tool use, independently of brain size, suggesting that these capacities have evolved together (Reader and Laland, 2002). While observational reports of social learning should be treated with caution (Galef, 1992; Reader and Laland, 2002; Lefebvre and Bouchard, in press), tool use is more easily characterized, and again we have evidence that the innovation measure correlates with alternative indices of cognition.

It is still an open question as to precisely what cognitive processes are involved in innovation. Reader and Laland (Chapter 1) outline some potential underlying processes. 'Innovation' could suggest an Edison-like breakthrough, but in fact many reported innovations are relatively simple behaviour patterns such as eating novel foods. Examples from the database include *Pan troglodytes* eating mangoes (Takahasi, 1983), *Cercopithecus mitis* eating a flying squirrel (Fairgrieve, 1997), *Alouatta caraya* eating marsh plant roots (Rodrigues and Maricho-Filho, 1995), *Saimiri sciureus* eating a bat (Sarsa *et al.*, 1997), and *Lemur catta* eating a chameleon (Oda, 1996). These examples contrast with reports of a food type known to be part of the diet, but eaten rarely, reports that would not qualify as innovations. Of course, the database also includes more vivid examples of innovation, such as lemurs (*L. catta*) immersing their tails in water and then drinking from the wet tail (an example of both innovation and social learning; Hosey *et al.*, 1997), and 'stepping-stick' use in chimpanzees: sticks were used to walk on, as protection against a spiny tree that bears edible fruit and flowers (described by Alp (1997) as a 'new type of tool', thus qualifying as both innovation and tool use. Note that only *novel* tool use was classified as innovation). A chimpanzee that repeatedly arose early and performed wild arboreal displays while others in the group were still in their nests, 'causing great confusion' provides another example, with the performance of the behaviour pattern coinciding with a successful bid for alpha position (Kummer and Goodall, Chapter 10). Reader and Laland (2001) give further examples. The variety of behaviour patterns illustrated above might suggest that a number of different psychological processes underlie these behaviour patterns, with particular processes important to a greater or lesser extent according to the particular circumstance.

Thus we have an apparent difficulty with the innovation measure. It appears to include a ragbag of diverse behaviour patterns that may not have similar underlying psychological mechanisms. A comparative psychologist testing animals in the laboratory on, say, reversal learning tasks would seem to have much greater control over the psychological processes under examination. Of course, the laboratory scientist must account for confounding contextual variables, a difficulty that may make interpretation of apparent species differences nigh impossible (Macphail, 1982). We argue that, at least at this early stage of investigation, the purported diversity of processes underlying innovation is unproblematic and can be ignored in comparative analyses. In fact, it may be best to avoid cognitive interpretations of innovation reports as such interpretations may be particularly susceptible to observer bias (Nicolakakis and Lefebvre, 2000). As research progresses it may become valuable, and feasible, to distinguish between kinds of innovation.

A distinction has been made, on the basis of the context of innovation, between foraging and nesting innovations in birds (Nicolakakis and Lefebvre, 2000). Foraging innovation frequency correlates with relative brain size, but nesting innovation frequency does not, a result that has two implications. First, if the patterns found for feeding were simply due to literature biases, then these should also affect reports of unusual nesting. Second, different neural substrates may underlie innovation in different contexts (or at least a subset of the innovations in each context). There is no presumed cognitive basis for changes in nest-building site or technique; nesting is viewed as predominantly pre-programmed (Hansell, 1984). Perhaps nesting innovations are simply chance occurrences, with accidental occurrences less costly in this context. Alternatively, errors in nesting may have more detrimental consequences than do foraging errors leaving little room for significant innovation, especially as the consequences of errors might not be immediately obvious, perhaps even delayed to the end of the breeding season. A focus on *learned* novel behaviour, a more restrictive definition of innovation (Reader and Laland, Chapter 1), could be particularly informative in distinguishing between these possibilities. This focus would exclude less significant or accidental happenings and so facilitate cross-context comparisons that are particularly relevant to studies of the distribution and causes of species differences in cognition (Johnston, 1982).

What does innovation rate measure?

Let us focus a little more on exactly what innovation frequency measures. Brain size has long been suggested as a proxy measure of behavioural flexibility, with numerous predicted links made to various ecological variables (see Sol, Chapter 3). It is an empirical question whether brain size indicates behavioural flexibility. The established relationship between relative brain size and innovation frequency would suggest that brain size does indicate behavioural flexibility, if the reasonable assumption is made that innovation rate estimates behavioural flexibility (Lefebvre *et al.*, 1997; Reader and Laland, 2002). However, behavioural flexibility is a concept that can lend itself to vagueness, and thus it is best to state clearly that it is innovation rate that is being measured (Reader and Laland, Chapter 1). It seems likely that the concept of behavioural flexibility will subsume a number of quite different processes, processes that may have quite different underlying causes, may not be correlated with one another, and may even be negatively correlated. Flexibility in behaviour could presumably be reached through several different routes. Innovation frequency may correlate with brain size, but this may say nothing about brain size's relationship with other proposed indicators of behavioural flexibility, such as niche breadth or dietary generalism. At this point, the ecological, evolutionary, and psychological significance of innovation itself is not clear. Comparative studies suffer from the problem that they usually rely upon correlation evidence (cf. Pagel, 1999). Thus relationships such as that between innovation rate and brain size may be due to some unknown intervening variable. It is difficult to say at this point in time whether innovation is an adaptation, an exaptation, or even an evolved trait. Conceivably, innovation could be an indicator of flexibility but not itself of evolutionary significance.

Is the relationship between brain size and innovation frequency robust?

Research effort

We now address more specific concerns regarding the relationship between brain size and innovation frequency and detail additional analyses that deal with these potential problems. Our aims are to test the robustness of the relationship as well as to illustrate general concerns relevant to comparative analyses of brains and behaviour.

The analysis hinges upon an appropriate correction for research effort. In the original analysis (Reader and Laland, 2002), research effort was computed by counting the number of publications on each species from a sample of around 950 papers. These were the same articles that were examined for reports of innovation. This measure has the advantage of estimating the most relevant aspect of research effort, that is, research in the fields from which the data were collected. However, it could be argued that a more extensive sample of research effort could give different results. Consequently, we estimated research effort from the 7144 primate studies listed in the Zoological Record 1993–2001 (v. 129–137). Like the original estimate of research effort, this new estimate correlates positively with both innovation frequency and relative brain size. In the original analysis species where no innovations were recorded were included. This is reasonable. For example, it seems equally informative that no innovations were recorded for *Saguinus oedipus*, with 153 studies listed in the Zoological Record, as that an innovation was noted for *Colobus kirkii*, with only seven articles in the Zoological Record. However, it might be argued that there was something different in the *way* that primates with innovations recorded were studied. If this is the case, it might be more informative to examine deviations from the relationship between research effort and innovation frequency only amongst the subset of 41 species where innovations are recorded. This analysis is described subsequently.

The original analyses took residuals from a plot of innovation frequency against research effort, and then took independent contrasts on these residuals. In this, Reader and Laland (2002) followed a number of studies that corrected for research effort in a similar manner (e.g. Byrne, 1993; Nicolakakis and Lefebvre, 2000). It seems likely, however, that research effort could be phylogenetically biased—for example, many more studies are devoted to the great apes than to bush babies—in which case it is sensible to take independent contrasts of research effort.[1] Consequently, the analysis presented below took independent contrasts of research effort, and then entered these contrasts into a multiple regression with those for innovation rate and neocortex volume. CAIC takes independent contrasts only

---

[1] Phylogeny should be considered regardless of the reasons behind similarity between close relatives (Purvis and Webster, 1999). It is easy to imagine that a well-studied, speciose clade that is also innovative could obscure the true relationship between research effort and innovation frequency, if species were treated as independent data points.

for species where data are available for all the variables of interest. Thus CAIC, like any multiple regression, would estimate the relationship between research effort and innovation frequency based upon the 17 species where brain data are also available. In comparison, if residual innovation frequency were taken before running CAIC, it would be possible to base the residuals upon all species where data on innovation and research effort were available (but see Darlington and Smulders, 2001). This consideration should be kept in mind when examining the results of independent contrast analyses, and provides a further reason to examine differences between independent contrast and across-species analyses (Price, 1997).

Measuring brains

Mammalian brain evolution has been of strong, continuing, and long-lasting interest (Count, 1947; Jerison, 1973; Harvey and Krebs, 1990; Finlay and Darlington, 1995; Barton and Harvey, 2000; Clark *et al.*, 2001; De Winter and Oxnard, 2001; Kaas and Collins, 2001). The benchmark studies of Stephan *et al.* (1970, 1981, 1988) measured volumes of the fundamental brain components of 44 species of non-human primate by taking serial sections (SS) through the brain. The importance of this data cannot be underestimated. Cited by 196 studies, it is the standard source (and until recently the only source) of brain area volumes for comparative studies of primate brain evolution. With the similar data of Zilles and Rehkämper (1988) on the orang-utan, Reader and Laland (2002) were able to analyse the relationship between brain volume and innovation frequency for 45 non-human primates. However, this represents only about a fifth of the 220 or so species of primate (Purvis, 1995; Rowe, 1996), and there are notable gaps in the database, such as the pygmy chimpanzee *Pan paniscus*. Moreover, many estimates of component brain part volumes rely on the measurement of only one or two brains. In only three of the 44 species that Stephan *et al.* measured does the sample size exceed 3, and many of man's closest relatives, such as the common chimpanzee, gorilla, and white-handed gibbon, are represented by a single sample. These small sample sizes may not be as problematic as they first appear because Stephan *et al.* (1981) correct brain volumes to species-typical means (see Appendix). Thus if the assumption holds that the ratio of one brain component to another is relatively constant between individuals of the same species, the volume estimates will not be biased by, for example, an atypically sized brain specimen.

These criticisms are not to decry the worth of the Stephan *et al.* (1981) data set. After an extensive search of the literature published since 1981, we were able to add data on the brain components of an additional 4 species, for a total of 49 species (Appendix).[2] This lack

---

[2] Most of the more recent brain volume data comes from magnetic resonance imaging (MRI). Rilling and Insel (1999) make a case that MRI is more accurate than the serial sectioning technique of Stephan *et al.* (1981). For example, measurements can be taken on living specimens, volumes do not need to be corrected for the effects of fixation, and it is possible to scan several animals and determine intraspecific variabilities. However it is difficult to compare the two techniques as only seven species have been measured by both MRI and SS techniques. The neocortex makes up the major part of the brain in non-human simians (mean ratio of neocortex to total brain volume is 0.68, based on Stephan *et al.*, 1981; Zilles and Rehkämper, 1988). Inspection of the N_C values (N_C = neocortex volume/[total brain−neocortex−cerebellum]) indicates that four species (all the great apes

of progress in the last 20 years demonstrates the significant amount of work involved in measuring brain components accurately, and the fact that primate brain specimens are not easily obtained. The orang-utan brain component measurements, for example, rely upon a 75-year-old brain specimen (Zilles and Rehkämper, 1988). That said, the data compendium presented in the Appendix includes valuable new data on the neocortex of all four great apes. The appendix also clarifies differences between various published sources and outlines the measurement techniques used.

In the analyses that follow, the relative brain size measurement used is the ratio of the neocortex to the rest of the brain, excluding the cerebellum (N_C). We focus on the neocortex because this is the brain area likely to be most involved in the neural processing underlying innovation (Jolicoeur *et al.*, 1984; Keverne *et al.*, 1996; Joffe and Dunbar, 1997). There are other methods of measuring relative brain size and accounting for the fact that large-bodied animals tend to have bigger brains than smaller animals. Many modern methods use the size of the brain itself as a reference variable, since alternative measures of body size such as body mass tend to be more subject to measurement error (Barton, 1999). The optimum measure of relative brain size is still an open question (Barton, 1999; Reader, in press). Deaner *et al.* (2000) examined currently used measures and concluded that there is little to chose between them on theoretical grounds. Different patterns of results can result from the different measures, particularly when comparing competing hypotheses of brain evolution using techniques such as multiple regression (Deaner *et al.*, 2000). Comparative scientists should be cautious when attempting to find the best predictor of brain size as findings may hinge on data quality or the precise measurement used, particularly when sample sizes are relatively small.

Our N_C measure differs from the more usual neocortex ratio measure, neocortex volume divided by the volume of the rest of the brain (Dunbar, 1992), and also from the executive brain ratio measure utilized by Reader and Laland (2002): [neocortex + striatum]/ [mesencephalon + diencephalon]. We focus on the neocortex here to harmonize with other studies, because there is controversy over the role of the striatum (Barton and Harvey, 2000), and because N_C can be calculated for more species. Exclusion of the cerebellum is simple, feasible, and vital for accurate relative neocortex measurements.[3]

---

with data available and humans) have considerably smaller N_C values using the MRI technique. Neocortex volume discrepancies between the two measurement techniques may be responsible for these differences. The Rilling and Insel neocortex volumes are on average 95 per cent of those of Stephan *et al.*, compared with 98 per cent for total brain volumes ($N = 7$, proportions excluding *Homo sapiens* are similar). It is a matter of some urgency to comparative studies of primate brain evolution that the relative accuracy of the two techniques is determined, and that data are gathered to cover both more species and more individuals from each species.

[3] Data on cerebellum volumes are widespread (Clark *et al.*, 2001), probably as it is relatively easily measured. The cerebellum makes up a significant proportion of the primate brain ($11.2 \pm 0.3\%$ of total brain volume in simians [mean $\pm$ SE, $N = 27$], $14.2 \pm 0.3\%$ in prosimians [$N = 18$]; Stephan *et al.*, 1981; Zilles and Rehkämper, 1988), and selection can act on the volume of the cerebellum independently of selection on overall brain size or neocortex volume (Rilling and Insel, 1998; Barton and Harvey, 2000; De Winter and Oxnard, 2001, cf. Clark *et al.*, 2001). For example, cerebellum volume has been linked to locomotion mode in primates, bats, and cetaceans (Stephan and Pirlot, 1970; Rilling and Insel, 1998; Marino *et al.*, 2000).

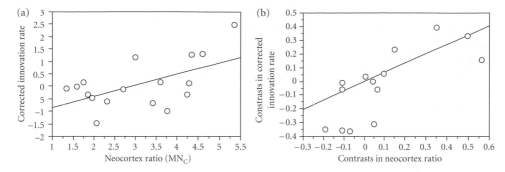

Figure 4.2 Residual innovation frequency vs neocortex ratio, MN_C (a) across-species and (b) independent contrasts. Innovation frequency was corrected for research effort by taking residuals from a natural log plot of innovation frequency against research effort. In the independent contrast analysis, contrasts in research effort were taken and entered into a multiple regression with the contrasts in innovation frequency and MN_C, though for illustration the residuals from a plot of innovation against research effort contrasts are shown.

Without taking the cerebellum into account there is a risk of confounding enlarged neocortex volumes with diminished cerebellum volumes.

In the analyses that follow, we take the unweighted mean of the N_C values calculated from the MRI and SS data, designated MN_C. We do not weight by sample size as this would tend to favour the MRI data, and at present it is not known which measurement technique is the more accurate. Analyses based solely on the SS data set designate the neocortex ratio as SN_C. Including the MRI data in analyses of brain size and innovation frequency will tend to be a conservative measure, as MRI N_C values tend to be smaller than SN_C values for the great apes, species where large numbers of innovations are reported.

The new analysis of innovation frequency and brain size

Reader and Laland (2002) demonstrated a positive correlation between innovation rate and relative brain size. In our revised analysis, the relationship between relative neocortex size and innovation frequency remains significant after (1) utilizing a new estimate of research effort, (2) incorporating research effort into an independent contrasts approach, (3) restricting the species examined to those where innovations have been recorded, and (4) using a different brain measure based upon an expanded data base (Figure 4.2). Figure 4.2(a) shows that there is a positive relationship between innovation frequency and relative neocortex size across species ($N = 17$, $r^2_{adj} = 0.28$, $F_{1,15} = 7.26$, $p < 0.02$). This relationship is maintained after taking independent contrasts (multiple regression with innovation frequency as the dependent variable, MN_C as the predictor variable, and research effort as a covariate: $N = 16$, $t_{14} = 2.14$, $p < 0.05$).[4]

---

[4] Analyses were conducted using the phylogeny of Purvis (1995), with some species added (Reader, in press) and branch lengths set equally. Multiple regression was through the origin (Garland et al., 1992) and analyses used CAIC version 2.6.8b (Purvis and Rambaut, 1995). The independent contrast analysis does not meet the assumptions of CAIC (Garland et al., 1992; Purvis and Rambaut, 1995; Diaz-Uriate and Garland, 1996; Purvis and

In addition to the measures described above, we have also addressed the effect of field vs captive studies, provisioning, and other human influences that could bias the innovation measure (Reader and Laland, 2002; Reader, in press). Furthermore, Lefebvre and colleagues (Lefebvre, 2000; Nicolakakis and Lefebvre, 2000) have demonstrated that the relationship between brain size and innovation rate in birds was not affected by the following potential confounding variables: mode of juvenile development, population size, journal source, editorial policy, research effort, observer interest, and common ancestry.

Thus the relationship between relative brain volume and innovation rate appears robust, supporting a link between behavioural flexibility and brain size. An obvious interpretation of this finding is that improvements in the ability to innovate may be the driving force behind the evolution of enhanced neocortex size. An additional or alternative explanation for the relationship with brain size is that some other selection pressure may have favoured brain enlargement, and this enlargement facilitated increased innovation rates. This link could be neurological: perhaps evolved neural architecture for, say, keeping track of social interactions (Dunbar, 1998) could also be applied to innovation in the ecological context. On the other hand, the link could be less direct. For example, perhaps the habitats or lifestyles of large-brained species tend to favour innovation, but an enlarged neocortex is not a necessary condition for innovation to occur. We discuss possible tests that would discriminate between these hypotheses below.

It is a common, but controversial, assumption that brain size and cognitive capacities are correlated (Macphail, 1982; Byrne, 1993), and we believe the link between neocortex size and innovation rate is evidence that this assumption is correct. The pervasive nature of the assumption that big brains equate with 'intelligence' creates a risk of circular reasoning. For example, it is easy to mistakenly argue that if a measure correlates with brain size then that measure provides an appropriate gauge of cognitive ability, while the hypothesis actually under test is whether brain size indicates cognitive capacity. Brain size and innovation frequencies correlate but should not be treated as equivalent measures. Innovation frequency measures current behaviour, and so, while the capacity to innovate may depend partly upon brain size, it is likely to be influenced by current conditions. For example, the 'necessity is the mother of invention hypothesis' (Laland and Reader, 1999a, b; Reader and Laland, 2001) argues that innovation is favoured by need, suggesting that innovative tendencies will only be revealed in limited circumstances. Macphail (1982, p. 4) makes a similar point from a quite different background: 'Intelligence . . . is held to manifest itself in all those situations in which subjects are required to adapt to novel circumstances', continuing to note that passive observation of animals in the wild with no unnatural demands will not reveal the full range of an animal's capabilities. Innovation rate may also

---

Webster, 1999). The following assumption checks were conducted: the absolute values of the three sets of contrasts were regressed against their nodal values and the standard deviation, and residuals from the multiple regression were standardised (Howell, 1997), converted to absolute values and plotted against MN_C, the predictor variable. There were violations ($p < 0.05$) for two of these seven assumption checks. However, removal of the two outliers in the regression (Purvis and Rambaut, 1995) resulted in an analysis that met the assumptions of CAIC, and the relationship between innovation frequency and MN_C remained significant (Figure 4.2(b), multiple regression: $N = 14$, $t_{12} = 4.46$, $p < 0.001$). The results are similar if residual innovation frequency is calculated prior to taking independent contrasts, or if SN_C is analysed instead of MN_C.

measure the extent to which an animal is prepared to discover and explore novel affordances of the environment, rather than just responding to changed circumstances. That is, innovation rate may partly measure the active role an animal takes in its own learning and development (Lewontin, 1983; Bateson, 1988).

The fact that innovation rate is influenced by current conditions could be one explanation why Nicolakakis *et al.* (2003) fail to find a relationship between avian innovation frequency and extinction risk. Species with the capacity to innovate may be particularly successful at coping with changed conditions and thus avoiding the risk of extinction (Sol, Chapter 3), whereas at risk species are likely to be living under the difficult circumstances that might favour innovation. Such a situation would tend to equalize the number of innovations reported for at risk and less-at-risk species.

There are a number of possible comparative tests of the necessity hypothesis that would also address the causes underlying the association of neocortex size with innovation rate. For example, comparisons could be made between different populations of the same species that were under varying degrees of stress. Populations on the edge of their species' geographical range, for instance, might be living under the difficult conditions thought to favour innovation. Comparisons between captive and wild populations are also likely to be informative. It is still somewhat of a mystery as to why tool use, for example, is relatively rare in the wild compared to the situation in captivity (Byrne, 1997). Population comparisons will also allow tests of the idea that innovation is favoured by a lack of environmental stresses (the 'spare time' hypothesis), a hypothesis diametrically opposed to the necessity hypothesis that may explain why innovation appears particularly common in captive and provisioned populations (Kummer and Goodall, Chapter 10).

Relative brain size is largely the result of a variety of selection pressures in the recent and ancient evolutionary past rather than a response to current conditions (Jerison, 1973; Martin, 1981; Barton and Harvey, 2000). The relative evolutionary lability of brain size and innovation rate has not been investigated in mammals. In birds innovatory capacities appear to have evolved more recently than brain size: the largest proportion of variance is located at the species level for innovation rate but at the parvorder level for brain size (Nicolakakis *et al.*, 2003; Sol, Chapter 3). Neocortex size correlates with several ecological and social variables (Deaner *et al.*, 2000), so if neocortex size is used as a proxy for behavioural flexibility known potential confounding variables should be included in the analysis (Sol *et al.*, 2002). Relative brain size is a less direct measure of behavioural flexibility than innovation rate, but unlike innovation rate it may estimate other components of behavioural flexibility, such as social learning or technical intelligence (Reader and Laland, 2002). Innovation rate and relative brain size have complementary roles in comparative studies.

Testing ecological hypotheses

Lee (1991) hypothesized that behavioural flexibility may radically affect the survival chances of animals under new conditions, an idea that has received empirical support from Sol (Chapter 3), who shows that innovative species are more successful invaders. Similarly, Gamble (1994) has argued that the ability to gather and communicate knowledge about the environment was key to hominid range expansion and to weathering environmental changes. A number of studies of hominid biogeography identified niche breadth and behavioural flexibility as characteristics that might be favoured during periods of increased climatic variability, and environmental deterioration has been argued as a key factor in the emergence of enlarged human brains (Vrba, 1985; Potts, 1998; Richerson and Boyd, 2000). In contrast, stable environments may favour the 'slow' life histories that tend to be associated with enlarged brains (Lewin, 1988). Fossil cranial volumes, fossil distributions and artefacts such as tools may provide important clues to the behavioural flexibility of extinct species, and techniques such as examination of lake sediments or ice cores allow past climatic variability to be determined (Calvin, 2002). However, hominid studies suffer from the fact that the sample size will be relatively small. An obvious place to look for support for the hypothesized link between behavioural flexibility, climatic variability, and species range size is in the extant primates.

The environment is heterogeneous in many dimensions (Ancel Meyers and Bull, 2002). Animals' environments are constructed of a large number of interrelated variables, including the distribution, numerosity and characteristics of conspecific competitors, food resources, refuges, predators, parasites, and competing species (Lee, 1991). Moreover, animals alter the world around them ('niche construction', Lewontin, 1983; Laland et al., 2000). All these variables can change cyclically through time, have consequences for the survivorship and reproduction of individuals, and thus pose problems of adaptation. The consequences of environmental variability will act at a variety of timescales and feedback processes may operate, such as interactions between food availability and community dynamics (Lee, 1991). Innovation may provide benefits only at particular timescales or in response to particular environmental changes; thus the choice of environmental variability measure is important.

Chivers (1991) argued that the central issue in considering primates tolerance to environmental change lies in the ability to locate, consume, and process adequate food. Most recorded observations of primate innovation are in the foraging context (Reader and Laland, 2001). Variation in climate, particularly rainfall, can be an important factor in causing changes in resource productivity (Janson and Chapman, 1999). Thus climate is likely to be a relevant proxy variable for aspects of environmental change likely to affect primates. Clark (1991) describes how particular rates of environmental change, scaled to an animals' developmental cycle and expected lifetime, may favour phenotypic or behavioural responses. Primates and other relatively long-lived animals will be exposed to change on a variety of timescales (Clark, 1991). In quite constant environments but also in environments that are alternating very rapidly, more rapidly than an animal can produce and utilize a new phenotype, individuals may develop and behave conservatively. More moderate rates of novelty and change may select for another attribute, 'flexibility' (Fagen, 1982): a common finding of theoretical models is that the evolution of both asocial and social learning is favoured at intermediate rates of environmental change (Boyd and Richerson, 1985; Stephens, 1991; Laland et al., 1996b; Shennan and Steele, 1999; Sibly, 1999). Temporal climatic fluctuation at scales shorter than the generation period for that species may thus be critical. In primates, it may be more appropriate to measure climatic change at seasonal or inter-annual scales than longer periods of time.

The geographic range of innovative or flexible species may be enlarged relative to other species due to (1) increased rates of invasion of new habitats (an example of 'inceptive' niche construction, Laland *et al.*, 2000), and/or (2) improved abilities to counter environmental change or habitat differentials (an example of 'counteractive' niche construction, Laland *et al.*, 2000). Speciation processes may complicate these relationships. Range expansion may be followed by speciation, effectively subdividing the range (Owens *et al.*, 1999). Innovative groups may exhibit less anatomical variation in response to geographic variation as they can cope with variation behaviourally (Lewontin, 1983; Laland *et al.*, 1996a; Laland *et al.*, 2000), and this may decrease the evolutionary subdivision of the taxa. Alternatively, innovative taxa may experience unusually high rates of speciation (Wyles *et al.*, 1983; Wcislo, 1989), a hypothesis for which there are some conflicting data (Lynch, 1990; Nicolakakis *et al.*, 2003).

Here, we examine the species range of extant primates and the temporal and spatial climatic variability that they encounter. Primates were selected for analysis for reasons of data availability, because of the interest in hominid evolution and range expansion, and to restrict the analysis to one well-known taxon. Like measures of behavioural flexibility, measures of climatic variability and range size will be informative only where relatively similar species are compared. Comparing very different taxa may give misleading results—seasonal variation in, say, rainfall is unlikely to influence whales and shrews in the same manner. We predict innovation frequencies and relative neocortex size will covary with both species range size and the climatic variability within the range, and predict that innovative species will utilize more different food types than less innovative species. Positive relationships would be consistent with the hypotheses that more innovative species can tolerate greater climatic variability, that innovation is a response to climatic change, or that climatic variability has selected for a propensity to innovate (Kummer and Goodall, Chapter 10; Lee, 1991; Reader and Laland, 2001).

Methods

Geographical ranges of primates were digitized from maps and their climatic variability analysed. Our spatial climatic variability measures were the coefficient of variation of rainfall and of temperature. Temporal variability was measured on seasonal timescales (temperature range throughout geographical range) and between years (inter-annual variation in rainfall). Dietary breadth data are from Eeley and Foley (1999), who estimated the number of food types utilized by simians on the African mainland. The maximum number of food types recorded was 10, so presumably broad dietary categories were utilized, such as gum, insects, and fruit. Full details on data sources are given below.

Species range size and climatic measurements

Primate species vary considerably in the area over which they are distributed. For example, the gelada baboon (*Theropithecus gelada*) is limited to highland Ethiopia while the vervet monkey (*Cercopithecus aethiops*) can be found throughout most of sub-Saharan Africa. Researchers have employed a number of alternative methods of estimating species range size in published literature because defining and mapping geographic ranges is inherently problematic, necessarily simplifying the complex spatial and temporal distribution patterns of individual organisms (Eeley and Foley, 1999). Given the global scale of the analyses and the problems with measures of latitudinal extent (Gaston *et al.*, 1998), extent of occurrence maps were utilized. We used maps of primate species ranges derived from a single source, *Primates of the World* (Wolfheim, 1983), so that range measurements were as comparable as possible. Wolfheim (1983) collated literature references and information from field biologists and produced maps based on the sum of location references for each primate species.

Primate range maps were scanned and digitized as vectors in AutoCAD map (Autodesk, 1997), imported into GRASS (1993) and converted to raster images. As the source maps were in an unknown projection, the raster images were rectified using a vector map of known projection and coordinates, and projected into equal area format using the ArcINFO (ESRI, 1999) PROJECT function. Three databases were produced for each of the main areas of primate distribution, Africa (excluding Madagascar), South America, and Asia. The African and South American databases were both created in Lambert's Azimuthal Equal Area projection. For the wider landmass of Asia, extending East–West rather than North–South, the Cylindrical Equal Area projection was used. African and South American maps were rectified into latitude–longitude format and then projected using ArcINFO; for the Asian maps one stage of this process was bypassed by rectifying the maps using a previously projected vector coastline map. Statistics were derived from the climatic variability maps with Imagine (Erdas, 1999).

Global surface climate data were obtained from two data sets: CRU05 0.5-degree 1961–1990 Mean Monthly Climatology and 1901–1995 Monthly Climate Time-Series (New *et al.*, 1999, 2000). Each map of climatic variability was based on a 30-year period (1961–1990) in order to attempt to include the full variation induced by various more short-term climatic cycles such as El Niño. Climatic cycles of longer duration exist but their effects have not been taken into account. The use of a 30-year period to calculate variables such as mean annual rainfall is standard practice in meteorology. A map of mean rainfall and temperature was produced using the 1961–1990 Mean Monthly Climatology data set (New *et al.*, 1999). The yearly mean was calculated for 1961–1990 based on monthly means, allowing for leap years. Secondly, the temperature range for the period 1961–1990 was used as an indicator of seasonality. Minimum and maximum yearly temperatures were calculated from monthly values and the difference taken for the final temperature range map. Finally, a map of inter-annual variability was based on variability in mean annual rainfall (New *et al.*, 2000). The mean was calculated for each year from 1961 to 1990, and new maps were produced based on standard deviation and coefficient of variation in annual rainfall for each 0.5-degree cell. Further processing was carried out to calculate the climatic variation within each species' range. Values of spatial variability tolerated by each species were calculated as the coefficient of variation of mean rainfall and temperature values within the species range. The seasonal and inter-annual variability tolerated by each species were calculated as mean values within the range. We used a principal components analysis (PCA) to analyse the climatic data and extract a factor that represented most of this variation while avoiding the problems associated with the analysis of large numbers of variables. However, this is a compromise solution, as it is conceivable, though we think unlikely, that

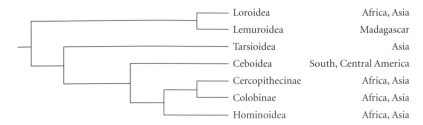

Loroidea	Africa, Asia
Lemuroidea	Madagascar
Tarsioidea	Asia
Ceboidea	South, Central America
Cercopithecinae	Africa, Asia
Colobinae	Africa, Asia
Hominoidea	Africa, Asia

Figure 4.3 A simple primate phylogeny outlining the links between phylogenetic affinity and geographic distribution. Modified from Purvis (1995), with distribution data from Rowe (1996). Branch lengths are arbitrary.

phylogenetic biases could modify the findings of a PCA of climatic variabilities. Thus we also analysed each climatic factor individually.

Comparative analyses
Range sizes are likely to be constrained by continent size, shape and habitat availability, and the historical patterns of climatic variability may have been quite different between different continents. There is thus a case to be made for separate analyses within each continent. However, as Figure 4.3 shows, primates' geographical distributions are clearly biased by phylogeny. This means that contrasts made within the same clade will generally be within the same continent and few contrasts will be made between continents, thus controlling for this potential confounding variable (Purvis and Webster (1999) detail how phylogenetic comparisons can control for the effects of unmeasured confounding variables). Pooling data across continents could obscure the detection of different correlation patterns on each continent, but we do not predict that such differences exist. Although the potential for range expansion may be greater in Africa than, say, the islands of south-east Asia, the same relationships are predicted within each area. More flexible species are predicted to have larger geographic ranges in both localities, even though absolute species range sizes may be smaller in south-east Asia than Africa.

Results

Dietary breadth
Relative neocortex volume (MN_C) correlates with dietary breadth (Figure 4.4, independent contrasts: $N = 11$, $F_{1,10} = 9.27$, $r^2_{adj} = 0.43$, $p < 0.02$). However, one assumption of CAIC was violated, and the situation was not improved by transformation of the data. The BRUNCH algorithm (Purvis and Rambaut, 1995) was utilized, and gave similar results without violating the assumptions of CAIC (independent contrasts: $N = 6$, $F_{1,5} = 7.13$, $r^2_{adj} = 0.51$, $p < 0.05$). Thus African anthropoid primates that eat a larger number of food types have larger neocortices. This result matches the existing evidence that particular diets are associated with enhanced brain size (Milton, 1988; Harvey and Krebs, 1990; Deaner et al., 2000).

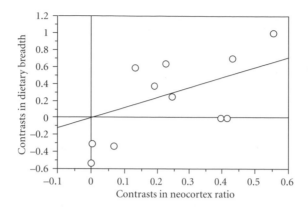

Figure 4.4 Independent contrasts in dietary breadth vs those for neocortex ratio, MN_C. Dietary breadth data are for African anthropoids, taken from Eeley and Foley (1999).

We examined the simulated effect of a data error by changing a single datum, the MN_C for *P. paniscus*, to an arbitrary value.[5] The shift in a single datum affects all contrasts ancestral to that datum as ancestral node values are estimated from daughter groups. Thus the points in an independent contrast plot are not independent from one another in the usual statistical sense (Björklund, 1994; Sokal and Rohlf, 1995). Comparisons between closely related taxa are particularly affected by the error, and such comparisons may have particularly high leverage. Thus data quality is crucial to the proper performance of CAIC (Purvis and Webster, 1999).

Corrected innovation frequency did not correlate with dietary breadth (multiple regression, independent contrasts: $N = 8$, $t_6 = 0.32$, $p = 0.76$, relationship unaltered by the removal of an outlier to meet the assumptions of CAIC: $N = 7$, $t_5 = 0.55$, $p = 0.61$, power = 0.20). In this and the following results, power was estimated based on $\alpha = 0.05$ and an effect size of $r = 0.45$ (Howell, 1997). Across-species analysis gave similar results to the independent contrasts (as in all analyses below). The lack of a correlation between dietary breadth and innovation rate was a surprise and may reflect the limited statistical power.[6] Power may be further limited by the ceiling of 10 food types set by Eeley and Foley (1999).

---

[5] The MN_C for *P. paniscus* was changed from the actual value of 4.25 to 5.0. The closest relative, *P. troglodytes*, has an MN_C of 5.36. The new data were analysed with CAIC as normal. Three data are shifted by the change in one datum: the comparison between *P. paniscus* and *P. troglodytes*, that between *Pan* and *Gorilla*, and that between the Cercopithecinae/Colobinae and Hominoidea (the oldest node at which contrasts are taken). Other contrasts are unaffected as they are conducted within the Cercopithecinae and Colobinae.

[6] Although this finding does not support our prediction, in one sense it is perhaps reassuring that innovation rate and diet breadth do not correlate strongly. Innovations were recorded in several behavioural contexts, the majority in the foraging context. One possible criticism of using foraging innovations to estimate behavioural flexibility is that rarely but regularly eaten foods could be falsely described as innovations, particularly in species with poorly known diets. Species with broad diets would then be incorrectly characterized as innovative.

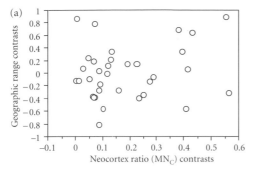
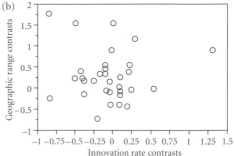

Figure 4.5 Independent contrasts in species range size and (a) neocortex ratio, MN$_C$, or (b) innovation rate, in African, Asian, and South American primates, excluding those dwelling in Madagascar. See legend, Figure 4.2 for notes on innovation rate.

Species range size

There was no relationship between species range size and relative neocortex volume (MN$_C$, Figure 4.5(a), independent contrasts: $N = 34$, $F_{1,33} = 0.62$, $r^2_{adj} = 0.00$, $p = 0.44$, power = 0.73). Similarly, corrected innovation frequency did not correlate with species range size (Figure 4.5(b), multiple regression, independent contrasts: $N = 32$, $t_{28} = 0.67$, $p = 0.51$, power = 0.71). These results are in contrast with the prediction that behaviourally flexible and large-brained species will have large geographic ranges.

This finding addresses two confounding effects that could influence innovation rate. First, species with large geographic ranges could have more innovations reported than those with small species ranges, because the behaviours of the former species are more likely to be observed, their population size is likely to be larger, and a limited range may have limited opportunities for innovation. Second, species inhabiting large geographic ranges may be more able to avoid the detrimental local conditions that might necessitate innovative responses by moving within the range. Neither of these two scenarios are supported by the data.

Climatic variability

Climatic variability does not correlate with relative neocortex size (MN$_C$, Figure 4.6(a), independent contrasts after removal of two outliers to meet the assumptions of CAIC: $N = 25$, $F_{1,24} = 0.35$, $r^2_{adj} = 0.00$, $p = 0.85$, power = 0.60). Corrected innovation frequency was also not correlated with climatic variability (Figure 4.6(b), multiple regression, independent contrasts: $N = 20$, $t_{18} = 0.42$, $p = 0.68$, power = 0.50). We raise possible methods for increasing the power of the analyses in the discussion. Similar results were found if African or South American primates were considered separately, and if the various climatic variables were analysed individually. Thus there seems to be little evidence that these measures of climatic variability and behavioural flexibility are linked.

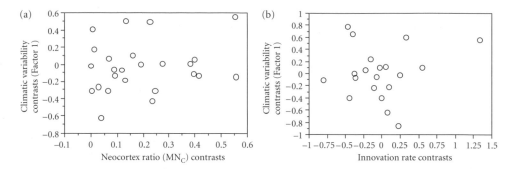

Figure 4.6 Independent contrasts in climatic variability versus those for (a) neocortex ratio, MN_C, or (b) innovation rate, in African and South American primates, excluding those dwelling in Madagascar. The measure of climatic variability was the first factor from a principal components analysis. Four variables were entered into the principal components analysis: spatial variation in temperature and rainfall across each species' range, seasonal variation in terms of temperature range, and inter-annual variation in terms of rainfall. Similar results are obtained if the individual climatic variables are analysed. See legend, Figure 4.2 for notes on innovation rate.

Discussion

Innovation rate correlates with relative neocortex size in primates, a result paralleled by similar analyses in birds (Lefebvre and Bolhuis, Chapter 2). Primate innovation rate correlates with measures of individual learning, with the reported variety of tool use, and with the reported frequency of social learning, suggesting that these cognitive capacities have evolved together. Contrary to predictions, we find little evidence that innovation rate or relative neocortex volume correlate with species range size or climatic variability.

   The unexpected results regarding species range size and climatic variability could be accounted for by the interesting possibility that the working hypotheses are flawed, or the less interesting possibility of a type II error. Statistical power was particularly low in certain cases, and ranged from 0.7 for the species range size analyses to 0.5 for some climatic variability analyses. One possible compromise route to augment power, by increasing the brain volume data set, would be to examine whole brain volumes rather than component volumes: a large proportion of the primate brain is neocortex. Whole brain volumes are available for a large number of primates, and intraspecific variabilities can also be determined. Information on population and temporal differences in innovation may also be valuable in avoiding type II errors. Primate populations make extensive shifts in their dietary behaviour at different times of the year, to the extent that would cause a species to be classified as a different type of dietary specialist, or as generalist as opposed to specialist (Chapman and Chapman, 1990). Recording innovation from just one population may also present a false picture of that species' innovatory capacities. Comparative students of innovation could also profitably pursue an experimental approach. It is relatively easy to present different species, or different populations of the same species, with novel foodstuffs or tasks and to observe their reaction (e.g. Webster and Lefebvre, 2001; Reader *et al.*, 2002; Day *et al.*, 2003). Experimental studies would complement the species innovation rate approach, allowing additional comparative questions to be addressed at different levels of analysis. For example, dynamic changes, functional payoffs or the proximate factors favouring innovation could be examined, informing evolutionary studies.

The possibility that the working hypotheses are wrong also deserves to be taken seriously. Disproving and comparing competing hypotheses is often difficult for the comparative student of brain evolution as there are frequently a large number of potential measures of the characteristics of interest (such as relative brain size or climatic variability), several potential alternative analytical methods, and the results can turn on the particular measure or analysis used (Dunbar, 1992; Deaner et al., 2000). For example, previous analyses have shown positive correlations between climatic variability and innovation rate within African primates (MacDonald, 2002). However, we feel that the innovation measure presented here, modified from Reader and Laland (2002), is likely to be more conservative, and are reluctant to assign significance to a finding restricted to one geographical region. Comparative analyses will be most informative when there are clear predictions to be tested, sample sizes are sufficiently high to allow conclusions to be drawn from negative results, and potential confounding variables are included in the analysis.

The finding that there was no significant relationship between innovation rate and species range size is perhaps not entirely a surprise considering that the common chimpanzee, the non-human primate most notable for its high innovation frequency and large brain, also has a relatively small species range. Moreover, the range size of African primates partly reflects climatic variation (Cowlishaw and Hacker, 1997; Eeley and Foley, 1999), and we also found no relationship between climatic variability and innovation frequency. At least three explanations may account for these patterns. First, other aspects of behavioural flexibility may correlate with species range size. For example, dietary niche breadth, a proposed measure of behavioural flexibility, correlates with species range size in African anthropoid primates (Eeley and Foley, 1999). However, relative neocortex size, a correlate of certain measures of primate behavioural flexibility, was not significantly associated with species range size. Second, some other species characteristic, such as dispersal ability, could account for differences between the ranges of species (Owens et al., 1999). Range size can be limited by a number of factors, potentially obscuring the role of one factor in predicting range sizes (Gaston, 1996). For example, large-brained, innovative species such as the apes might be particularly vulnerable to human persecution because they are larger-bodied, less arboreal, diurnal, live in larger groups, have long inter-birth intervals, and rely more on patchily distributed foods. Hunting can eliminate larger species, but have a positive effect on smaller species' population densities (Peres, 1999). Species vary in their responses to human activities (Owens and Bennett, 2000; Purvis et al., 2000), and current geographic ranges may not reflect those previous to human activities. Anthropogenic changes could have destroyed any relationship between brain size or innovation and species range size. Third, geographic range size in primates may be a particularly labile trait, as in New World birds, and so evolutionary patterns might be swamped by short-term fluctuations in species range size (Gaston and Blackburn, 1997).

If true, the lack of a relationship between species range size and innovation rate has a number of implications for the interpretation of hominid range expansion. While the primate analysis does not disprove the hypothesis that hominid range expansion was related to an increase in brain size and behavioural flexibility, it raises the possibility that it may not have been part of a general primate trend. If behavioural flexibility was important in hominid range expansion, it may have been on a different scale, or involved behaviours absent in other primates that were particularly useful in encountering new environments. Alternatively, some other process may have been critical to hominid range expansion.

It is a valid question as to whether the scales of analysis we utilized here were appropriate; perhaps animals respond to more fine-grained climatic changes, or to other environmental changes. Moreover, variation in itself may be less important than unpredictability in climatic conditions. As noted above, we utilized two timescales of climatic variability relevant to primate life histories, and also measured spatial variability, our aim being to measure variation across the species range rather than within a particular individual's habitat. In sum, we believe our choice of climatic measures and measurement scales are appropriate to investigations of environmental variability, but accept the possibility that other measures at other scales may also be informative.

If climatic variability and innovative tendencies are not related, we need to consider what processes may underlie the evolution of innovative capabilities. An obvious candidate would be the demands of social living. The 'social intelligence' hypothesis argues that enhanced brain size in primates is a response to increases in the complexity of group life (Jolly, 1966; Humphrey, 1976; Byrne and Whiten, 1988; Whiten and Byrne, 1997; Dunbar, 1998). Social innovation could be a major component of such social intelligence (Byrne, Chapter 11). A large number of reported innovations are in the foraging domain, a finding that would seem to conflict with social pressures being the driving force behind innovation (Reader and Laland, 2002). However, ecological innovations could also have social payoffs, and innovative tendencies are unlikely to be restricted to one context (Seyfarth and Cheney, 2002). A related candidate process that might have favoured the evolution of innovative propensities is sexual selection (Slater and Lachlan, Chapter 5; Miller, 2000). Innovative males, for example, may be able to attract the attention of potential mates with novel courtship displays, circumvent the restrictions of a dominance hierarchy, or intimidate potential rivals for females (Kummer and Goodall, Chapter 10; Goodall, 1986; Parnell and Buchanan-Smith, 2001; Reader and Laland, 2001). Comparative analyses, using innovation rate to operationally quantify innovative propensities, will clearly have an essential role in testing these hypotheses and understanding the evolution of innovation.

Summary

Flexibility in behaviour, particularly innovation, is predicted to allow animals to cope with environmental change and to inhabit wide geographical ranges. An operational measure of innovation, the frequency of reports of novel behaviour, allows us to address the hypothesis that innovation, species range size, and climatic variability are linked. We collated 533 records of non-human primate innovation from the published literature and calculated species innovation indexes, accounting for differences in research effort. Species' ranges were digitized from maps and their spatial and temporal climatic variability analysed. Temporal variability was measured on seasonal and inter-annual timescales. After controlling for phylogeny, innovation frequency correlated positively with species' relative neocortex volumes. Innovation frequencies also correlated with laboratory measures of learning, increasing our confidence in the innovation measure, and with social learning frequencies, suggesting that innovation and social learning propensities have evolved together. Species range size did not correlate with innovation frequency or neocortex volume. If innovation frequency is a measure of behavioural flexibility, this finding contrasts with the prediction that flexible species will have large geographic ranges. Innovation frequencies and neocortex size did not correlate with climatic variability, inconsistent with the hypotheses that more innovative species can tolerate greater climatic variability, that innovation is a response to climatic change, or that climatic variability has selected for a propensity to innovate.

Acknowledgements

This chapter benefited from discussions with and comments from Louis Lefebvre, Daniel Sol, James Steele, Kevin Laland, and two anonymous referees. Heiko Frahm, Samuel Wang, and Rob Deaner were kind enough to provide or suggest sources of brain volume data, or to clarify discrepancies between different data sources. SMR was funded by a Royal Society postdoctoral fellowship, KM by NERC.

Appendix

Data on body masses, brain masses, and volumes of total brain and some component parts in primates are given in Table 4.A1. Various ratio measures are calculated from component volumes: the neocortex ratio (Dunbar, 1992), executive brain ratio ([neocortex + striatum]/[mesecephalon + diencephalon]; Reader and Laland, 2002), and the neocortex ratio taking cerebellum into account (see this chapter). Some problematic data are presented for information only, these data are in italics and marked by an asterisk in the notes column. The data should not be used 'as it is' without reference to the notes below.

Species   Prosimians and simians are listed in alphabetical order.

Sources   1. Stephan *et al.* (1970, 1981, 1988). 2. Stephan *et al.* (1970). 3. Rilling and Insel (1998, 1999). 4. Hopf and Claussen (1971). 5. Semendeferi and Damasio (2000). 6. Zilles and Rehkämper (1988). Stephan *et al.*'s papers are referred to in the notes by the year of publication.

Measurement methods   MRI, Magnetic resonance imaging; FM, fresh mass; SS, Serial sections. See main text for notes on the comparability of the various measures.

Body and brain masses   Body and brain masses presented by Stephan *et al.* (1970, 1981, 1988) are 'standard' or 'species typical' values, modified from data presented in 1966 and 1969 (Bauchot and Stephan, 1966, 1969) that combined data from the literature with Bauchot *et al.*'s own data. Hence the sample size for these values frequently exceeds the n quoted here, the sample size for the brain volume measurements. Rilling and Insel (1998, 1999) quote body masses that are based upon captive specimens, and so these data are not listed here. Smith and Jungers (1997) provide a recent compendium of primate body masses.

Notes on sources   Stephan *et al.* (1970, 1981, 1988) took serial sections through fixed brains, and corrected the calculated volumes for shrinkage and to species-typical brain

Table 4.A1 Data on body masses, brain masses, and volumes of total brain and some component parts in primates

Species	Body mass (g)	Brain mass (g)	Total brain volume (mm³)	Mesencephalon volume (mm³)	Medulla oblongata volume (mm³)	Cerebellum volume (mm³)	Striatum volume (mm³)
Prosimians							
Avahi l. laniger	1285.0	10.49	9798	343.0	553.0	1489.0	524.0
Avahi l. occidentalis	860.0	9.67	9124	345.0	508.0	1383.0	478.0
Cheirogaleus major	450.0	6.80	6373	241.0	382.0	947.0	327.0
Cheirogaleus medius	177.0	3.14	2961	145.0	202.0	437.0	145.0
Daubentonia madagascariensis	2800.0	45.15	42,611	897.0	1517.0	6461.0	2765.0
Galago senegalensis	186.0	4.80	4512	205.0	254.0	672.0	210.0
Galagoides demidoff	81.0	3.38	3203	135.0	169.0	413.0	169.0
Indri indri	6250.0	38.30	36,285	978.0	1342.0	5504.0	1855.0
Lepilemur ruficaudatus	915.0	7.60	7175	260.0	449.0	1165.0	366.0
Loris tardigradus	322.0	6.60	6269	220.0	233.0	728.0	351.0
Microcebus murinus	54.0	1.78	1680	85.2	99.1	234.0	85.7
Nycticebus coucang	800.0	12.50	11,755	345.0	528.0	1310.0	760.0
Otolemur crassicaudatus	850.0	10.30	9668	384.0	540.0	1414.0	556.0
Perodicticus potto	1150.0	14.00	13,212	391.0	680.0	1699.0	712.0
Petterus fulvus	1400.0	23.30	22,106	590.0	909.0	3328.0	1215.0
Propithecus verreauxi	3480.0	26.70	25,194	780.0	1223.0	3957.0	1411.0
Tarsius sp.	125.0	3.60	3393	197.0	207.0	428.0	133.0
Tarsius syrichta	*87.5*	*3.63*	*3416*	*171.0*	*185.0*	*422.0*	*133.0*
Varecia variegata	3000.0	31.50	29,713	1018.0	1420.0	4286.0	1591.0
Simians							
Alouatta sp.	6400.0	52.00	49,009	986.0	1593.0	5699.0	2829.0
Aotus trivirgatus	830.0	17.10	16,195	409.0	686.0	1873.0	862.0
Ateles geoffroyi	8000.0	108.00	101,034	1482.0	1834.0	12,438.0	4950.0
Callicebus moloch	650.0	15.50	14,434	469.0	683.0	1287.0	721.0
...	*900.0*	*19.00*	*17,944*	*530.0*	*787.0*	*1622.0*	*920.0*
Callimico goeldii	480.0	11.00	10,510	340.0	460.0	1240.0	493.0
Callithrix jacchus	280.0	7.60	7241	295.0	318.0	757.0	372.0
Cebuella pygmaea	140.0	4.50	4302	192.0	201.0	468.0	174.0
Cebus apella			66,500			6500.0	
Cebus sp.	3100.0	71.00	66,939	1221.0	1738.0	7871.0	3258.0
Cercocebus albigena	7900.0	104.00	97,603	1770.0	2708.0	10,726.0	4146.0
Cercocebus torquatus atys			98,900			9200.0	
Cercopithecus aethiops			62,748				
Cercopithecus ascanius	3400.0	67.00	63,505	1162.0	1632.0	5828.0	2827.0
Cercopithecus mitis	6300.0	75.00	70,564	1354.0	1999.0	6758.0	2733.0
Colobus badius	7000.0	78.00	73,818	1333.0	2007.0	8648.0	3217.0
Erythrocebus patas	7800.0	108.00	103,167	1621.0	2616.0	8738.0	3624.0
Gorilla gorilla			397,300			64,700.0	
...						*69,300.0*	
...	105,000.0	500.00	470,359	4352.0	7509.0	69,249.0	14,567.0

Neocortex volume (mm³)	n (M = Male; F = Female)	Neocortex/ (Total brain– Neocortex)	Executive/ Brainstem	Neocortex/ (Total brain– Neocortex– Cerebellum)	Data reference	Measurement method	Notes
4813.0	2	0.97	5.96	1.38	1	SS	1
4443.0	2	0.95	5.77	1.35	1	SS	2
2938.0	2	0.86	5.24	1.18	1	SS	
1221.0	2	0.70	3.94	0.94	1	SS	
22,127.0	1	1.08	10.31	1.58	1	SS	
2139.0	1	0.90	5.12	1.26	1	SS	
1568.0	2	0.96	5.71	1.28	1	SS	
20,114.0	2	1.24	9.47	1.89	1	SS	
3282.0	3	0.84	5.15	1.20	1	SS	
3524.0	1	1.28	8.55	1.75	1	SS	10
740.0	6	0.79	4.48	1.05	1	SS	
6192.0	2	1.11	7.96	1.46	1	SS	
4723.0	2	0.96	5.71	1.34	1	SS	
6683.0	2	1.02	6.90	1.38	1	SS	
12,207.0	3	1.23	8.95	1.86	1	SS	
13,170.0	2	1.10	7.28	1.63	1	SS	
1768.0	2	1.09	4.71	1.48	1	SS	3c
1890.0	1	*1.24*	*5.68*	*1.71*	2	SS	*3c
15,293.0	1	1.06	6.93	1.51	1	SS	
31,660.0	2	1.82	13.37	2.72	1	SS	3b
9950.0	5	1.59	9.87	2.28	1	SS	
70,856.0	1	2.35	22.86	3.99	1	SS	
8973.0	1	1.64	8.41	2.15	2	SS	4
11,163.0	2	*1.65*	*9.17*	*2.16*	1	SS	*4
6476.0	1	1.61	8.71	2.32	1	SS	
4371.0	4	1.52	7.74	2.07	1	SS	
2535.0	2	1.43	6.89	1.95	1	SS	9
45,000.0	2M 2F	2.09		3.00	3	MRI	12
46,429.0	2	2.26	16.79	3.67	1	SS	3a
68,733.0	1	2.38	16.27	3.79	1	SS	
63,900.0	2M 2F	1.83		2.48	3	MRI	12
27,268.0	12M 41F	*0.77*			4	FM	*16
45,166.0	1	2.46	17.18	3.61	1	SS	
49,933.0	1	2.42	15.71	3.60	1	SS	
50,906.0	2	2.22	16.20	3.57	1	SS	13
77,141.0	2	2.96	19.06	4.46	1	SS	6
246,900.0	1M 1F	1.64		2.88	3	MRI	12
	2				5	MRI	*15
341,444.0	1	2.65	30.02	5.72	1	SS	

(Continued)

Table 4.A1 (*Continued*)

Species	Body mass (g)	Brain mass (g)	Total brain volume (mm³)	Mesencephalon volume (mm³)	Medulla oblongata volume (mm³)	Cerebellum volume (mm³)	Striatum volume (mm³)
Homo sapiens			1,298,900			134,100.0	
...						155,100.0	
...	65,000.0	1330.00	1,251,847	8087.0	9622.0	137,421.0	28,689.0
Hylobates lar			83,000			10,900.0	
...	5700.0	102.00	97,505	1459.0	2251.0	12,078.0	4784.0
Hylobates sp.						10,700.0	
Lagothrix lagotricha	5200.0	101.00	95,503	1582.0	2110.0	11,268.0	4947.0
Macaca mulatta			79,100			7100.0	
...	7800.0	93.00	87,896	1380.0	1992.0	8965.0	4032.0
Miopithecus talapoin	1200.0	40.00	37,776	826.0	1035.0	3374.0	1908.0
Nasalis larvatus	14,000.0	97.00	92,797	1556.0	2945.0	12,113.0	3735.0
Pan paniscus			311,200			41,300.0	
...						45,800.0	
Pan troglodytes			337,300			46,400.0	
...						41,200.0	
...	46,000.0	405.00	382,103	3739.0	5817.0	43,663.0	12,246.0
Papio anubis	25,000.0	201.00	190,957	2711.0	5297.0	18,683.0	7182.0
Papio cynocephalus			143,300			13,700.0	
Papio hamadryas			150,965				
Pithecia monachus	1500.0	35.00	32,867	745.0	1009.0	3908.0	1918.0
Pongo pygmaeus			406,900			46,000.0	
...						52,000.0	
...	54,000.0	333.00	308,500	4000.0	5500.0	47,200.0	11,500.0
Pygathrix nemaeus	7500.0	77.00	72,530	1345.0	2206.0	8063.0	3166.0
Saguinus midas	340.0	10.30	9569	332.0	428.0	1061.0	471.0
Saguinus oedipus	380.0	10.00	9537	333.0	413.0	984.0	453.0
Saimiri sciureus			23,100			2000.0	
...	660.0	24.00	22,572	526.0	722.0	2260.0	1042.0
Theropithecus gelada gelada			113,658				

Neocortex volume (mm³)	n (M = Male; F = Female)	Neocortex/ (Total brain– Neocortex)	Executive/ Brainstem	Neocortex/ (Total brain– Neocortex– Cerebellum)	Data reference	Measurement method	Notes
980,400.0	3M 3F	3.08		5.32	3	MRI	*11,12
	10				5	MRI	*11,15
1,006,525.0	1	4.10	58.46	9.33	1	SS	*11
56,600.0	2M 2F	2.14		3.65	3	MRI	12
65,800.0	1	2.08	19.03	3.35	1	SS	
	4				5	MRI	*15
65,873.0	3	2.22	19.18	3.59	1	SS	
55,700.0	2M 2F	2.38		3.42	3	MRI	12
63,482.0	1	2.60	20.02	4.11	1	SS	
26,427.0	2	2.33	15.23	3.31	1	SS	8
62,685.0	1	2.08	14.76	3.48	1	SS	5
218,500.0	2M 2F	2.36		4.25	3	MRI	12
	3				5	MRI	*15
238,000.0	3M 3F	2.40		4.50	3	MRI	12
	6				5	MRI	*15
291,592.0	1	3.22	31.80	6.22	1	SS	
140,142.0	2	2.76	18.40	4.36	1	SS	
100,100.0	2M 0F	2.32		3.39	3	MRI	12
78,098.0	5M 7F	1.07			4	FM	*16
21,028.0	2	1.78	13.08	2.65	1	SS	
287,700.0	3M 1F	2.41		3.93	3	MRI	12
	4				5	MRI	*15
219,800.0	1–2	2.48	24.35	5.30	6	SS	14
48,763.0	1	2.05	14.62	3.11	1	SS	
5883.0	2	1.60	8.36	2.24	1	SS	7
5894.0	3	1.62	8.51	2.22	1	SS	
16,500.0	3M 1F	2.50		3.59	3	MRI	12
15,541.0	1	2.21	13.29	3.26	1	SS	
41,631.0	4	0.58			4	FM	*16

masses. Stephan *et al.* calculated total fresh brain volume (including ventricles and meninges etc.) from fresh brain mass divided by 1.036, the density of fresh brain tissue in grams per cubic centimetre. Zilles and Rehkämper (1988) follow a similar procedure, also collating data from the literature to estimate a species typical brain mass and correcting measured volumes to this species typical datum. Like Stephan *et al.* and Zilles and Rehkämper, here total brain volume refers to the volume excluding ventricles, meninges etc. For additional details on neocortex components, see Frahm *et al.* (1982). Stephan *et al.* (1988) present body mass, brain mass, and N, along with various indices calculated from the Stephan *et al.* (1981) data set. The 1970 data are based upon smaller sample sizes, and so the 1981 data are typically listed here.

Notes
1. One of the *Avahi laniger laniger* specimens listed in 1981 was listed as *Hapalemur simus* in 1970, but was subsequently reclassified on the basis of tooth morphology (Frahm, personal communication). This specimen was also listed as *Hapalemur simus* in 1966.
2. In 1970 data was presented for '*Avahi laniger*' ($n = 1$). Body and brain mass match the 1966, 1981, and 1988 data for *Avahi laniger occidentalis*. Small differences in the brain volumes cited in the 1970 and 1981 papers are presumably the result of the addition of a second specimen.
3. Malagasy prosimians, African prosimians, and African simians were collected and prepared by Stephan and colleagues in the field. However, material from South America and south-east Asia was more heterogeneous, partly collected from zoos or the animal trade, which made species identity more difficult, especially where species of the same genus have similar brain and body masses (Frahm, personal communication). There are thus doubts over the identity of the *Cebus*, *Alouatta*, and *Tarsius* specimens, leading to differences between the species names in 1970, 1981, and 1988, detailed below. Hence species (but not genus) identities in these three cases should be treated with caution.
(3a) *Cebus.* In 1970, data are presented for *Cebus* sp. ($n = 1$) and *Cebus albifrons* ($n = 1$). In 1981, data are presented for *Cebus* sp. only ($n = 2$). Though the 1981 brain volumes do not match the mean of the 1970 data, note that the brain volumes are corrected to different brain masses so this would not be expected to be the case. The mean ratio of neocortex to total brain volume calculated from the 1970 data matches that calculated to the 1981 data to four decimal places. In 1988, the same brain data are presented, but the species is described as *C. albifrons* ($n = 2$). For the analyses described here, we thus assumed the species concerned was *C. albifrons*.
(3b) *Alouatta.* Data is presented for *Alouatta seniculus* ($n = 1$) in 1970, but in 1981 there is reference to *Alouatta* sp. only ($n = 2$). In 1988, the same brain data are presented, but the species is described as *A. seniculus* ($n = 2$). The 1970 body mass and brain mass data match 1969 data for '*Alouatta* sp.', where an average was calculated for four species of *Alouatta* (Bauchot and Stephan, 1969). Here, we assumed the species concerned was *A. seniculus*.
(3c) *Tarsius.* The same data are presented as *Tarsius* sp. ($n = 2$) in 1981 and 1988. In 1970, *Tarsius syrichta* ($n = 1$) is described, the body mass and brain mass matching that

quoted as *T. syrichta* in 1966 and the mean of the three female *T. syrichta* in the Stephan collection (Stephan, 2002). *Tarsius spectrum*, the other tarsier detailed in 1966, is rather different in size (body mass = 170 g, brain mass = 4.65 g, $n = 2$) to *T. syrichta* (body mass = 87.5 g, brain mass = 3.63 g, $n = 3$; Bauchot and Stephan, 1969). However, the mean of male and female body masses quoted by Smith and Jungers (1997) are more similar to one another and close to the body mass quoted for *Tarsius* sp. in 1981 (*T. spectrum* = 117 g, $n = 39$, *T. syrichta* = 126 g, $n = 27$). We did not attempt to assign the *Tarsius* sp. data to one species, and so *Tarsius* was excluded from the analyses presented here.

4. *Callicebus moloch*. As noted by Clark *et al.* (2001), the components fail to sum to the whole in the 1981 data. The sum of telencephalon components checks with that of telencephalon volume, but total brain volume neither matches the sum of components nor the expected brain volume calculated from brain mass as detailed above. Frahm (personal communication) notes this is probably an error due to correction for an enlarged hydrocephalus in one specimen. Hence data for *C. moloch* from 1970 is also presented, and was utilized in the analyses presented here.

5. Does not match some previously published data. There is a typographical error in the neocortex ratio of *Nasalis* in Dunbar (1992).

6. There is a typographical error in the neocortex ratio of *Erythrocebus patas* in Dunbar (1998, figure 3).

7. Listed as *Saguinus tamarin* in 1981, *Saguinus midas* in 1988.

8. Listed as *Cercopithecus talapoin* in 1981, *Miopithecus talapoin* in 1988.

9. *Cebuella pygmaea* body mass was 120 g and brain mass 4.15 g in 1988.

10. Listed as *Loris gracilis* in 1970.

11. *Homo sapiens* was not included in these analyses.

12. Rilling and Insel (1998, 1999) do not correct their volume measurements to the mean total brain volume of the species. Ratio measures may provide an advantage over other relative brain measures if the assumption is made that the ratio of one brain component to another is relatively constant between individuals of the same species, whereas the absolute volumes of components may be quite different (e.g. between males and females in species with significant sexual dimorphism in body size). The Rilling and Insel (1998, 1999) data appear to be based on the same specimens. There are differences between the two papers in the data for two species. *Gorilla gorilla* brain volume is 397.3 cc and body mass 61.7 kg in 1999, but 383.5 cc and 85.0 kg in 1998, with one male and one female measured in both years. *Cercocebus atys* (listed as *C. torquatus atys* in the text in 1999) brain volume is 98.7 cc and body mass 8.8 kg in 1999, 99.7 cc and 10.5 kg in 1998; two males and two females were measured in 1999, three males and one female in 1998. The most recent brain volume data (tabulated) were used in both cases. Cerebellum volumes are from Rilling and Insel (1998). Rilling and Insel (1999) calculate neocortex volume from the sum of the neocortical grey matter and the cerebral white matter. Whole brain volume was calculated by separating brain tissue from surrounding tissues (cerebrospinal fluid, meninges, blood vessels, muscle, fat, and bone; Rilling and Insel, 1999). It may thus not be exactly comparable to the net volume presented by Stephan *et al.* (see above). In five species, one of the specimen was subadult, but Rilling and

Insel (1999) note that these subjects' brains were probably adult-sized or nearly adult-sized. Sample sizes differ for some brain components, see Rilling and Insel (1999).

13. *Colobus badius*. Neocortex volume incorrectly given as 50905 mm^3 in Clark *et al.* (2001).

14. Two brains were measured, but the components of the telencephalon were measured in only one brain. Cerebellum and pons volume were quoted separately, these have been summed here.

15. For comparison, cerebellum volumes from Semendeferi and Damasio (2000) are presented. Semendeferi and Damasio present whole brain volumes, but this is not comparable with the data presented here, as it excludes the medulla, pons, and the 'greater part of the midbrain' (p. 318). The species of gibbon is not identified.

16. Brain volume is calculated from brain mass using the formula above. The unweighted mean of the two sexes was taken, and we assume an equal sex ratio in *Teropithecus gelada gelada*, where the sex ratio was not stated. Hopf and Claussen (1971) give the ratio of the neocortex to the spinal cord, presumably based upon mass as they present spinal cord mass data. It should thus be possible to calculate neocortex mass, and, from this, neocortex volume. However, though mass data are presented separately for males and females, it is not clear how the neocortex ratios were calculated. It was not possible to calculate the brain mass: spinal cord mass ratios quoted by Hopf and Claussen (1971) using weighted or unweighted mean of both sexes, or examining males and females individually. The neocortex volumes presented (calculated using an unweighted average of the two sexes) are for information only, and should not be utilized: note that the neocortex ratios seem unusually low.

References

Alp, R. (1997). 'Stepping-sticks' and 'seat-sticks': New types of tools used by wild chimpanzees (*Pan troglodytes*) in Sierra Leone. *American Journal of Primatology*, **41**, 45–52.

Ancel Meyers, L. and Bull, J. J. (2002). Fighting change with change: Adaptive variation in an uncertain world. *Trends in Ecology and Evolution*, **17**, 551–7.

Autodesk, Inc. (1997). *AutoCAD Map Release 2*.

Barton, R. (1999). The evolutionary ecology of the primate brain. In *Comparative primate socioecology* (ed. P. C. Lee), pp. 167–94. Cambridge: Cambridge University Press.

Barton, R. A. and Harvey, P. H. (2000). Mosaic evolution of brain structure in mammals. *Nature*, **405**, 1055–8.

Bateson, P. P. G. (1988). The active role of behaviour in evolution. In *Evolutionary processes and metaphors* (ed. M.-W. Ho and S. W. Fox), pp. 191–207. Chichester: Wiley.

Bauchot, R. and Stephan, H. (1966). Données nouvelles sur l'encephalisation des insectivores et des prosimiens. *Mammalia*, **30**, 160–96.

Bauchot, R. and Stephan, H. (1969). Encéphalisation et niveau évolutif chez les simiens. *Mammalia*, **33**, 225–75.

Björklund, M. (1994). The independent contrast method in comparative biology. *Cladistics*, **10**, 425–33.

Boyd, R. and **Richerson, P. J.** (1985). *Culture and the evolutionary process.* Chicago, IL: University of Chicago.

Byrne, R. W. (1992). The evolution of intelligence. In *Behaviour and evolution* (ed. P. J. B. Slater and T. R. Halliday), pp. 223–65. Cambridge: Cambridge University Press.

Byrne, R. W. (1993). Do larger brains mean greater intelligence? *Behavioral and Brain Sciences,* **16**, 696–7.

Byrne, R. W. (1997). The technical intelligence hypothesis: An additional evolutionary stimulus to intelligence? In *Machiavellian intelligence II* (ed. A. Whiten and R. W. Byrne), pp. 289–311. Cambridge: Cambridge University Press.

Byrne, R. W. and **Whiten, A.** (1988). *Machiavellian intelligence: Social expertise and the evolution of intellect in monkeys, apes and humans.* Oxford: Oxford University Press.

Calvin, W. H. (2002). *A brain for all seasons: Human evolution and abrupt climate change.* Chicago, IL: University of Chicago.

Chapman, C. A. and **Chapman, L. J.** (1990). Dietary variability in primate populations. *Primates,* **31**, 121–8.

Chivers, D. J. (1991). Species differences in tolerance to environmental change. In *Primate responses to environmental change* (ed. H. O. Box), pp. 5–38. London: Chapman & Hall.

Clark, A. B. (1991). Individual variation in responsiveness to environmental change. In *Primate responses to environmental change* (ed. H. O. Box), pp. 91–114. London: Chapman & Hall.

Clark, D. A., Mitra, P. P., and **Wang, S. S.-H.** (2001). Scalable architecture in mammalian brains. *Nature,* **411**, 189–93.

Count, E. W. (1947). Brain and body weights in man: Their antecedents in growth and evolution. *Annals of the New York Academy of Sciences,* **46**, 993–1122.

Cowlishaw, G. and **Hacker, J. E.** (1997). Distribution, diversity and latitude in African primates. *American Naturalist,* **150**, 505–12.

Darlington, R. B. and **Smulders, T. V.** (2001). Problems with residual analysis. *Animal Behaviour,* **62**, 599–602.

Day, R. L., Coe, R. L., Kendal, J. R., and **Laland, K. N.** (2003). Neophilia, innovation and social learning: A study of intergeneric differences in callitrichid monkeys. *Animal Behaviour,* **65**, 559–71.

De Winter, W. and **Oxnard, C. E.** (2001). Evolutionary radiations and convergences in the structural organization of mammalian brains. *Nature,* **409**, 710–14.

Deaner, R. O., Nunn, C. L., and **van Schaik, C. P.** (2000). Comparative tests of primate cognition: Different scaling methods produce different results. *Brain, Behavior and Evolution,* **55**, 44–52.

Diaz-Uriate, R. and **Garland, T.** (1996). Testing hypotheses of correlated evolution using phylogenetically independent contrasts: Sensitivity to deviations from Brownian motion. *Systematic Biology,* **45**, 27–47.

Dunbar, R. I. M. (1992). Neocortex size as a constraint on group size in primates. *Journal of Human Evolution,* **20**, 469–93.

Dunbar, R. I. M. (1998). The social brain hypothesis. *Evolutionary Anthropology,* **6**, 178–90.

Eeley, H. A. C. and **Foley, R. A.** (1999). Species richness, species range size and ecological specialization among African primates. *Biodivesity and Conservation,* **8**, 1033–56.

Erdas, Inc. (1999). *Erdas Imagine version 8.4*: Erdas worldwide headquarters, 2801 Buford Highway, NE, Atlanta, Georgia 30329-2137, USA.

ESRI, Inc. (1999). *ArcInfo version 8.0.1*: 380 New York Street, Redlands, CA 92373-8100, USA.

Fagen, R. (1982). Evolutionary issues in development of behavioral flexibility. In *Perspectives in ethology* (ed. P. G. Bateson and P. H. Klopfer), Vol. 5, pp. 365–83. New York: Plenum Press.

Fairgrieve, C. (1997). Meat-eating by blue monkey (*Cercopithecus mitis stuhlmanni*): Predation of a flying squirrel (*Ananchanus derbianus jacksani*). *Folia Primatologica*, **68**, 354–6.

Felsenstein, J. (1985). Phylogenies and the comparative method. *American Naturalist*, **125**, 1–15.

Finlay, B. L. and **Darlington, R. B.** (1995). Linked regularities in the development and evolution of mammalian brains. *Science*, **268**, 1578–84.

Frahm, H. D., Stephan, H., and **Stephan, M.** (1982). Comparison of brain structure volumes in insectivora and primates. I. Neocortex. *Journal für Hirnforschung*, **23**, 375–89.

Galef, B. G. Jr. (1992). The question of animal culture. *Human Nature*, **3**, 157–78.

Gamble, C. (1994). *Timewalkers: The prehistory of global colonization.* Cambridge, MA: Harvard University Press.

Garland, T. Jr., Harvey, P. H., and **Ives, A. R.** (1992). Procedures for the analysis of comparative data using phylogenetically independent contrasts. *Systematic Biology*, **41**, 18–31.

Gaston, K. J. (1996). Species-range-size distributions: Patterns, mechanisms and implications. *Trends in Ecology and Evolution*, **11**, 197–201.

Gaston, K. J. and **Blackburn, T. M.** (1997). Age, area and avian diversification. *Biological Journal of the Linnean Society*, **62**, 239–53.

Gaston, K. J., Blackburn, T. M., and **Spicer, J. I.** (1998). Rapoport's rule: Time for an epitaph? *Trends in Ecology and Evolution*, **13**, 70–4.

Gibson, K. R. (1999). Social transmission of facts and skills in the human species: Neural mechanisms. In *Mammalian social learning: Comparative and ecological perspectives* (ed. H. O. Box and K. R. Gibson), pp. 351–66. Cambridge: Cambridge University Press.

Goodall, J. (1986). *The chimpanzees of Gombe: Patterns of behaviour.* Cambridge, MA: Belknap Press.

GRASS (1993). *GRASS 4.1 Reference manual.* Champaign, IL: US Army Corps of Engineers, Construction Engineering Laboratories.

Hansell, M. H. (1984). *Animal architecture and building behaviour.* London: Longman.

Harvey, P. H. and **Krebs, J. R.** (1990). Comparing brains. *Science*, **249**, 140–6.

Harvey, P. H. and **Pagel, M. D.** (1991). *The comparative method in evolutionary biology.* Oxford: Oxford University Press.

Hopf, A. and **Claussen, C.-P.** (1971). Comparative studies on the fresh weights of the brains and spinal cords of *Theropithecus gelada, Papio hamadryas* and *Cercopithecus aethiops*. *Proceedings of the Third International Congress of Primatology, Zurich*, **1**, 115–21.

Hosey, G. R., Jacques, M., and **Pitts, A.** (1997). Drinking from tails: Social learning of a novel behaviour in a group of ring-tailed lemurs (*Lemur catta*). *Primates*, **38**, 415–22.

Howell, D. C. (1997). *Statistical Methods for Psychology*, 4th edn. Belmont: Duxbury.

Humphrey, N. K. (1976). The social function of intellect. In *Growing points in ethology* (ed. P. P. G. Bateson and R. A. Hinde), pp. 303–17. Cambridge: Cambridge University Press.

Janson, C. H. and **Chapman, C. A.** (1999). Resources and primate community structure. In *Primate communities* (ed. J. G. Fleagle, C. Janson, and K. E. Reed), pp. 237–67. Cambridge: Cambridge University Press.

Jerison, H. J. (1973). *Evolution of the brain and intelligence.* New York: Academic Press.

Joffe, T. H. and **Dunbar, R. I. M.** (1997). Visual and socio-cognitive information processing in primate brain evolution. *Proceedings of the Royal Society of London, Series B*, **264**, 1303–7.

Johnston, T. D. (1982). The selective costs and benefits of learning: An evolutionary analysis. *Advances in the Study of Behaviour*, **12**, 65–106.

Jolicoeur, P., Pirlot, P., Baron, G., and **Stephan, H.** (1984). Brain structure and correlation patterns in insectivora, chiroptera, and primates. *Systematic Zoology*, **33**, 14–29.

Jolly, A. (1966). Lemur social behavior and primate intelligence. *Science*, **153**, 501–6.

Kaas, J. H. and **Collins, C. E.** (2001). Evolving ideas of brain evolution. *Nature*, **411**, 141–2.

Keverne, E. B., Martel, F. L., and **Nevison, C. M.** (1996). Primate brain evolution: Genetic and functional considerations. *Proceedings of the Royal Society of London, Series B*, **262**, 689–96.

Laland, K. N. and **Reader, S. M.** (1999a). Foraging innovation in the guppy. *Animal Behaviour*, **57**, 331–40.

Laland, K. N. and **Reader, S. M.** (1999b). Foraging innovation is inversely related to competitive ability in male but not in female guppies. *Behavioral Ecology*, **10**, 270–74.

Laland, K. N., Odling-Smee, F. J., and **Feldman, M. W.** (1996a). The evolutionary consequences of niche construction: A theoretical investigation using two-locus theory. *Journal of Evolutionary Biology*, **9**, 293–316.

Laland, K. N., Richerson, P. J., and **Boyd, R.** (1996b). Developing a theory of animal social learning. In *Social learning in animals: The roots of culture* (ed. C. M. Heyes and B. G. Galef, Jr.), pp. 129–54. London: Academic Press.

Laland, K. N., Odling-Smee, J., and **Feldman, M. W.** (2000). Niche construction, biological evolution and cultural change. *Behavioural and Brain Sciences*, **23**, 131–75.

Lee, P. (1991). Adaptations to environmental change: An evolutionary perspective. In *Primate responses to environmental change* (ed. H. O. Box), pp. 39–56. London: Chapman & Hall.

Lefebvre, L. (2000). Feeding innovations and their cultural transmission in bird populations. In *The evolution of cognition* (ed. C. Heyes and L. Huber), pp. 311–28. Cambridge, MA: MIT Press.

Lefebvre, L. and **Bouchard, J.** (in press). Social learning about food in birds. In *The biology of traditions: Models and evidence* (ed. D. M. Fragaszy and S. Perry). Cambridge: Cambridge University Press.

Lefebvre, L. and **Giraldeau, L.-A.** (1996). Is social learning an adaptive specialization? In *Social learning in animals: The roots of culture* (ed. C. M. Heyes and B. G. Galef, Jr.), pp. 107–28. London: Academic Press.

Lefebvre, L., Whittle, P., Lascaris, E., and **Finkelstein, A.** (1997). Feeding innovations and forebrain size in birds. *Animal Behaviour*, **53**, 549–60.

Lewin, R. (1988). Living in the fast track makes for small brains. *Science*, **242**, 513–14.

Lewontin, R. C. (1983). Gene, organism and environment. In *Evolution from molecules to men* (ed. D. S. Bendall), pp. 273–85. Cambridge: Cambridge University Press.

Lynch, M. (1990). The rate of morphological evolution in mammals from the standpoint of the neutral expectation. *The American Naturalist*, **136**, 727–41.

MacDonald, K. (2002). Statistical analysis of the distribution of modern primates: A comparative approach to the spatial analysis of the Palaeolithic. In *Archaeological informatics: Pushing the envelope CAA 2001. BAR International Series 1016* (ed. G. Burenhult and J. Arvidsson), pp. 105–12.

Macphail, E. M. (1982). *Brain and intelligence in vertebrates.* Oxford: Clarendon Press.

Marino, L., Rilling, J. K., Lin, S. K., and **Ridgway, S. H.** (2000). Relative volume of the cerebellum in dolphins and comparison with anthropoid primates. *Brain, Behavior and Evolution*, **56**, 204–11.

Martin, R. D. (1981). Relative brain size and basal metabolic rate in terrestrial vertebrates. *Nature*, **293**, 57–60.

Miller, G. (2000). *The mating mind: How sexual choice shaped the evolution of human nature.* London: William Heinemann.

Milton, K. (1988). Foraging behaviour and the evolution of primate intelligence. In *Machiavellian intelligence: Social expertise and the evolution of intellect in monkeys, apes and humans* (ed. R. W. Byrne and A. Whiten), pp. 271–84. Oxford: Oxford University Press.

New, M. G., Hulme, M., and Jones, P. D. (1999). Representing 20th century space–time climate variability. I: Development of a 1961–1990 mean monthly terrestrial climatology. *Journal of Climate,* **12**, 829–56.

New, M. G., Hulme, M., and Jones, P. D. (2000). Representing 20th century space–time climate variability. II: Development of 1901–1996 monthly terrestrial climate fields. *Journal of Climate,* **13**, 2217–38.

Nicolakakis, N. and Lefebvre, L. (2000). Forebrain size and innovation rate in European birds: Feeding, nesting and confounding variables. *Behaviour,* **137**, 1415–29.

Nicolakakis, N., Lefebvre, L., and Sol, D. (2003). Behavioural flexibility predicts species richness in birds, but not extinction risk. *Animal Behaviour,* **65**, 445–52.

Oda, R. (1996). Predation on a chameleon by a ring-tailed lemur (*Lemur catta*) in the Berenty Reserve, Madagascar. *Folia Primatologica,* **67**, 40–3.

Owens, I. P. F. and Bennett, P. M. (2000). Ecological basis of extinction risk in birds: Habitat loss versus human persecution and introduced predators. *PNAS,* **97**, 12144–8.

Owens, I. P. F., Bennett, P. M., and Harvey, P. H. (1999). Species richness among birds: Body size, life history, sexual selection or ecology? *Proceedings of the Royal Society of London, Series B,* **266**, 933–9.

Pagel, M. (1999). Inferring the historical patterns of biological evolution. *Nature,* **401**, 877–84.

Parnell, R. J. and Buchanan-Smith, H. M. (2001). Animal behaviour: An unusual social display by gorillas. *Nature,* **412**, 294.

Peres, C. A. (1999). Effects of subsistence hunting and forest types on the structure of Amazonian forest communities. In *Primate communities* (ed. J. G. Fleagle, C. Janson, and K. E. Reed), pp. 268–83. Cambridge: Cambridge University Press.

Potts, R. (1998). Variability selection in hominid evolution. *Evolutionary Anthropology,* **7**, 81–96.

Price, T. (1997). Correlated evolution and independent contrasts. *Philosophical Transactions of the Royal Society of London,* **352**, 519–29.

Purvis, A. (1995). A composite estimate of primate phylogeny. *Philosophical Transactions of the Royal Society of London, Series B,* **348**, 405–21.

Purvis, A. and Rambaut, A. (1995). Comparative analysis by independent contrasts (CAIC): An Apple Macintosh application for analysing comparative data. *Computer Applications in the Biosciences,* **11**, 247–51.

Purvis, A. and Webster, A. J. (1999). Phylogentically independent comparisons and primate phylogeny. In *Comparative primate socioecology* (ed. P. C. Lee), pp. 44–70. Cambridge: Cambridge University Press.

Purvis, A., Gittleman, J. L., Cowlishaw, G., and Mace, G. M. (2000). Predicting extinction risk in declining species. *Proceedings of the Royal Society of London, Series B, Biological Sciences,* **267**, 1947–52.

Reader, S. M. (in press). Relative brain size and the distribution of innovation and social learning across the nonhuman primates. In *The biology of traditions: Models and evidence* (ed. D. M. Fragaszy and S. Perry). Cambridge: Cambridge University Press.

Reader, S. M. and Laland, K. N. (2001). Primate innovation: Sex, age and social rank differences. *International Journal of Primatology,* **22**, 787–805.

Reader, S. M. and Laland, K. N. (2002). Social intelligence, innovation and enhanced brain size in primates. *Proceedings of the National Academy of Sciences, USA,* **99**, 4436–41.

Reader, S. M., Nover, D., and Lefebvre, L. (2002). Locale-specific sugar packet opening by Lesser Antillean bullfinches in Barbados. *Journal of Field Ornithology,* **73**, 82–5.

Richerson, P. J. and Boyd, R. (2000). Climate, culture and the evolution of cognition. In *The evolution of cognition* (ed. C. Heyes and L. Huber), pp. 329–46. Cambridge, MA: MIT Press.

Riddell, W. I. and Corl, K. G. (1977). Comparative investigation of the relationship between cerebral indices and learning abilities. *Brain, Behavior and Evolution,* **14**, 385–98.

Rilling, J. K. and Insel, T. R. (1998). Evolution of the cerebellum in primates: Differences in relative volume among monkeys, apes and humans. *Brain, Behavior and Evolution*, **52**, 308–14.

Rilling, J. K. and Insel, T. R. (1999). The primate neocortex in comparative perspective using magnetic resonance imaging. *Journal of Human Evolution*, **37**, 191–223.

Rodrigues, F. H. G. and Maricho-Filho, J. (1995). Feeding on a marsh-living herbaceous plant by black howler monkeys (*Alouatta caraya*) in central Brazil. *Folia Primatologica*, **65**, 115–17.

Rowe, N. (1996). *The pictorial guide to the living primates.* New York: Pogonias Press.

Sarsa, L. L., Ferrari, S. F., and Dina, A. L. C. B. (1997). Feeding behaviour and predation of a bat by *Saimiri sciureus* in a semi-natural Amazonian environment. *Folia Primatologica*, **68**, 194–8.

Semendeferi, K. and Damasio, H. (2000). The brain and its main anatomical subdivisions in living hominoids using magnetic resonance imaging. *Journal of Human Evolution*, **38**, 317–32.

Seyfarth, R. M. and Cheney, D. L. (2002). What are big brains for? *Proceedings of the National Academy of Sciences, USA*, **99**, 4141–2.

Shennan, S. J. and Steele, J. (1999). Cultural learning in hominids: A behavioural ecological approach. In *Mammalian social learning: Comparative and ecological perspectives* (ed. H. O. Box and K. R. Gibson), pp. 367–88. Cambridge: Cambridge University Press.

Sibly, R. M. (1999). Evolutionary biology of skill and information transfer. In *Mammalian social learning: Comparative and ecological perspectives* (ed. H. O. Box and K. R. Gibson), pp. 57–71. Cambridge: Cambridge University Press.

Smith, R. J. and Jungers, W. L. (1997). Body mass in comparative primatology. *Journal of Human Evolution*, **32**, 523–59.

Sokal, R. R. and Rohlf, F. J. (1995). *Biometry*, 3rd edn. New York: Freeman.

Sol, D., Lefebvre, L., and Timmermans, S. (2002). Behavioural flexibility and invasion success in birds. *Animal Behaviour*, **63**, 495–502.

Stephan, H. (2002). Stephan collection (http://turing.commtechlab.msu.edu). *Housed at the Institut für Neuroanatomie und C. und O. Vogt Institut für Hirnforschung, University of Düsseldorf, Germany.*

Stephan, H. and Pirlot, P. (1970). Volumetric comparisons of brain structures in bats. *Zietschrift Zoologiste Systematik Evolutionforschung*, **8**, 200–36.

Stephan, H., Bauchot, R., and Andy, O. J. (1970). Data on the size of the brain and of various parts in insectivores and primates. In *The primate brain* (ed. C. R. Noback and W. Montagna), pp. 289–97. New York: Appleton-Century-Crofts.

Stephan, H., Frahm, H., and Baron, G. (1981). New and revised data on volumes of brain structure in insectivores and primates. *Folia Primatologica*, **35**, 1–29.

Stephan, H., Baron, G., and Frahm, H. (1988). Comparative size of brain and brain components. In *Comparative primate biology, vol. 4, Neuroscience* (ed. J. Erwin and H. D. Steklis), pp. 1–38. New York: Alan R. Liss.

Stephens, D. W. (1991). Change, regularity, and value in the evolution of animal learning. *Behavioural Ecology*, **2**, 77–89.

Takahasi, K. (1983). Mahale chimpanzees taste mangoes—toward acquisition of a new food item? *Primates*, **24**, 273–5.

Timmermans, S., Lefebvre, L., Boire, D., and Basu, P. (2000). Relative size of the hyperstriatum ventrale is the best predictor of feeding innovation rate in birds. *Brain, Behavior and Evolution*, **56**, 196–203.

Vrba, E. S. (1985). Ecological and adaptive changes associated with early hominid evolution. In *Ancestors: The hard evidence* (ed. E. Delson), pp. 63–71. New York: Alan R. Liss.

Wcislo, W. T. (1989). Behavioral environments and evolutionary change. *Annual Review of Ecology and Systematics*, **20**, 137–69.

Webster, S. J. and **Lefebvre, L.** (2001). Problem solving and neophobia in a Columbiforme—Passeriforme assemblage in Barbados. *Animal Behaviour*, **62**, 23–32.

Whiten, A. and **Byrne, R. W.** (1997). *Machiavellian intelligence II. Extensions and evaluations.* Cambridge: Cambridge University Press.

Wolfheim, J. H. (1983). *Primates of the World: Distribution, abundance and conservation.* Seattle: Washington University Press.

Wyles, J. S., Kunkel, J. G., and **Wilson, A. C.** (1983). Birds, behaviour, and anatomical evolution. *Proceedings of the National Academy of the USA*, **80**, 4394–7.

Zilles, K. and **Rehkämper, G.** (1988). The brain, with special reference to the telencephalon. In *Orang-utan biology* (ed. J. H. Schwartz), pp. 157–76. Oxford: Oxford University Press.

CHAPTER 5

IS INNOVATION IN BIRD SONG ADAPTIVE?

PETER J. B. SLATER AND ROBERT F. LACHLAN

Introduction

Innovation can be defined in a variety of different ways. The definition in the introduction to this volume is a broad one: 'a process that results in new or modified learned behaviour and that introduces novel behavioural variants into a population's repertoire'. Such a definition may include four different phenomena that it might help to differentiate between:

1. *Immigration.*   Behaviour patterns may only be novel from the geographical view-point, having been developed elsewhere but introduced by the movement of animals from one population to another. From the point of view of changes in behaviour at the level of the individual, this is relatively trivial, though it may be a very important reason why novel traits appear in a particular population. The 'immigration' of new types of song by accurate copying between populations (rather than by movement of individuals) is also a logical possibility; perhaps the most frequent example here is inter-specific mimicry.
2. *Innovation.*   Deriving their ideas from the vocal learning literature, Janik and Slater (2000) define innovation as where new behaviour patterns are derived by modifications of pre-existing ones.
3. *Invention.*   Janik and Slater (2000) suggest this as a term to describe the appearance of a behaviour pattern that is totally novel, not obviously derived from one that an animal has been exposed to.
4. *Improvisation.*   In line with its dictionary meaning, or that in music, this term refers to novel behaviour that is 'offhand' (Shorter Oxford English Dictionary) or 'according to the inventive whim of the moment' (Oxford Dictionary of Music). In other words, new patterns that are not necessarily based on previous ones, are made up on the spot and transient.

In discussing bird song, while innovation in this strict sense may be a dominant influence in leading to changes with time, as we shall see each of these other three influences may have a part to play.

Bird song is the classic case where individual animals learn their behaviour from others, usually young birds from adults, so that particular songs are passed on from one individual to another through cultural transmission. Learning can be remarkably accurate, but even in such cases new variants arise not infrequently and, in birds with large repertoires of songs, it is usual for the precise combination of songs that a bird has in its repertoire to be unique.

Much of bird song learning is accurate, leading the songs of young birds to conform with those round them, so that traditions may arise that last for many generations (Janik and Slater, 2003). This may be beneficial in various ways, for example through matching the form of song to the transmission qualities of the habitat, or allowing matched counter-singing between neighbours. So, is there a benefit to generating new song types or do these arise from copying errors leading to a breach in conformity? Whether novelties are in any sense beneficial to the individual that introduces them is the main question to be considered here. They may be novel simply because the new variant is different and 'stands out from the crowd' (in which case they are under negative frequency dependent selection). This situation might arise if, for example, females are attracted to novel sounds as a result of a sensory bias. On the other hand, a new variant may be adaptive in some absolute sense (i.e. be directly favoured by natural or cultural selection), so that the new song would, like an advantageous mutant gene, benefit any individual possessing it. This might be the case if, for example, a new song possesses acoustic qualities that allow it to be heard in a given environment at a greater distance from its source. We will discuss whether there is evidence for such processes and whether they lead to the spread of novel types of song within the population. Where we use the word 'adaptive' here it is in its conventional biological sense of whether the trait enhances the inclusive fitness of the individual possessing it.

While we will explore these issues here, and consider them in relation to a wide range of species, one thing we will not be able to do is provide a single definitive answer. As with so much else in bird song, the remarkable variety between species makes generalizations hard to come by. The commonly held view is that song learning is accurate in the great majority of instances and that, where it fails to be so, this is because errors have crept in and these are either neutral or deleterious. Our main question is: could these ever be beneficial to the individuals that generate them?

How changes take place

Vocal learning has so far been found in three bird groups: the parrots, the hummingbirds, and the oscine passerines or songbirds (Catchpole and Slater, 1995). The last of these comprises more than half the bird species and is the most extensively studied. In every songbird case that has been examined, learning has been found to play a role in song development but, within that generalization, there lies a huge variety of patterns.

In species with large song repertoires, in which individuals may have hundreds or even thousands of distinct song phrases, it is easy to gain the impression that songs are improvised, for two songs of the same type are seldom produced in close proximity and the bird may appear to be generating variety as it goes along. However, sonagraphic analysis usually reveals that adult birds have a stable and finite repertoire size, certainly as far as elements or syllables are concerned, and often of song types as well. There is a difficulty in generalizing here, because of the different singing styles adopted by different species. In many species, particularly those with small repertoires, song types are more or less fixed in structure and distinct from each other, so the number of song types is the best measure of repertoire. In other cases, the basic unit of the repertoire is the element, a variable subset of which are used to construct each song. Here, as for example in the sedge warbler (*Acrocephalus schoenobaenus*, Catchpole, 1976), a relatively limited element repertoire can be used to construct an almost infinite number of song types. Between these two extremes there are many intermediate possibilities although, because changes in them are easier to identify and quantify, much of the more detailed work on cultural change in bird song has been on species with a small number of rather fixed song types.

The standard method of plotting the number of new elements or song types found against the number analysed tends to give a logistic graph that rises steeply at first but then approaches an asymptotic level, which represents the animal's repertoire size. Of course, such a plot only fits a logistic model if the units are produced at random, and this is seldom the case, but most such plots do tend to level-off. An interesting exception is the study by Luschi (1993) of the Sardinian warbler (*Sylvia melanocephala*), in which new element types became scarcer as a single recording session progressed, but old ones were dropped and new ones continued to appear when the same bird was recorded in a subsequent bout (Figure 5.1). This may be because each bird has a very large repertoire and only uses a subsample of it in any one singing session or because there is genuine turnover, in a way more akin to improvisation, with elements being continuously dropped from the repertoire and new ones developed. Whichever is the case here, continuous repertoire

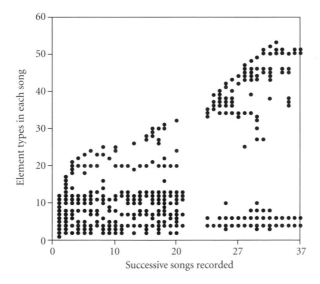

Figure 5.1 Two successive song bouts as sung by a male Sardinian warbler. Thirty-seven songs were produced and these were composed from 53 different elements, the black dots indicating the exact elements that occurred in each song. New elements continued to appear throughout the series, and a variety of those common in the first bout were rare or absent in the second. Redrawn from Luschi (1993) with permission.

change in adulthood is a phenomenon that certainly deserves further study. It occurs in thrush nightingales (*Luscinia luscinia*, Sorjonen, 1987), in which adjacent males learn from each other and come to share an increasing proportion of their repertoire during the course of the breeding season. The effect is so strong that a male's repertoire in a particular year is more like that of his neighbours than his own in the previous year. Interestingly, a very similar phenomenon occurs among mammals, in humpback whales (*Megaptera novaeangliae*, Payne and Payne, 1985), with songs being shared among individuals within a population but changing with time in concert with one another (see below).

In most species with well-defined song types, the repertoire appears to be learnt early in life and is then fixed. However, it is necessary to be a little cautious with this generalization as better conditions for learning, such as having a live tutor to copy rather than a tape-recording, may often lead birds to learn outside what has been thought to be the conventional sensitive phase for song learning (see e.g. Baptista and Petrinovich, 1984). Nevertheless, though a small amount of change may occur later, there is no doubt that in many species song learning is more or less 'age-limited' (see e.g. Jones *et al.*, 1996). An exception to this, found in some species with small repertoires, is where birds move territories and modify their songs to match those of their newly acquired neighbours (e.g. village indigobird, *Vidua chalybeata*, Payne, 1985). But in most cases the repertoire is fixed once the young bird reaches adulthood and does not change regardless of whether the songs in it match those of neighbours. However, a great deal of sharing of song types between nearby birds does occur and young birds entering the population very often copy songs present in it with a high degree of fidelity, resulting in stability in song traditions over several generations. Accurate copying is at the basis of these examples, but novel songs must also often be introduced despite this accuracy simply because song varies over relatively short distances. Where a novel song type does arise, this will be by means of one of the mechanisms mentioned at the beginning of this chapter. Here we shall discriminate between immigration (which does not require an individual to change a song type) and the other three mechanisms.

Immigration

The bird singing a song type may have learnt it accurately, but then moved some distance away into an area in which it is not normally sung. This is probably a relatively common source of novelty. While adult birds usually remain in the same area, and commonly on the same territory, between breeding seasons, young birds often disperse some distance before breeding and learning of song prior to dispersal is common. In the common chaffinch (*Fringilla coelebs*), for example, song may be learnt in the first summer, before dispersal, or the following spring, as the bird sets up his territory. A male that was the only one in a population in the Orkney islands singing types G and H might have been imagined to have created these types anew, had we not found them to be the commonest types in another population some 12 km away from which he had doubtless dispersed (Slater and Ince, 1979).

Innovation/invention/improvisation

While occasional cases of possible improvisation (e.g. Luschi, 1993) or invention (e.g. Kroodsma *et al.*, 1999) have been described, innovation as defined above is the major source of novelty in songs arising within a population rather than being introduced from elsewhere. We will therefore use the word innovation from here on to refer to this generation of novelty within a population. New song types may be formed during the learning process because all the song types in a bird's repertoire are not accurate copies of pre-existing ones. In his long-term study of a small closed population of saddlebacks (*Philesturnus carunculatus*) on an isolated island, Jenkins (1978) was able to exclude immigration and could therefore be sure that new types he found were generated in this way. Other studies with marked populations of birds have also found young birds that learn their songs within the population producing song types that differ in systematic ways from those already present (e.g. Payne, 1996). Even where birds are not marked the relationship between new types and pre-existing ones may suggest how changes have taken place, though it cannot be certain exactly where and when this occurred (e.g. Slater and Ince, 1979).

Most studies have looked at changes in individual song types within a population. We will consider this first, and then look at the elements that make up these types and the repertoires to which they contribute.

Song types

Song types are often complex sequences of elements, so that the possibility that two might be the same by chance rather than through cultural inheritance is negligible. Using the analogy between social learning and genetic transmission, population genetic models can be adapted to become models of cultural evolution (Cavalli-Sforza and Feldman, 1981). Since in most of the best studied species, the song type appears to be the unit of cultural inheritance, with whole song types tending to be copied from one individual to another (e.g. Slater and Ince, 1979; Baker, 1996), models of the cultural evolution of bird song have been developed to interpret the distribution of song types in time and space, as recorded in natural populations. The distribution of song types within a population depends on a number of factors:

1. Where the song was learnt, relative to the singer's territory (including whether an individual learns before or after natal dispersal).
2. Which songs an individual chooses to learn. For example, individuals might exhibit a 'conformist bias' (Boyd and Richerson, 1985), tending to learn the commonest song types, or might end up learning some songs that they hear more than others because they transmit better through the habitat (see below).
3. Ecological factors. For example small, isolated populations also tend to have a higher rate of song sharing (e.g. Baker, 1996).
4. The history of the population. If the population was founded recently by a small number of individuals then it might be expected to exhibit a smaller number of song types and a higher level of song sharing (e.g. Schottler, 1995).
5. The rate of innovation.

In reviewing various examples of cultural change in bird song, Lynch (1996) estimated that immigration and innovation were about equally important in the introduction of new types to an area. By adapting pre-existing population genetics models, he estimated the rate of introduction of new types from either source as 0.12 for the great tit (*Parus major*, data from McGregor and Krebs, 1982) and 0.13 for the chaffinch (data from Slater *et al.*, 1980), a figure slightly lower than the 0.15 estimate of the original authors (see Figure 5.2).

However, as Lynch (1996) pointed out, the many simplifying assumptions needed when adapting population genetics models mean that they are less than ideal. An alternative approach is to build a model of bird song 'from scratch'. In order to incorporate the important spatial component of song learning, one solution is a spatially explicit simulation model (a cellular automaton), based on individuals (Goodfellow and Slater, 1986; Williams and Slater, 1990). Lachlan and Slater (in press) designed an accurate spatial simulation of chaffinch song learning, and fitted it to field recordings. Using this method, they estimated an innovation rate of between 1 and 3 per cent in samples of chaffinch song from four locations in Scotland, with an upper confidence limit of around 6 per cent. The principle reason why previous estimates were so much higher was probably because they could not differentiate between novelty due to immigration and that due to genuine innovation within the population. Lachlan and Slater (in press) partially removed the effects of immigration in two different ways: first, by sampling song sharing on the Isle of Lewis, a very small, isolated population, in which song dispersal into the population was probably much lower; and second by including song dispersal explicitly in their spatial simulations of the mainland populations. The results are lent some credence by the similarity of the estimates using these two techniques. In summary, our results show that the great majority of song type copying by chaffinches is accurate, and only in a small proportion of cases, under 5 per cent once immigration is allowed for, do new types arise through innovation.

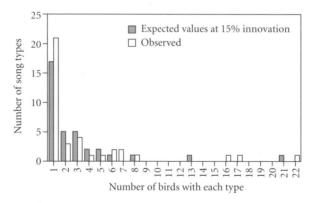

Figure 5.2 The frequencies of song types in a population of 42 chaffinches (white bars) in which each male sang 1–5 song types. Twenty-one songs were unique to the individual singing them, while the commonest type was sung by 22 birds. The black bars are based on a computer simulation that gave the best match to the data: in this young birds setting up their territories copied accurately from others in the population on 85% of occasions and introduced new song types on 15%. Simulations involving greater accuracy gave more birds sharing, while poorer accuracy led to more unique song types. These studies do not differentiate between immigration and innovation; recent work suggests that much of the novelty arising in a population is due to the former, with less than 5% of copying inaccurate. Data from Slater *et al.* (1980).

What form does innovation leading to a new song type take? Examination of published sonagrams suggests several main ways in which it may arise (see, e.g., Slater and Ince, 1979; Payne, 1996). The commonest changes appear to be where a phrase is added to or deleted from a song, where one type of element is substituted for another (or where a pre-existing one is modified in a substantial way), or where sequences that were previously only found in two different song types are recombined to form a new type. In all these cases the change involved is relatively minor, fitting in with the notion that copy errors are involved rather than the generation of difference for some functional reason.

In addition to the frequency with which innovations occur, and the form that these take, the consequences of innovations for the songs present in a population depend to an important extent on dispersal patterns. As we have discussed, species vary in whether they modify their songs after dispersal, but young males usually settle close to where they hatch so that song types also tend not to move far. For example male chaffinches appear to disperse 700 m on average (Paradis *et al.*, 1998), and this fits well with a peak in song sharing that occurs between chaffinches located 600 m apart (Lachlan and Slater, in press). Alternatively, young males may move between populations, as in the indigo bunting (*Passerina cyanea*), where only 20 per cent of new territory owners had hatched in the population but where song is learnt after dispersal (Payne, 1996). In either case, song is learnt within the population and, if accuracy of copying is high, then song types can persist for many cultural generations, but there is gradual turnover so that those present within a population tend to differ, at an extreme leading to the existence of dialects.

Elements

Studies of cultural mutation rates suggest, perhaps not surprisingly, that these are higher at the sequence (or song type) level than at that of the element (e.g. Lynch and Baker, 1994; Burnell, 1998). Indeed it has been argued for some species that the whole song repertoire is composed of a limited species-universal number of elements. In the indigo bunting, for example, the elements may be split into 127 different types (Baker and Boylan, 1995), and the authors argue that these are well defined and do not blend with each other so that it is relatively easy to recognize them repeatedly. Songs commonly consist of sequences of five such elements so that, even if all songs were exactly that length and no element could occur at two points in the sequence, this would allow for a species song-type repertoire of over 30 billion!

The swamp sparrow also has a very limited number of element types, and the study by Marler and Pickert (1984) suggests that there are distinct physical constraints on the sorts of sounds that can be produced. They classified 1307 notes and found that they fell into 96 types, which in turn could be grouped into only 6 broad categories. Each bird produces many different types in subsong, but most of these are discarded so that the final song incorporates only a small number (Marler and Peters, 1982). Again the number of possible songs that can be composed is enormous but, with such a small number of element types of very clearly defined form, it is perhaps questionable whether the elements themselves need to be learnt at all. The fact that deafened individuals in this and other species do not produce normal elements and that birds with abnormal experience often produce elements outside the normal species range does, however, suggest that learning has a role at the level of element acquisition as well as that of the song. In cardinals (*Richmondena cardinalis*) isolated birds have been found to develop elements within the normal range, but the elements of tutored birds still achieve a precise match to those they are exposed to (Yamaguchi, 2001), implying a process at least of selection if not of memorization.

A particularly fascinating case, which appears to involve genuine invention, as defined at the start of this chapter, is in the grey catbird (*Dumetella carolinensis*, Kroodsma *et al.*, 1997). This species is a mimic thrush, and each individual produces a large number of brief song phrases, often of a single element or syllable. Hand-reared individuals produce large and varied repertoires of apparently normal units whether or not they were tutored, and even in tutored birds there was little evidence of imitation. To date, this is the only evidence of invention in the generation of large repertoires.

Repertoires

Just as elements may combine to produce a repeatable song type, so may song types combine to generate a repertoire. In some species the whole song repertoire is shared between individuals, so that some birds may have A and B and others C and D, but no birds have combinations of songs from those two groupings (e.g. corn bunting, *Miliaria* (formerly *Emberiza*) *calandra*, McGregor, 1980, though not in all populations, see Latruffe *et al.*, 2000). This probably arises where a young bird learns all his songs from a single individual. More commonly, however, all possible combinations of song types may occur in different birds' repertoires, and here young birds are clearly learning songs from different tutors. Especially where they have many song types, they may well end up with a repertoire unique to themselves, an example of innovation but not involving new elements or song types. For example, in three samples of 36, 43, and 32 male chaffinches (which have a repertoire of between one and six song types) within each of which territories were contiguous or near so, Lachlan and Slater (in press), found only three pairs of individuals that shared their entire repertoire.

Interestingly, in species with song types that are fixed in their sequence of elements, such as the chaffinch, it seems usual for a bird to learn each of its song types from a single tutor (an example of particulate inheritance like that of genes) rather than blending characteristics of several. Thielcke (1987) argues for blending in the case of short-toed tree-creepers (*Certhia brachydactyla*). If such blending does occur it would have a conservative influence, leading the variability of song to be greatly reduced.

We can conclude from this section that novelty in bird song may arise for a variety of different reasons, from a pre-existing song being introduced from elsewhere (immigration) to a singing bird recombining elements to produce new songs continuously (improvisation). It can also happen at different levels, ranging from elements of new form being generated to new repertoires arising because song types have been recombined. But all these things may be of little consequence, functionless by-products of the vocal learning process. Is there evidence that they may sometimes be of adaptive significance?

Benefits and costs of innovation?

In discussing the benefits and costs of innovation it is worth separating species with large and varied repertoires, especially those without fixed song types, from those with small repertoires in which the song types tend to be more stereotyped. Where repertoires are large, each individual may have a repertoire unique to itself. The endless recombination of element types may also generate an effectively infinite variety of types. Yet greater variety may stem from the importing of the phrases of alien species through mimicry. For example, the song of the starling (*Sturnus vulgaris*) consists of a mixture of species-typical and mimicked elements put together in a complex pattern (Eens, 1997). Both the species-typical elements and the patterning leave little doubt, at least to the human ear, that the singer is a starling, even though the mimicked sounds are, or have been at some stage in the past, imported from other species. As a way of introducing novelty, mimicry, which is often startlingly accurate, is more akin to immigration than to the other mechanisms discussed earlier. Invention, as in the grey catbird, may be yet another way of bolstering variety. Good evidence from several species with large repertoires points to females preferring males whose songs are more varied. In various experiments females have been found to be more attracted to loudspeakers playing larger repertoires (Lampe and Saetre, 1995), to respond to such playbacks with more copulation solicitation or to pair earlier with large repertoire males (e.g. sedge warbler, Catchpole *et al.*, 1984; Buchanan and Catchpole, 1997). The high level of innovation and individuality involved here seems to be a means of boosting variety, as favoured by sexual selection.

There are also several potential costs to innovation. Arguments that it may be costly to generate variety are undercut by the fact that during early development many species pass through a stage of 'plastic' singing in which a very wide variety of different songs are produced, apparently with considerable innovation. The final repertoire of the individual is then selected from within this diversity. Instead a cost of innovation is likely to lie with the risk that communication may be impaired. For example, innovation may lead to the production of a song outside the normal species-typical range, with obvious ramifications. In at least some species this cost may be avoided by constraints on the sort of sounds that can be mastered (e.g. Marler and Pickert, 1984; Podos, 1997). Another cost of innovation may involve the loss of conformity. Several studies suggest benefits of song type sharing between neighbours to survival and territory retention (Wilson *et al.*, 2000) and in reproductive success (Payne, 1982; Payne *et al.*, 1988). Several explanations for this have been proposed, ranging from more efficient communication occurring when individuals share song types, through sharing being a conventional signal, to sharing being a way for individuals to identify and exclude intruders that do not conform (see, e.g. Bertram, 1970). A third possibility has recently been described by Nowicki *et al.* (2002), and this is that females may, for some reason, prefer the songs of males that copy accurately. They found that female song sparrows (*Melospiza melodia*) responded more to the songs of males that more accurately matched those of the tutors to which they had been exposed. They suggest that accurate song learning may be an indication that a male is of high quality as he has successfully resisted developmental stress. This is in line with earlier findings in which they suggested that developmental stress may influence song in other ways and so song may give an indication of male quality (Nowicki *et al.*, 1998, 2000). But the suggestion that accuracy of learning might be such a measure is a novel idea and one that would certainly constrain innovation.

Podos (1996) demonstrated an interesting constraint on swamp sparrow song by training young birds with recordings in which the trill rate had been artificially increased.

While they copied this new pattern they were unable to sustain the increased trill rate, but introduced periodic silent gaps into their songs, a phenomenon he referred to as 'broken syntax'. In the field adult males respond to playback of such songs as strongly as to normal song, but females in the laboratory show less copulation solicitation to them (Nowicki *et al.*, 2001). This suggests that it is female choice that is acting as the constraint and stopping innovation along this particular axis.

Innovation is much less common in species with small repertoires of fixed song types than in ones with large numbers of elements that can be recombined in various ways. If, novel song types are generated at a rate of less than 6 per cent, as Lachlan and Slater (in press) suggest, it is hard to see how this could benefit the individuals possessing them unless the two strategies of novelty and conformity comprised a mixed ESS. No theory to this effect has been put forward. Instead, the evidence from several studies is that song type innovation is of no selective advantage, but arises at this low level through the inevitable errors in copying that occur in any learning process. While such errors may occasionally occur as single major changes, as where a bird reverses the order of elements or merges two song types, a small degree of error is probably introduced in every instance of copying. This may be imperceptible, but mount over cultural generations until we recognize that a new song type has arisen.

Payne *et al.* (1981) favoured such a 'neutral model' to account for changes over time in indigo buntings, calculating that the average song type had a half-life of 3.8 years. In chaffinches, both changes with time (Ince *et al.*, 1980) and the distribution of song types in a population (Slater *et al.*, 1980) suggested a pattern in which song types arose and became extinct by chance. Most notably, the studies by Lynch and his collaborators, also on chaffinches (Lynch *et al.*, 1989; Lynch and Baker, 1993, 1994), using a 'population memetics'[1] approach, suggest that the processes of mutation, migration, and drift account for the observed changes without it being necessary to assume that innovations are selected for or against. Changes with time in white-crowned sparrow (*Zonotrichia leucophrys*) song dialects (Chilton and Lein, 1996) and in a population of Eurasian tree sparrows (*Passer montanus*) introduced into North America (Lang and Barlow, 1987) point in the same direction. The latter case is interesting as only 12 pairs were introduced from Germany in 1870, yet the diversity of songs present currently matches that found in Germany, suggesting that mutation was originally greater than extinction, but that the two became balanced as the new world population rose to its present level.

While these results on cultural change are clear, only a fairly small range of species is concerned. It will be interesting to see whether they are confirmed by subsequent studies on other species. For the moment, both this work and that on vocal learning point to most

---

[1] The term memetics is derived from genetics, following the meme/gene analogy, and several authors of papers on bird song have used the word meme to refer to a unit of cultural inheritance. Its usefulness is limited by the many different levels at which changes in song may arise. It is perhaps at its most appropriate in reference to species with clear song types, which may be copied precisely between individuals, as in the chaffinches studied by Lynch *et al.* On the other hand, even here it may obscure important processes going on at the level of elements or repertoires. There are dangers in arguing by analogy, and this term does not seem to us very helpful in studies of song.

bird song innovation being at a low level, occurring as a by-product of the vocal learning process, and either neutral in its effect or possibly disadvantageous. The exception may be in species with large repertoires where sexually selected variety may be achieved through innovation, invention, or improvisation.

No learning process is 100 per cent accurate so, whether or not novelty is favoured, it is bound to arise. If novel songs arise, what conditions might lead them to spread, and is there evidence that they do so?

Do novel songs spread?

If a novel song appears within a population it may not be copied, and so become extinct when its possessor dies, or be copied so that more than one bird comes to sing it. In a population of chaffinches it is not unusual for the commonest song type present to be sung by more than half the birds in a wood. However, while this might suggest some process of selection, computer simulations point to this outcome resulting simply from random processes: many songs become extinct but the occasional one spreads, purely by chance, until it occurs in the repertoires of many birds (Slater *et al.*, 1980). The question we need to address is whether some song types spread to a greater extent than such random processes would account for.

The most obvious way for an advantageous new behaviour pattern to spread in a population is, as with mutant genes, where it is passed down within a family. In bird song this would be (at least in temperate regions, where song tends to be a male preserve), where sons learnt from their fathers. Thus, where a song type confers a reproductive advantage, and it is passed from father to son, then it would be expected to become commoner generation by generation. This is, however, unlikely to be important in bird song as, with some notable exceptions (e.g. Darwin's finches, *Geospiza fortis* and *Geospiza scandens*, Millington and Price, 1985; Grant and Grant, 1996, stripe-backed wrens, *Campylorhynchus nuchalis*, Price, 1998, 1999), there is little evidence that vocalizations are passed down within families. Song learning in most temperate zone species, where seasonality is a strong influence, seems largely to occur in the first summer after independence from the parents and from territorial neighbours next spring (Slater, 1989). Given this, it is not surprising, first, that changes in the frequency of song types tend to be random rather than progressive and, second, that one of the best examples of a progressive change is that described by Gibbs (1990) in *G. fortis*. He found that the commonest song type in a population became progressively less so over the course of 6 years, while the three less common types became commoner (Figure 5.3). Furthermore, this arose because males singing the rarer types survived better and produced more male offspring that survived to join the breeding population singing their fathers' songs. However, it is not clear why possessing a rare song should lead to greater longevity and fecundity. Gibbs suggests that, in the dense population he studied, there may be advantage to being individually recognizable, as birds with a rare song will be. However, this raises the question of why most birds copy accurately. In any case an advantage of having a less common song type cannot be the whole story as the effect would lead all song types to become equal in frequency, and this was certainly far from the case at the start of Gibbs' study. It is perhaps most likely that new songs are being generated

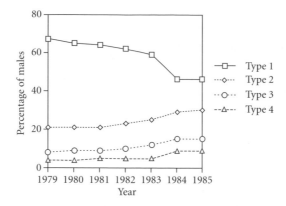

Figure 5.3 Changes in the frequency of four song types in the population of Darwin's medium ground finch on Daphne Major over 6 years. During this period the commonest song became progressively rarer, while the three less common songs rose in frequency. Redrawn from Gibbs (1990) with permission from Elsevier.

occasionally (though Gibbs reports none in the 6 years of his study), and that these spread in the population at the expense of pre-existing ones. But, if rarity itself is an advantage, it is not clear how any song would become extinct, as any song that was about to die out would become favoured; indeed it would seem most likely that the number of types in the population would grow progressively.

If song types are not passed from father to son, the main way in which new songs may spread is if they are more attractive as models than other types. Anything that makes them more likely to be copied will lead them to spread, even if they do not confer a reproductive advantage, so that this sort of 'cultural selection' (Cavalli-Sforza and Feldman, 1981) need not parallel natural selection. However, in most descriptions of cases where a song has spread within an area, reasons have been suggested for the song being more beneficial than the alternatives.

It was Hansen (1979) who originally suggested that songs may become matched to habitat because young birds, learning at a distance, will be more prone to learn song types that travel undistorted through the particular habitat they are occupying. This may account for the fact, for example, that great tits (Hunter and Krebs, 1979) and rufous-collared sparrows (*Zonotrichia capensis*, e.g. Handford, 1988) tend to have slower trills in denser woodland where fast trills would be distorted by echoes off the trees. In Finland, the songs of great tits were also found to have reduced in complexity between 1947 and 1981, especially in urban areas, with 4 phrase syllables disappearing, 3 becoming rarer, 2 increasing strongly and a new 1 syllable phrase appearing (Bergman, 1980; Lehtonen, 1983). This change has been suggested as perhaps due to the increasing number of birds living in noisier urban environments where long songs would be more difficult to learn, and where there are fewer coal tits (*Parus ater*) whose short and simple songs might lead to confusion (Bergman, 1980). The amplitude of a song alone may make it more likely to be learnt, as louder songs will cover a wider area; if songs are used in advertising, louder songs are also more likely to attract a mate, so again the chance of the song being learnt is likely be linked to its effectiveness as a signal. In the thrush nightingale, a species in which there is turnover of songs between seasons, Sorjonen (1987) found that loud and simple songs, which he referred to aptly as 'well penetrating', were more likely to be copied, whereas weak and complex ones were abandoned.

A very interesting case of an innovation spreading, though admittedly not in a bird, is that described by Noad *et al.* (2000) in humpback whales (Figure 5.4). The songs sung by the population off the east coast of Australia showed an almost total switch to match those of animals off the west coast between the southwards migrations in 1996 and 1997. It is suggested that the new songs were introduced by a small number of immigrants and then spread rapidly through the population. Like thrush nightingales, the songs of these animals are well matched within the population but change in concert over time (Payne and Payne, 1985). It appears from this latest work that novel songs introduced from outside may have a particular attractiveness as models for copying. If animals gain from singing novel or different songs, perhaps that advantage would be gained by others also by copying those songs, though they would then become common in the population so that the advantage would be progressively lost.

An equivalent example to the last, though less dramatic, is in another species where songs of groups of animals change in concert both within and between seasons, the yellow-rumped cacique (*Cacicus cela*). All males in a colony share 5–8 song types and these change gradually, apparently in constant directions, suggesting that the change results from a benefit in adopting certain songs rather than from accumulating error or drift (Trainer, 1989). Following an idea put forward by Payne (1985) to account for directional changes in village indigobird song, Trainer suggests that dominant and successful males may change their songs slightly to diverge from that of the group, and it may then pay others to match that song because of its association with dominance, leading to a progressive shift.

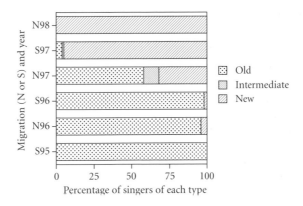

Figure 5.4 Changes in humpback whale songs off the east coast of Australia over 3 years between their southwards migration in 1995 (S95) and that northwards in 1998 (N98). The new song that first appeared in 1996, and was sung by all members of the population by 1998, was that common off the west coast in 1996. It was probably introduced by a small number of animals moving from that population to the other coast. Redrawn from Noad *et al.* (2000) with permission from Nature.

To summarize, while purely random processes may account for a certain amount of spreading of song types within a population, a number of cases have been described where progressive change has taken place. While the reasons for such changes are not yet clear, the most likely are that the songs are being modified to match either a changing environment or the songs of changing social companions. Such cases are, however, few compared with the large number of bird species that have been studied.

Innovation and diversity

One of the most general statements that can be made about bird song is that it tends to identify an individual as a member of a species. As mentioned above, even mimics usually sing in such a way that their own species identity is clear. Females, and male competitors, respond preferentially to species-typical songs, or even local variants of them. An obvious possibility, therefore, is that innovation in song could play an important part in the development of prezygotic reproductive isolation. If a novel type of song were to spread through a population, it could lead to the isolation of that population from others. Is this likely to occur?

For an innovation made by a singing male to cause prezygotic isolation, local females must show a clear preference for that new song form. The possibility of such assortative mating has been examined in the white-crowned sparrow, a 'dialect' singer in which all males within one area tend to sing similar song types, which are different from those some distance away. However, evidence from experiments suggests that females have, if any, only a weak preference for that dialect (Chilton *et al.*, 1990), and dialect boundaries also seem not to limit dispersal (Zink and Barrowclough, 1984). There is certainly more than enough gene flow between dialect groups to prevent the formation of new species. Moreover, in many species that do not show dialects, there tends to be a good deal of song dispersal between local populations. All in all, it thus seems very unlikely that innovation could cause speciation through the formation of a new dialect. Nevertheless, from a theoretical perspective, innovation due to song learning could lead to more rapid divergence of song in an isolated population than if the mating signal was not learned (Lachlan *et al.*, unpublished results). If so, allopatric speciation might occur more readily in species that learn their songs and occasionally innovate.

What level of innovation would have the largest effect on population divergence? In a recent model, Ellers and Slabbekoorn (in press) demonstrate that higher levels of innovation lead to greater phenotypic diversity between local populations. However, Baptista and Trail (1992), while arguing against the role of song learning in speciation, pointed out that species in which innovation is most common, such as mimics, are not very great in number. The logic behind this observation is that, if innovation is too common, either the signal provides little information about species identity or the high rate of novelty serves as a homogenizing factor between populations (because novel song types become superseded rather than spreading through the population).

Concluding remarks

Bird song is learnt by copying from other individuals, and this is a matching process, leading individuals to behave similarly to one another. Yet, by its very nature, transcription errors are bound to arise quite frequently in any learning process, and the key question is whether this is all there is to the innovations we see arising. The opposite to innovation is conformity, and there is obviously a tension between the two processes when it comes to communication. A male with a different song may be recognizable and may stand out from the crowd so that he is more attractive to prospective mates. On the other hand he may be less able to interact with territorial neighbours through matched countersinging, or fail to produce the correct 'password' in a colony in which individuals share their songs (Feekes, 1977). The relative levels of innovation and conformity will depend on how such conflicts are resolved. Much of the evidence at the moment is in favour of song learning leading to conformity, with innovation occurring through copying errors rather than being selected for as such. This should be the null hypothesis, though hopefully it will be disproved in an increasing number of interesting cases. At present, in only a few cases, such as catbirds and mimics, does innovation appear to be very common, and in these it seems to serve simply to generate variety. There is perhaps a useful distinction to be made between these cases and innovation in other domains: rather than innovation here leading to new functional specializations, innovation is itself the functional specialization.

Bird song is a cultural phenomenon in the rather loose sense that its development involves copying by one animal of the behaviour of another. Why learning has such a key role in song development remains perplexing (see Slater *et al.*, 2000), as there is rather little evidence for the cultural transmission involved enabling individuals to produce more effective songs or to benefit by learning from the experience of others. Which particular songs birds learn seems more often to be selectively neutral. It is nevertheless fascinating to observe the complex patterns in space and time that have resulted from the simple fact that song is culturally transmitted. But it is a cultural phenomenon only in a very weak sense: while it is clearly social learning, it is in many ways very similar to individual learning except that it happens to be from other animals. On the other hand, perhaps that is altogether too confusing!

Summary

Song in songbirds is a classic case of social learning, but that learning is not always accurate so that song changes in both time and space. We outline the sorts of changes that may take place, at the level of elements, song types and repertoires, and also several distinct ways in which novel songs may arise in a population (immigration, innovation, invention, and improvisation). Much of the evidence is that such changes are random, and that their consequences are either neutral or negative, but some cases have been described where there have been progressive changes in song within a population. These are likely to stem from individuals adapting their songs to match a changing physical or social environment. The individual that generates variety and stands out from the crowd may also benefit through sexual selection, with a tension between innovation and conformity.

Acknowledgements

We are grateful to Vincent Janik, Nigel Mann, the editors and two anonymous referees for helpful comments on the manuscript.

References

Baker, M. C. (1996). Depauperate meme pool of vocal signals in an island population of singing honeyeaters. *Animal Behaviour,* **51**, 853–8.

Baker, M. C. and Boylan, J. T. (1995). A catalog of song syllables of indigo and lazuli buntings. *Condor,* **97**, 1028–40.

Baptista, L. F. and Petrinovich, L. (1984). Social interaction, sensitive phases and the song template hypothesis in the white-crowned sparrow. *Animal Behaviour,* **32**, 172–81.

Baptista, L. F. and Trail, P. W. (1992). The role of song in the evolution of passerine diversity. *Systematic Biology,* **41**, 242–7.

Bergman, G. (1980). Die Veranderung der Gesangmelodie der Kohlmeise *Parus major* in Finnland und Schweden. *Ornis Fennica,* **57**, 97–111.

Bertram, B. (1970). The vocal behaviour of the Indian hill mynah. *Animal Behaviour Monographs,* **3**, 81–192.

Boyd, R. and Richerson, P. J. (1985). *Culture and the evolutionary process.* Chicago, IL: University of Chicago Press.

Buchanan, K. L. and Catchpole, C. K. (1997). Female choice in the sedge warbler, *Acrocephalus schoenobaenus:* Multiple cues from song and territory quality. *Proceedings of the Royal Society of London, Series B,* **264**, 521–6.

Burnell, K. (1998). Cultural variation in savannah sparrow, *Passerculus sandwichensis,* songs: An analysis using the meme concept. *Animal Behaviour,* **56**, 995–1003.

Catchpole, C. K. (1976). Temporal and sequential organisation of song in the sedge warbler (*Acrocephalus schoenobaenus*). *Behaviour,* **59**, 226–46.

Catchpole, C. K. and Slater, P. J. B. (1995). *Bird song: Biological themes and variations.* Cambridge: Cambridge University Press.

Catchpole, C. K., Dittami, J., and Leisler, B. (1984). Differential responses to male song repertoires in female songbirds implanted with oestradiol. *Nature,* **312**, 563–4.

Cavalli-Sforza, L. L. and Feldman, M. W. (1981). *Cultural transmission and evolution: A quantitative approach.* Princeton, NJ: Princeton University Press.

Chilton, G. and Lein, M. R. (1996). Long-term changes in songs and song dialect boundaries of Puget sound white-crowned sparrows. *Condor,* **98**, 567–80.

Chilton, G., Lein, M. R., and Baptista, L. F. (1990). Mate choice by female white-crowned sparrows in a mixed dialect population. *Behavioral Ecology and Sociobiology,* **27**, 223–7.

Eens, M. (1997). Understanding the complex song of the European starling: An integrated ethological approach. *Advances in the Study of Behavior,* **26**, 355–434.

Ellers, J. and Slabbekoorn, H. (in press). Song dispersal and male dispersal among bird populations: A spatially explicit model testing the role of vocal learning. *Animal Behaviour.*

Feekes, F. (1977). Colony specific song in *Cacicus cela* (Icteridae, Aves): The password hypothesis. *Ardea,* **65**, 197–202.

Gibbs, H. L. (1990). Cultural evolution of male song types in Darwin's medium ground finches, *Geospiza fortis. Animal Behaviour,* **39**, 253–63.

Goodfellow, D. J. and Slater, P. J. B. (1986). A model of bird song dialects. *Animal Behaviour,* **34**, 1579–80.

Grant, B. R. and Grant, P. R. (1996). Cultural inheritance of song and its role in the evolution of Darwin's finches. *Evolution,* **50**, 2471–87.

Handford, P. (1988). Trill rate dialects in the rufous-collared sparrow, *Zonotrichia capensis*, in northwestern Argentina. *Canadian Journal of Zoology*, **66**, 2658–70.

Hansen, P. (1979). Vocal learning: Its role in adapting sound structures to long-distance propagation and a hypothesis on its evolution. *Animal Behaviour*, **27**, 1270–1.

Hunter, M. L. and **Krebs, J. R.** (1979). Geographical variation in the song of the great tit (*Parus major*) in relation to ecological factors. *Journal of Animal Ecology*, **48**, 759–85.

Ince, S. A., Slater, P. J. B., and **Weismann, C.** (1980). Changes with time in the songs of a population of chaffinches. *Condor*, **82**, 285–90.

Janik, V. M. and **Slater, P. J. B.** (2000). The different roles of social learning in vocal communication. *Animal Behaviour*, **60**, 1–11.

Janik, V. M. and **Slater, P. J. B.** (2003). Traditions in mammalian and avian vocal communication. In *The biology of traditions* (ed. S. Perry and D. M. Fragaszy), pp. 213–35. Cambridge: Cambridge University Press.

Jenkins, P. F. (1978). Cultural transmission of song patterns and dialect development in a free-living bird population. *Animal Behaviour*, **26**, 50–78.

Jones, A. E., ten Cate, C., and **Slater, P. J. B.** (1996). Early experience and plasticity of song in adult male zebra finches (*Taeniopygia guttata*). *Journal of Comparative Psychology*, **110**, 354–69.

Kroodsma, D. E., Houlihan, P. W., Fallon, P. A., and **Wells, J. A.** (1997). Song development in grey catbirds. *Animal Behaviour*, **54**, 457–64.

Kroodsma, D. E., Sanchez, J., Stemple, D. W., Goodwin, E., da Silva, M. L., and **Vielliard, J. M. E.** (1999). Sedentary lifestyle of neotropical sedge wrens promotes song imitation. *Animal Behaviour*, **57**, 855–63.

Lachlan, R. F. and **Slater, P. J. B.** (in press). Song learning in chaffinches: How accurate, and from where? *Animal Behaviour.*

Lampe, H. M. and **Saetre, G.-P.** (1995). Female pied flycatchers prefer males with larger song repertoires. *Proceedings of the Royal Society of London, Series B*, **262**, 163–7.

Lang, A. L. and **Barlow, J. C.** (1987). Syllable sharing among North American populations of the Eurasian tree sparrow. *Condor*, **89**, 746–51.

Latruffe, C., McGregor, P. K., Tavares, J. P., and **Mota, P. G.** (2000). Microgeographic variation in corn bunting (*Miliaria calandra*) song: Quantitative and discrimination aspects. *Behaviour*, **137**, 1241–55.

Lehtonen, L. (1983). The changing song patterns of the great tit *Parus major*. *Ornis Fennica*, **60**, 16–21.

Luschi, P. (1993). Improvisation of new notes during singing by male Sardinian warblers. *Bioacoustics*, **4**, 235–44.

Lynch, A. (1996). The population memetics of birdsong. In *Ecology and evolution of acoustic communication in birds* (ed. D. E. Kroodsma and E. H. Miller), pp. 181–97. Ithaca and London: Comstock Publishing Associates.

Lynch, A. and **Baker, A. J.** (1993). A population memetics approach to cultural evolution in chaffinch song: Meme diversity within populations. *American Naturalist*, **141**, 597–620.

Lynch, A. and **Baker, A. J.** (1994). A population memetics approach to cultural evolution in chaffinch song: Differentiation among populations. *Evolution*, **48**, 351–9.

Lynch, A., Plunkett, G. M., Baker, A. J., and **Jenkins, P. F.** (1989). A model of cultural evolution of chaffinch song derived with the meme concept. *American Naturalist*, **133**, 634–53.

Marler, P. and **Peters, S.** (1982). Developmental overproduction and selective attrition: New processes in the epigenesis of bird song. *Developmental Psychobiology*, **15**, 369–78.

Marler, P. and **Pickert, R.** (1984). Species-universal microstructure in a learned birdsong: The swamp sparrow (*Melospiza georgiana*). *Animal Behaviour*, **32**, 673–89.

McGregor, P. K. (1980). Song dialects in the corn bunting (*Emberiza calandra*). *Zeitschrift für Tierpsychologie*, **54**, 285–97.

McGregor, P. K. and Krebs, J. R. (1982). Song types in a population of great tits (*Parus major*): Their distribution, abundance and acquisition by individuals. *Behaviour*, **79**, 126–52.

Millington, S. J. and Price, T. D. (1985). Song inheritance and mating patterns in Darwin's finches. *Auk*, **102**, 342–6.

Noad, M. J., Cato, D. H., Bryden, M. M., Jenner, M.-N., and Jenner, K. C. S. (2000). Cultural revolution in whale songs. *Nature*, **408**, 537.

Nowicki, S., Peters, S., and Podos, J. (1998). Song learning, early nutrition and sexual selection in songbirds. *American Zoologist*, **38**, 179–90.

Nowicki, S., Hasselquist, D., Bensch, S., and Peters, S. (2000). Nestling growth and song repertoire size in great reed warblers: Evidence for song learning as an indicator mechanism in mate choice. *Proceedings of the Royal Society of London, Series B*, **267**, 2419–24.

Nowicki, S., Searcy, W. A., Hughes, M., and Podos, J. (2001). The evolution of bird song: Male and female response to song innovation in swamp sparrows. *Animal Behaviour*, **62**, 1189–95.

Nowicki, S., Searcy, W. A., and Peters, S. (2002). Quality of song learning affects female response to male bird song. *Proceedings of the Royal Society of London, Series B*, **269**, 1949–54.

Paradis, E., Baillie, S. R., Sutherland, W. J., and Gregory, R. (1998). Patterns of natal and breeding dispersal in songbirds. *Journal of Animal Ecology*, **67**, 518–36.

Payne, R. B. (1982). Ecological consequences of song matching: Breeding success and intraspecific song mimicry in indigo buntings. *Ecology*, **63**, 401–11.

Payne, R. B. (1985). Behavioral continuity and change in local song populations of village indigobirds *Vidua chalybeata*. *Zeitschrift für Tierpsychologie*, **70**, 1–44.

Payne, R. B. (1996). Song traditions in indigo buntings: Origin improvisation, dispersal, and extinction in cultural evolution. In *Ecology and evolution of acoustic communication in birds* (ed. D. E. Kroodsma and E. H. Miller), pp. 198–220. Ithaca and London: Comstock Publishing Associates.

Payne, K. and Payne, R. S. (1985). Large scale changes over 19 years in songs of humpback whales in Bermuda. *Zeitschrift für Tierpsychologie*, **68**, 89–114.

Payne, R. B., Thompson, W. L., Fiala, K. L., and Sweany, L. L. (1981). Local song traditions in indigo buntings: Cultural transmission of behavior patterns across generations. *Behaviour*, **77**, 199–221.

Payne, R. B., Payne, L. L., and Doehlert, S. M. (1988). Biological and cultural success of song memes in indigo buntings. *Ecology*, **69**, 104–17.

Podos, J. (1996). Motor constraints on vocal development in a songbird. *Animal Behaviour*, **51**, 1061–70.

Podos, J. (1997). A performance constraint on the evolution of trilled vocalizations in a songbird family (Passeriformes: Emberizidae). *Evolution*, **5**, 537–51.

Price, J. J. (1998). Family- and sex-specific vocal traditions in a cooperatively breeding songbird. *Proceedings of the Royal Society of London, Series B*, **265**, 497–502.

Price, J. J. (1999). Recognition of family specific calls in stripe-backed wrens. *Animal Behaviour*, **57**, 483–92.

Schottler, B. (1995). Songs of blue tits *Parus caeruleus palmensis* from La Palma (Canary Islands)—a test of hypotheses. *Bioacoustics*, **6**, 135–52.

Slater, P. J. B. (1989). Bird song learning: Causes and consequences. *Ethology, Ecology & Evolution*, **1**, 19–46.

Slater, P. J. B. and Ince, S. A. (1979). Cultural evolution in chaffinch song. *Behaviour*, **71**, 146–66.

Slater, P. J. B., Ince, S. A., and Colgan, P. W. (1980). Chaffinch song types: Their frequencies in the population and distribution between the repertoires of different individuals. *Behaviour*, **75**, 207–18.

Slater, P. J. B., Lachlan, R. F., and Riebel, K. (2000). The significance of learning in signal development: The curious case of the chaffinch. In *Animal signals: signalling and signal design in animal communication* (ed. Y. Espmark, T. Amundsen and G. Rosenqvist), pp. 341–52. Trondheim, Norway: Tapir Publishers.

Sorjonen, J. (1987). Temporal and spatial differences in traditions and repertoires in the song of the thrush nightingale (*Luscinia luscinia*). *Behaviour*, **102**, 196–212.

Thielcke, G. (1987). Langjährige Dialektronstanz beim Gartenbaumläufer (*Certhia brachydactyla*). *Journal für Ornithologie*, **128**, 171–80.

Trainer, J. M. (1989). Cultural evolution in song dialects of yellow-rumped caciques in Panama. *Ethology*, **80**, 190–204.

Williams, J. M. and Slater, P. J. B. (1990). Modelling bird song dialects: The influence of repertoire size and numbers of neighbours. *Journal of Theoretical Biology*, **145**, 487–96.

Wilson, P. L., Towner, M. C., and Vehrencamp, S. L. (2000). Survival and song-type sharing in a sedentary subspecies of the song sparrow. *Condor*, **102**, 355–63.

Yamaguchi, A. (2001). Sex differences in vocal learning in birds. *Nature*, **411**, 257–8.

Zink, R. M. and Barrowclough, G. F. (1984). Allozymes and song dialects: A reassessment. *Evolution*, **38**, 444–8.

CHAPTER 6

SOCIAL LEARNING: PROMOTER OR INHIBITOR OF INNOVATION?

BENNETT G. GALEF, JR.

Introduction

During the 130-year history of the scientific study of animal behaviour, there have been two periods of relatively intense interest in the role of social learning in behavioural development. The first of these episodes occurred in the latter part of the nineteenth century at a time when instinct and imitation were considered to be the main sources of adaptive behaviour in animals. The second began some 30 years ago with the publication of Ward and Zahavi's (1973) classic paper on information centres, and continues to the present day. Below, I discuss contrasting views of the interaction of social learning and innovation characteristic of these two periods.

Social learning as a conservative force

At the end of the nineteenth century, social learning was seen as a way in which the normal behaviours of a population of animals were conserved and transmitted intact from one generation to the next. Consequently, social learning was viewed primarily as interfering with acquisition of novel patterns of behaviour.

I quote from the work of but three of the several nineteenth century behavioural scientists with an interest in the role of imitation in behavioural development. First, a paragraph from a wonderful essay by Alfred Russell Wallace (1870) entitled 'The philosophy of birds' nests.' In this brief paper, Wallace argues that constancy across generations in the structure of nests built by various avian species, like constancy across generations in the shape of human habitations, results from social learning.

'No one' Wallace asserts, 'imputes [the] stationary condition of domestic architecture among...savage tribes to instinct, but to simple imitation from one generation to another, and the absence of any sufficiently powerful stimulus to change or improvement. When once a particular mode of building has become confirmed by habit and by hereditary custom, it will be long retained, even when its utility has been lost through changed conditions, or through migration to a different region....These characteristics of the abode of savage man will be found exactly paralleled by the nests of birds' (Wallace, 1870, p. 212–5).

C. L. Morgan, writing in 1896, expressed a similar view, 'The conservative tendency of imitation, bringing the newly born members of the animal community into line with the average behaviour of the species is probably its most important office. The young bird or mammal...is born into a community where certain behaviour is constantly exhibited before its eyes. Through imitation it falls in with the traditional habits...' (Morgan, 1896, p. 183–4). Or, from an earlier monograph by the same author, 'Where the young animal is surrounded during the early plastic and imitative period of life by its own kith and kin, imitation will undoubtedly have a conservative tendency. The education of young animals by their parents has also a conservative tendency' (Morgan, 1890, p. 455).

Like Morgan, James Baldwin (1895, p. 298), the philosophically minded social psychologist, felt that the role of imitation in acquisition of novel behaviours had been over-emphasised, that 'many of the most "innate" powers of the animals are brought out, perfected and constantly kept efficient, by imitation of their own species.' Thus, imitation was to be viewed as playing an important, if sometimes underestimated, role in conservation of species-typical behaviours.

Of course others, for example Darwin's protege in matters behavioural, George Romanes (1882), saw imitation (especially of human behaviour by animals) as a source of behavioural innovations. Still, the prevailing view of imitation was that it played an essentially conservative role in behavioural development.

Social learning as a progressive force

During the past 30 years, animal behaviourists have more often been concerned with the role of social learning in diffusing novel or innovative patterns of behaviour through a population than with its possible role in maintaining existing patterns of behaviour. The list of innovative behaviours exhibited by animals that have been attributed to social learning of one kind or another is long indeed. It ranges from sweet-potato-washing by Japanese macaques on Koshima Island (Kawai, 1965) and termite fishing by chimpanzees at Gombe (Goodall, 1986) to pinecone stripping by roof rats in Israel (Terkel, 1996) and diving for molluscs among Norway rats living along the Po River in Italy (Gandolfi and Parisi, 1973).

Why such different views of social learning?

Several possible reasons for the change in perspective on the role of social learning come to mind. First, at the end of the nineteenth century, those engaged in the first scientific studies of animal behaviour were struggling to explain the observation that all the members of any given species tend to engage in similar patterns of behaviour. Today's understanding of behavioural consistency within a species rests largely on work carried out in mid-twentieth century by a large group of psychologists and biologists (e.g. Tinbergen, Lorenz, Lehrman, Kuo, etc.), and that work largely ignored the possibility that social learning might contribute to development of species-typical behaviours.

Results of Kasper–Hauser experiments convinced ethologists that behavioural transmission across generations was relatively unimportant in the development of instincts. Lehrman's (1953) reinterpretation of the results of isolation-rearing studies, focussed as it was on the importance of the interaction of individual and environment in behavioural development, did little to undermine that conclusion. In the late 1800s, on the contrary, discussions on contributions of experience to development of species-typical behaviours focussed on social learning, particularly imitation, rather than on individual trial and error.

Second, scientists working in the late 1890s did not have access to the myriad field observations collected during the last 50 years indicating that, especially in primates, there are systematic differences in the behaviour of allopatric populations of a single species. Now that existence of such population- or locale-specific patterns of behaviour has been clearly established, these animal traditions require explanation, and social learning provides an obvious potential source of differences in the behaviour of allopatric populations of a species.

Third, scientists working in the 1890s had not been through the nature–nurture controversy of the 1960s that so clearly revealed the difficulty of determining causes of similarities in behaviour, and the comparative ease with which analyses of sources of differences in behaviour can be carried out. With the general shift in emphasis from study of sources of constancy to study of sources of variability, social learning was increasingly used to explain the latter rather than the former.

Last, we live in an era when many behavioural scientists, particularly primatologists, seek evidence of human-like performance in animals. Diffusion of technical innovations through human populations is part of the everyday experience of those of us fortunate enough to live in the twenty-first century. If, as anthropomorphic approaches to the study of behaviour require, behavioural capacities of animals in general and of primates in particular are fundamentally like those of humankind, then primates would be expected to transmit behavioural novelty.

Such increased acceptance of anthropomorphism as a heuristic (if not as an explanation) may be a necessary response to the rigid Behaviorism of the first half of the twentieth century. In any case, when anthropomorphic speculations are acceptable, the role of social learning in behavioural development is likely to be viewed differently than it was late in the nineteenth century, when naturalists were intent on rejecting the excessive anthropomorphism that characterised the work of their predecessors (Galef, 1996c).

Is there a resolution?

So, who got it right? Is social learning in animals a force for conservation of the old ways or a force for change, spreading innovative behaviours through populations? Or, are both views correct?

At least part of the answer to such questions lies in results of experiments undertaken to determine just how information is transmitted socially from one individual to another. Understanding how animals learn socially and how social learning interacts with both individual learning and unlearned predispositions of animals should provide some insight into the role of social learning in both promoting and inhibiting the spread of behavioural innovations.

Food choices of wild and laboratory rats as model systems

My students and I have used the feeding behaviour of both a common laboratory animal, the domesticated Norway rat (*Rattus norvegicus*), and of its wild progenitor, as model systems in which to study: (1) behavioural mechanisms supporting social learning and (2) interactions of socially learned behaviours with other influences on behavioural development. Results of such studies of social feeding in Norway rats, described very briefly below (and reviewed more extensively in Galef, 1976, 1982, 1988, 1996a,b), suggest that social learning is inherently neither progressive nor conservative in its impact on behaviour. Rather, social learning acts in concert with an animal's behavioural proclivities and individual experiences, sometimes to maintain old habits in new recruits to a social group, sometimes to diffuse novel patterns of behaviour through a population.

Social learning as a conservative force in the food choices of Norway rats

Experiments with wild rats

Many years ago, my co-workers and I (Galef and Clark, 1971a) took wild Norway rats, first and second generation descendants of animals that we had trapped on garbage dumps in southern Ontario, and established them in small groups in 1×2 m enclosures. Using taste-aversion learning, we taught all the members of each of our colonies to eat only one of the two foods that we made available to each colony for 3 h each day. Our wild rats learned rapidly to avoid eating the adulterated diet placed in their cages each day and continued, for months, to avoid that diet even when later offered uncontaminated samples of it.

We then waited impatiently until young ones were born to colony members, grew to weaning age and began to eat solid food. By watching on closed-circuit television throughout daily 3-h feeding periods, we could observe and record every mouthful of food that the weaning juveniles in each enclosure ate.

We found, invariably, that the young members of each colony ate only the food that the adults of their colony were eating, and never even sampled the alternative food that adult members of their colony had learned to avoid. For weeks, the young wild rats remained faithful to the food preference we had taught to the adult members of their colony even though both adults and young were presented only with uncontaminated samples of both diets (Figures 6.1(a) and (b)).

Such avoidance of bait by young rats after adult members of their colony have been poisoned on it and learned to avoid eating it is no mere laboratory artefact. Applied ecologists trying to exterminate pest populations of wild Norway rats have reported, as we found in captive animals, that if members of an adult population learn to avoid ingesting a poison bait, their young ones will also avoid all contact with that bait for some time (Steiniger, 1950).

The reason for the socially induced conservatism in food choice seen in wild Norway rats is easy to understand. Wild rats, unlike their domesticated conspecifics, are extremely hesitant to eat any food that they have not previously eaten. For example, in the laboratory, wild Norway rats that are used to eating one food, and are then offered access only to an unfamiliar food, will often starve themselves for days before starting to eat the unfamiliar food, even if the unfamiliar food is highly nutritious and palatable (Galef, 1970). Domesticated rats placed in a similar situation will begin eating the unfamiliar food in a matter of minutes or hours.

Results of our experiments have shown that young wild rats living with older conspecifics are biased in a variety of different ways to begin eating the same food that the

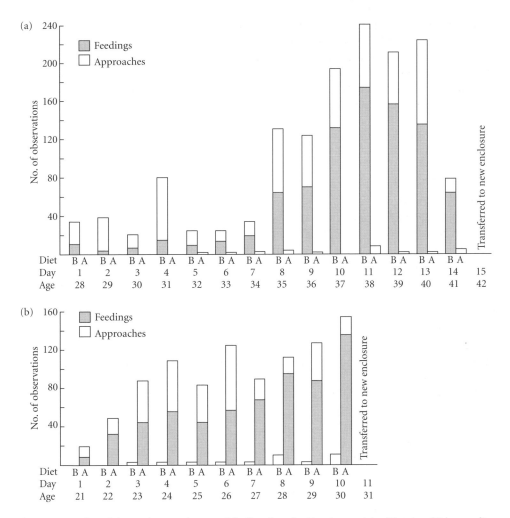

Figure 6.1 Number of observed approaches to and feedings from food bowls containing Diets A and B by weanling wild rat pups the adults of whose colonies had been trained to avoid ingesting: (a) Diet A or (b) Diet B. Reprinted from Galef and Clark (1971a) with permission. Copyright © 1971 by the American Psychological Association.

adults of their colony are eating (Galef, 1996b): (1) Young rats detect the flavour of their mother's diet in her milk and, when weaning, prefer foods having a flavour they experienced while suckling (Galef and Henderson, 1972; Galef and Sherry, 1973). (2) When seeking their first meals of solid food outside the nest, weaning rats approach adults feeding at a distance from the nest site and begin to feed close to those adults, often crawling up under an adult's belly and starting to eat under its chin (Galef and Clark, 1971a, b). (3) Young rats prefer to eat both foods and at feeding sites that have been scent-marked by adults of their species (Galef and Heiber, 1976; Galef and Beck, 1985; Laland and Plotkin, 1993). (4) Rats both young and old can detect the odour of a food on the breath of a conspecific and show enhanced preferences for foods experienced in that way (Galef and Wigmore, 1983;

Galef, 1996b). (5) Young rats show enhanced preferences for foods that they have previously stolen directly from the mouths of conspecifics (Galef *et al.* 2001), and (6) young rats follow scent trails that adults deposit when travelling from feeding sites back to their nest (Galef and Buckley, 1996).

Once weaning wild rats have been biased by interaction with adults of their colony (or with their peers) to begin eating one food rather than available alternatives, the young ones will ignore those alternatives because of their inherent reluctance to ingest unfamiliar substances. Greenberg (Chapter 8) discusses the role of neophobia, and its converse 'neophilia', in development of novel behaviours. Here we will be more concerned with the role of neophobia in maintenance of behaviours once they have been introduced into a population.

We have used comparative methods to test directly the hypothesis that the conservative nature of social influences on the food choices of young wild rats depends on an interaction between the social biasing of initial food choices and the reluctance of wild rats to ingest unfamiliar potential foods. As mentioned earlier, members of domesticated strains of Norway rat are far more willing to eat unfamiliar foods than are wild Norway rats. Consequently, although weaning domesticated rats, like weaning wild rats, might initially eat the same food that adults of their colony are eating, we predicted that domesticated rats should soon sample available foods other than the food to which adults of their colony have introduced them.

Our findings supported these predictions. Domesticated rats, like their wild forebears, initially eat the same food that adults of their colony are eating, but unlike wild rats domesticated rats soon begin first to sample and then to eat, available alternatives (Figure 6.2; Galef and Clark, 1971a). Thus, the 'neophobia' of young wild rats (their tendency to avoid

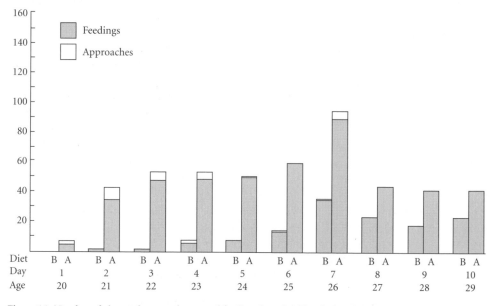

Figure 6.2 Number of observed approaches to and feedings from food bowls containing Diets A and B by weanling domestic rat pups the adults of whose colony had been trained to avoid ingesting the more palatable Diet B. Adapted from Galef and Clark (1971a) with permission. Copyright © 1971 by the American Psychological Association.

contact with unfamiliar objects or foods (Barnett, 1963)) together with social biasing of weaning rats' initial food choices, results in a highly conservative social influence on feeding behaviour.

Experiments with domesticated rats

Conservative influences of social learning are, however, not confined to neophobic wild rats and can be seen even in their domesticated, relatively neophilic descendants. As illustrated below, in domesticated rats, environmental factors can determine whether social influences conserve feeding patterns in a social group.

Evaluating consequences of ingesting various foods takes time because foods must be sampled and evaluated individually, if their relative value is to be determined accurately. Consequently, domestic rats might be expected to show stronger conservative effects of social learning on their food choices the shorter the time that they have available to determine for themselves the relative values of available foods.

We established groups of four domesticated Norway rats in 1×2-m enclosures and again trained all members of each colony to eat one or the other of two foods placed in each enclosure, this time for 1 h/day. Once all the subjects in each colony had learned to avoid eating the adulterated food, we offered them uncontaminated samples of both foods, and they continued to avoid whichever base diet had previously contained toxin.

Once each day, we replaced one of the trained colony members with a naive conspecific. After four days of such replacement, we had a new generation, a colony of four rats none of which we had taught directly to avoid one of the foods available in the enclosure each day. After all four original colony members had been removed, we replaced each day the member of a colony that had been longest in an enclosure, and we continued this replacement process generation after generation until we exhausted our supply of naive rats.

Even after four generations of replacements, we still saw profound impact on the food choices of the last generation of the training that members of the first generation had received. Fourth-generation rats introduced into colonies whose founding members had been trained to eat cayenne-pepper flavoured diet ate far more of that diet than did fourth-generation rats introduced into colonies whose founding members had been trained to eat the alternative available diet flavoured with Japanese horseradish (Figure 6.3; Galef and Allen, 1995).

However, we observed this conservative function of social learning only when we sharply restricted the time that subjects had to sample the two foods available in their enclosures, thus denying our subjects opportunity to learn for themselves about the relative worth of the two foods. The conservative role of social learning on the food choices of successive generations of rats was far greater in rats that had access to food for 2 h/day than in rats that had access to food for 24 h/day (Figure 6.4; Galef and Whiskin, 1997). So, although domestic rats are not particularly neophobic, there are circumstances in which social learning can play a largely conservative role in behavioural development, as Wallace (1870), Morgan (1890, 1896) and Baldwin (1895) suggested was generally the case.

Social learning as a progressive force in the food preferences of Norway rats

There are also circumstances in which social learning acts progressively, to diffuse novel behaviours through a population. For example, after a naive rat (an 'observer' rat) interacts

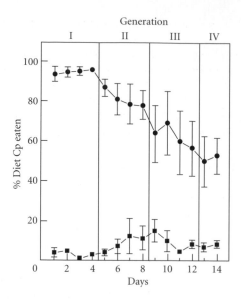

Figure 6.3 Mean amount of Diet Cp (cayenne-pepper flavoured), as a percentage of total amount eaten by domesticated rats housed in floor enclosures that contained founding colonies trained to eat either Diet Cp (circles) or Diet Jh (squares, Japanese horseradish flavoured). Day 1: enclosures contained only members of the founding colony; Days 2–4: enclosures contained both original colony members and replacement subjects; Days 5–14: colonies contained only replacement subjects. Flags ± 1 SEM. Reprinted from Galef and Allen (1995) with permission from Elsevier.

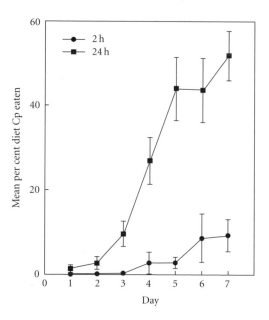

Figure 6.4 Mean amount of Diet Cp eaten, as a percentage of total intake, by founding colony members trained to eat Diet Cp and their replacements that had access to food for either 2 h (circles) or 24 h (squares) each day. Flags ± 1 SEM. Reprinted from Galef & Whiskin (1997) with permission from Elsevier.

for a few minutes with a recently fed conspecific (a 'demonstrator' rat), the observer shows substantial enhancement of its preference for whatever food its demonstrator ate (Galef and Wigmore, 1983). In many of our experiments on such social transmission of food preference from demonstrator to observer rats, independent groups of observer rats first interacted with demonstrator rats fed either cinnamon-flavoured diet or cocoa-flavoured diet and the observers then chose between cinnamon- and cocoa-flavoured diets. Observers that had interacted with demonstrators fed cinnamon-flavoured diet preferred cinnamon-flavoured diet, whereas observers that had interacted with demonstrators fed cocoa-flavoured diet preferred that diet (Figure 6.5; Galef and Wigmore, 1983).

Of course, such a social learning process could act either conservatively to bring new recruits to a population into line with their colony's established food preferences, or progressively cause individual colony members to increase their probability of eating any unfamiliar foods being eaten by other members of their social group. Which way is this social learning process most likely to act?

To look at progressive and conservative functions of this type of social learning about foods we examined three groups of rats that were once again offered a choice between cinnamon- and cocoa-flavoured diets (Galef, 1993). Before testing for food preference, one group of subjects (whose data are depicted at the left of Figure 6.6) had no experience whatsoever of either cocoa-flavoured diet or demonstrators fed cocoa-flavoured diet. The food choices of this group of subjects provided a baseline measure of preference for the two diets. Members of a second group (whose data are presented in the middle of Figure 6.6) had eaten cocoa-flavoured diet for 3 days before interacting with the demonstrator rats that had been fed cocoa-flavoured diet. Members of a third group (whose data are depicted to the right in Figure 6.6) had never seen cocoa-flavoured diet before they interacted with a demonstrator rat fed cocoa-flavoured diet.

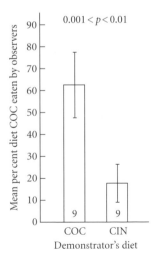

Figure 6.5 Mean amount of cocoa-flavoured diet(COC) eaten, as a percentage of total amount ingested, by observers whose demonstrators had eaten either cocoa- or cinnamon(CIN)-flavoured diet. Flags ±1 SEM. Reprinted from Galef & Wigmore (1983) with permission from Elsevier.

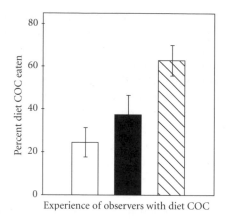

Figure 6.6 Mean amount of cocoa flavoured diet(COC), as a percentage of total amount eaten, ingested by observers when cocoa diet was totally unfamiliar (open bar), a familiar diet eaten by a demonstrator (closed bar), and an unfamiliar diet eaten by a demonstrator (hatched bar). Flags = ±1 SEM. Reprinted from Galef (1993) with permission from Elsevier.

Comparison of the food choices of observers that were either familiar (middle bar) or unfamiliar (right bar) with cocoa-flavoured diet before they interacted with a demonstrator that had eaten cocoa-flavoured diet, shows that social-learning had a greater impact when inducing intake of an unfamiliar than of a familiar food. Such data suggest that this particular mechanism for social learning is more likely to induce population members to introduce a new food into their feeding repertoires than to continue eating a familiar food.

Commentary

The message to be extracted from the laboratory data of the sort reviewed above is that social learning can act either to conserve existing patterns of behaviour or to facilitate diffusion of novel patterns of behaviour, depending on other behavioural processes acting in an animal. Of course, social learning is only one way in which behavioural innovations can be introduced into an individual's behavioural repertoire. Individuals, especially those individuals whose established behaviours are failing to produce a desirable density of rewards, may try new ways to achieve their goal. Indeed, such individual trial-and-error must be the source of any innovative behaviours that subsequently diffuse through a population of non-human animals by social learning.

So far as we know, socially learned behaviours are neither more nor less resistant to effects of reward and punishment than are individually learned behaviours (Galef, 1995; Heyes, 1993, 1994). Consequently, frequency of expression of innovations learned socially, like that of novel behaviour patterns learned individually, will be largely determined by the relative frequency and magnitude of rewards that result from engaging in the innovative behaviour and any existing alternatives.

On such a view, when we see an innovative behaviour persevering in a population over many generations, it is reasonable to assume that it is more frequently or better rewarded than alternatives present in the behavioural repertoires of population members. For example, Aisner and Terkel (1992) have described populations of roof rats (*Rattus rattus*) living in the pine forests of Israel that subsist on a diet consisting entirely of seeds extracted from pinecones (Zohar and Terkel, 1992). Extraction of pine seeds has been a stable tradition in these relatively short-lived animals for many decades.

Laboratory studies of development of the energetically efficient method of stripping of scales from pinecones to access pine seeds has shown that only rats reared by dams that strip pinecones efficiently learn reliably to strip pinecones in a similar way. I would suggest that, although the efficient pattern of pinecone stripping is socially learned, its maintenance in populations of forest-dwelling roof rats is a consequence of its providing greater rewards than alternative methods of attacking pinecones.

Consistent with this view is the finding that when socially learned behaviours are less efficient than available alternatives, they rapidly disappear. For example, Giraldeau and his co-workers (Giraldeau and Lefebvre, 1986, 1987; Giraldeau and Templeton, 1991) found that 'observer' pigeons tested after watching conspecifics peck open paper-covered food wells and recover seed, learned rapidly to open such wells for themselves, whereas pigeons without opportunity to learn the behaviour socially acquired it very slowly. However, when observer pigeons that had learned socially to open food wells were tested in flocks, rather than individually, some pigeons ('producers') continued to open wells, whereas others stopped exhibiting the socially learned behaviour and, instead, scrounged seed from the wells opened by producers. When Giraldeau and Lefebvre subsequently removed all producers from the flock, scroungers began, once again, to open food wells for themselves, and when producers were returned to the flock, these same birds went back to scrounging. Clearly, expression of socially learned behaviour depended on the absence of alternative routes to reward, and the longevity of the socially learned producer behaviour in individuals depended on the environmental situation, not the fact that producing was socially learned.

In sum, laboratory research with both wild and domesticated animals suggests that effects of social learning on a population can either be conservative or progressive, long lasting or ephemeral. The type and duration of social effects on behaviour of population members appears to depend on the behavioural proclivities of the social learners and on environmental contingencies, not on the fact that social processes initiated a behavioural variant.

Of course, as Laland and his co-workers have made clear in their innovative experimental work on innovation, the social environment, as well as the ecological environment, can provide rewards (e.g. Laland and Williams, 1997). If, for example, threat of predation is high and that threat is reduced substantially by remaining close to conspecifics, then ecological rewards may be foregone in the interests of social defence against predation. However, even in such cases, contingencies maintain the behaviour, not its social origins.

Implications of laboratory studies for interpretation of field observations

Controlled studies of the spread and maintenance of socially learned behaviours in convenient species should inform discussion of the origins of stable differences in behaviour observed in allopatric populations living free in their natural habitat. Failure to consider information about both the nature of social learning and the role of relative reward in maintenance of socially acquired responses revealed by laboratory experiments, has led to some curious proposals concerning causes of differences in behaviour of allopatric populations.

For example, contemporary accounts of origins of behavioural traditions in chimpanzees require that social learning play both progressive and conservative roles. An innovative behaviour is assumed to spread through a population by social learning and then to become fixed in that population because of the presumed highly conservative nature of social learning.

In particular, Whiten et al. (1999), in a thought-provoking paper, described as 'cultural' two methods of tool use that chimpanzees employ when dipping for driver ants. At Gombe in Tanzania (East Africa) ant-dipping chimpanzees hold a long wand in one hand, introduce it into an underground nest of driver ants, and then quickly withdraw the wand from the nest as ants stream up the wand to attack. The feeding chimpanzee then sweeps the wand with its free hand, collecting the ants in a loose mass that it then pops into its mouth and chews rapidly to avoid being bitten (McGrew, 1992). In a second method, used by chimpanzees in the Tai forest in the Ivory Coast (West Africa), a short stick is held in one hand and used to collect a small number of ants, which are then transferred directly to the mouth by sweeping the stick through the mouth. The method of ant dipping used by chimpanzees in the Tai forest, results in far fewer ants being consumed per unit time spent ant dipping than does the technique used at Gombe (Whiten et al., 1999).

If the less efficient technique used at Tai is, as Whiten et al. (1999) proposed, 'cultural', then, in the past, an innovator discovered the inefficient technique of ant dipping currently used by Tai chimpanzees, and that technique diffused through the Tai population by social learning. During this diffusion of behaviour, social learning had a progressive role, spreading the novel behaviour. On the culture hypothesis, either no Tai chimpanzee ever discovered the more efficient technique for ant dipping that is currently used by chimpanzees living at Gombe or learning the more efficient method of ant dipping was in some way inhibited by the socially learned, inefficient technique. In the latter case social learning would have played, and is playing, a conservative role.

It seems unlikely, for reasons indicated below, that the inefficient foraging technique continues to be used at Tai because no member of the Tai population ever discovered the more efficient technique used at Gombe. First, chimpanzees at Bossou in Guinea (West Africa), like those at Gombe, have learned to use slender wands and to gather driver ants with their hands, so, obviously, discovery of the Gombe technique occurs with some frequency. Further, and unexpectedly on the cultural explanation for the difference in ant-dipping seen at Tai and Gombe, chimpanzees at Bossou not only use both Tai and Gombe ant-dipping techniques, but also use the Tai technique (the one human observers consider relatively inefficient) more frequently than they use the Gombe technique (the one human observers consider relatively efficient). How are such data to be interpreted from the cultural perspective?

If traditions in chimpanzees are highly conservative, then a socially learned inefficient method of ant dipping might persist even if individuals at Bossou occasionally discovered the more efficient ant-dipping technique for themselves. However, as discussed previously, laboratory data suggest that socially learned behaviours are as easily modified by individual experience of their consequences as are individually learned behaviours (Galef, 1995; Heyes, 1993, 1994). Alternatively, it is possible, though statistically unlikely, that diffusion of the more efficient method of foraging for ants is currently in progress in the Bossou population of chimps. The reason I suggest that such an explanation is unlikely is that if one examines the many behaviour patterns that Whiten *et al.* label as cultural, nearly half are habitual or common in some populations and only occasional in others. Some general explanation is required for the varying frequency of expression of population-specific behaviours in different social groups. Given the number of such cases, recency of introduction of a behaviour pattern into a population seems unlikely to be a general cause of observed variation in frequency of expression of 'cultural' behaviours.

On the other hand, and as some proponents of the 'cultural' interpretation of variability in chimpanzee behaviour have suggested: 'differences [in behaviour] could result from biotic or physical factors acting directly in transaction with the individual chimpanzee... we can never rule out unknown (to us) environmental factors' (McGrew, 1992, p. 166). Maintenance of different techniques of ant dipping in different areas might, for example, be due to differences in soil conditions, the size of ant nests, or the behaviour of nest occupants in response to intruding probes.

Of course, as the editors of the present volume suggested to me, the cultural explanation of the mixed ant-dipping behaviour seen at Bossou can be strengthened by ad hoc elaboration. For example, low-status troop members may have initiated the Gombe technique at Bossou, as low-status individuals may be more likely than high-status individuals to be innovators (Reader and Laland, 2001), and low-status individuals may be less likely than high-status individuals to serve as models for imitation (Rogers, 1962). However, a simpler explanation consistent with available data is that ant-dipping behaviours are not cultural, but instead, reflect variation in the consequences of using long and short wands in different situations.

The suggestion that ant dipping with short sticks and with long wands are efficient in different situations may be testable (See Humle and Matsuzawa (2002) for recent evidence consistent with this view). A human experimenter would first become proficient at using long wands at Gombe and short sticks at Tai. Then he or she would have to conduct an experiment using both long wands and short sticks to secure driver ants at nests exploited by chimpanzees in each location. If, in the Tai forest, short sticks caught more ants than long wands, and at Gombe, long wands were more productive of ants than short sticks, then the cultural explanation (Whiten *et al.*, 1999) for observed differences in ant dipping techniques at Gombe and Tai could be rejected.

Concluding remarks

It is, perhaps, appropriate to conclude with a quotation from Morgan (1896, p. 184). 'Often we are unable to say in the present condition of our knowledge whether the performance of certain activities is due to heredity or tradition...' To make the quote thoroughly modern, we would have to add to Morgan's proposal only 'or to individual learning about environmental contingencies.'

We know that social learning acts sometimes to spread innovation through a population and sometimes to conserve established patterns of behaviour. Possibly, social learning can act first to introduce a novel pattern of behaviour into a population of animals and then to sustain it there without environmental support (Laland, 1996). However, demonstration under controlled conditions of this last feature of social learning is needed before it is accepted as an explanation for observed differences in the behaviour of allopatric populations living in natural habitat.

Summary

Historically, social learning has been seen as playing both conservative and progressive roles in the development of behaviour, acting both to maintain current patterns of behaviour and to spread novel behaviours through a population. Experiments investigating social influences on the food choices of both wild and domesticated laboratory rats indicate that social learning can, in fact, play either a conservative or progressive role in behavioural development, depending on environmental circumstances and the unlearned behavioural proclivities of subjects.

Consideration, from the perspective provided here, of field data describing the distribution of behavioural variants in allopatric populations of chimpanzees across Africa suggests that some purported 'cultural' differences in behavioural repertoires of chimpanzee troops may, in fact, reflect subtle differences in ecology, rather than effects of social learning. Perhaps the most pressing open question regarding social learning in animals concerns whether and under what conditions social learning can act, first, progressively to introduce a novel pattern of behaviour into a population, then conservatively to maintain the behaviour in the population in the absence of environmental support.

Acknowledgements

Preparation of this paper was facilitated by a grant from the Natural Sciences and Engineering Research Council of Canada. I thank Simon Reader, Kevin Laland and two anonymous reviewers for their thoughtful comments on an earlier draft.

References

Aisner, R. and Terkel, J. (1992). Ontogeny of pine-cone opening in the black rat (*Rattus rattus*). *Animal Behaviour*, **44**, 327–36.

Baldwin, J. M. (1895). *Mental development in the child and in the race.* New York: Macmillan.

Barnett, S. A. (1963). *The rat: A study in behavior.* Chicago, IL: University of Chicago Press.

Galef, B. G. Jr. (1970). Aggression and timidity: Responses to novelty in feral Norway rats. *Journal of Comparative and Physiological Psychology*, **70**, 370–81.

Galef, B. G. Jr. (1977). Mechanisms for the social transmission of food preferences from adult to weaning rats. In *Learning mechanisms in food selection* (ed. L. M. Barker, M. Best, and M. Domjan), pp. 123–48. Waco: Baylor University Press.

Galef, B. G. Jr. (1982). Studies of social learning in Norway rats: A brief review. *Developmental Psychobiology*, **15**, 279–95.

Galef, B. G. Jr. (1988). Communication of information concerning distant diets in a social, central-place foraging species (*Rattus norvegicus*). In *Social learning: Psychological and biological perspectives* (ed. T. R. Zentall and B. G. Galef, Jr.), pp. 119–40. Hillsdale, NJ: Erlbaum.

Galef, B. G. Jr. (1993). Functions of social learning about foods in Norway rats: A causal analysis of effects of diet novelty on preference transmission. *Animal Behaviour*, **46**, 257–65.

Galef, B. G. Jr. (1995). Why behaviour patterns that animals learn socially are locally adaptive. *Animal Behaviour*, **49**, 1325–34.

Galef, B. G. Jr. (1996a). Social influences on food preferences and feeding behavior of vertebrates. In *Why we eat what we eat* (ed. E. D. Capaldi), pp. 207–32. Washington, DC: American Psychological Association.

Galef, B. G. Jr. (1996b). Social enhancement of food preferences in Norway rats: A brief review. In *Social learning in animals: The roots of culture* (ed. C. M. Heyes and B. G. Galef, Jr.), pp. 49–60. San Diego, CA: Academic Press.

Galef, B. G. Jr. (1996c). The making of a science. In *Foundations of animal behavior* (ed. L. D. Houck and L. C. Drickamer), pp. 5–12. Chicago, IL: University of Chicago Press.

Galef, B. G. Jr. and Allen, C. (1995). A new model system for studying animal tradition. *Animal Behaviour*, **50**, 705–17.

Galef, B. G. Jr. and Beck, M. (1985). Aversive and attractive marking of toxic and safe foods by Norway rats. *Behavioral and Neural Biology*, **43**, 298–310.

Galef, B. G. Jr. and Buckley, L. L. (1996). Use of foraging trails by Norway rats. *Animal Behaviour*, **51**, 765–71.

Galef, B. G. Jr. and Clark, M. M. (1971a). Social factors in the poison avoidance and feeding behavior of wild and domesticated rat pups. *Journal of Comparative and Physiological Psychology*, **25**, 341–57.

Galef, B. G. Jr. and Clark, M. M. (1971b). Parent-offspring interactions determine time and place of first ingestion of solid food by wild rat pups. *Psychonomic Science*, **25**, 15–16.

Galef, B. G. Jr. and Heiber, L. (1976). The role of residual olfactory cues in the determination of feeding site selection and exploration patterns of domestic rats. *Journal of Comparative Psychology*, **90**, 727–39.

Galef, B. G. Jr. and Henderson, P. W. (1972). Mother's milk: A determinant of feeding preferences of weaning rat pups. *Journal of Comparative and Physiological Psychology*, **78**, 213–19.

Galef, B. G. Jr. and Sherry, D. F. (1973). Mother's milk: A medium for the transmission of cues reflecting the flavor of mother's diet. *Journal of Comparative and Physiological Psychology*, **83**, 374–8.

Galef, B. G. Jr. and Whiskin, E. E. (1997). Effects of social and asocial learning on longevity of food preference traditions. *Animal Behaviour*, **53**, 1313–22.

Galef, B. G. Jr. and Wigmore, S. W. (1983). Transfer of information concerning distant foods: A laboratory investigation of the 'information-centre' hypothesis. *Animal Behaviour*, **31**, 748–58.

Galef, B. G. Jr., Marczinski, C. A., Murray, K. A., and Whiskin, E. E. (2001). Studies of food stealing by young Norway rats. *Journal of Comparative Psychology*, **115**, 16–21.

Gandolfi, G. and Parisi, V. (1973). Ethological aspects of predation by rats, *Rattus norvegicus*, (Berkenhout) on bivalves, *Unio pictorum* L. and *Cerastoderma lamarckii* (Reeve). *Bolletino di Zoologia*, **40**, 69–74.

Giraldeau, L.-A. and Lefebvre, L. (1986). Exchangeable producer and scrounger roles in a captive flock of feral pigeons: A case for the skill pool effect. *Animal Behaviour*, **34**, 797–803.

Giraldeau, L.-A. and Lefebvre, L. (1987). Scrounging prevents cultural transmission of food-finding behaviour in pigeons. *Animal Behaviour*, **35**, 387–94.

Giraldeau, L.-A. and Templeton, J. J. (1991). Food scrounging and diffusion of foraging skills in pigeons (*Columbia livia*): The importance of tutor and observer rewards. *Ethology*, **89**, 63–72.

Goodall, J. (1986). *The chimpanzees of gombe: Patterns of behavior.* Cambridge, MA: Belknap Press.

Heyes, C. M. (1993). Imitation, culture and cognition. *Animal Behaviour,* **46**, 99–110.

Heyes, C. M. (1994). Imitation and culture: Longevity, fecundity and fidelity in Social transmission. In *Behavioral aspects of feeding* (ed. by B. G. Galef, Jr, M. Mainardi, and P. Valsecchi), pp. 271–8. Chur, Switzerland: Harwood Academic.

Humle, T. and Matsuzawa, T. (2002). Ant dipping among the chimpanzees of Boussou, Guinea, and some comparisons with other sites. *American Journal of Primatology,* **58**, 133–48.

Kawai, M. (1965). Newly acquired pre-cultural behavior of the natural troop of Japanese monkeys on Koshima Inlet. *Primates,* **6**, 1–30.

Laland, K. N. (1996). Is social learning always locally adaptive? *Animal Behaviour,* **52**, 637–40.

Laland, K. N. and Plotkin, H. C. (1993). Social transmission of food preferences among Norway rats by marking of food sites and gustatory contact. *Animal Learning & Behavior,* **21**, 35–41.

Laland, K. N. and Williams, K. (1997). Shoaling generates social learning of foraging information in guppies. *Animal Behaviour,* **53**, 1161–9.

Lehrman, D. S. (1953). A critique of Konrad Lorenz's theory of instinctive behavior. *Quarterly Review of Biology,* **28**, 337–63.

McGrew, W. C. (1992). *Chimpanzee material culture: Implications for human evolution.* Cambridge: Cambridge University Press.

Morgan, C. L. (1890). *Animal life and intelligence.* London: Edward Arnold.

Morgan, C. L. (1896). *Habit and instinct.* London: Edward Arnold.

Rogers, E. M. (1962). *Diffusion of innovations.* New York: Free Press.

Romanes, G. J. (1882). *Animal intelligence.* London: Kegan, Paul, Trench.

Steiniger, von F. (1950). Beitrage zur Soziologie und sonstigen Biologie der Wanderratte. *Zeitschrift fur Tierpsychologie,* **7**, 356–79.

Terkel, J. (1996). Cultural transmission of feeding behavior in the black rat (*Rattus rattus*). In *Social learning in animals: The roots of culture* (ed. C. M. Heyes and B. G. Galef, Jr.), pp. 17–47. San Diego, CA: Academic Press.

Wallace, A. R. (1870). The philosophy of birds' nests. In *Contributions to the theory of natural selection* (ed. A. R. Wallace), pp. 211–30. New York: Macmillan & Co.

Ward, P. and Zahavi, A. (1973). The importance of certain assemblages of birds as 'information centres' for food finding. *Ibis,* **115**, 517–34.

Whiten, A., Goodall, J., McGrew, W. C., Nishida, T., Reynolds, V., Sugiyama. Y. *et al.* (1999). Culture in chimpanzees. *Nature,* **399**, 682–5.

Zohar, O. and Terkel, J. (1992). Acquisition of pine cone stripping behaviour in black rats (*Rattus rattus*). *International Journal of Comparative Psychology,* **5**, 1–6.

PATTERNS AND CAUSES OF ANIMAL INNOVATION

CHAPTER 7

EXPERIMENTAL STUDIES OF INNOVATION IN THE GUPPY

KEVIN N. LALAND AND YFKE VAN BERGEN

Introduction

Innovation refers to a process that results in new or modified learned behaviour and that introduces novel behavioural variants into a population's repertoire (see Chapter 1). Virtually all relevant data on behavioural innovation in animals come from observations of natural populations, which frequently have an anecdotal quality. In several instances, the novel behaviour has been observed only once, and in a single individual (Kummer and Goodall, 1985; Chapter 10). There are a host of complications with many such reports (see McGrew, 2002). For example, the report may require clarification before it can be certain that the behaviour concerned is more than a random or accidental event, or that it serves the function attributed to it. The same reservations apply to the diffusion of innovations in animal populations, where the processes that determine whether an innovation spreads to others, at what rate, and by which pathways are masked by the idiosyncratic characteristics of each example. While there are numerous experimental studies demonstrating that particular individuals, or species, are capable of solving a novel problem, each of which could arguably be regarded as an example of innovation, the vast majority of such experiments have not investigated within-species variation in problem-solving ability, and there are surprisingly few experimental studies of problem solving that focus on sex, age, or dominance-rank differences, or try to determine the principal sources of variation (Hutt, 1973). What is needed is a systematic experimental investigation of how innovation occurs in animal populations under controlled conditions, in order to integrate field, laboratory, and theoretical findings.

Here we summarize the results of a comprehensive laboratory experimental investigation into innovation and social learning in the guppy (*Poecilia reticulata*), a small tropical freshwater fish endemic to South America and the Lesser Antilles. In this context, the innovation involves exploration, learning, and problem solving of simple mazes to locate a novel food source. We concentrate on the guppy for two reasons. First, we wish to present an overview of the research that has been carried out in our laboratory, where the guppy has proved a valuable model system for experimental research into animal innovation. Second, while there are countless studies on the behaviour of other fishes that may have implications for the study of innovation, innovation has not been their principal focus. While we describe how innovation operates within a single species, we believe that similar processes will underpin innovation in other taxonomic groups, and indeed in the final section of this chapter we argue that the explanations we offer for guppy innovation may also apply to primates and other animals.

The guppy is an excellent model species for research into animal innovation for several reasons. First, social learning of foraging information, predator avoidance behaviour, and mate choice has been demonstrated in guppies (Sugita, 1980; Dugatkin and Godin, 1992, 1993; Laland and Williams, 1997; Brown and Laland, 2002). Moreover, novel learned foraging behaviour can diffuse through laboratory populations of these fish (Reader and Laland, 2000), and social learning processes can mediate behavioural traditions (Laland and Williams, 1997, 1998), in which population-specific behaviour patterns are maintained in spite of changes in the composition of a population. Traditional behaviour is a feature of many natural fish populations (Helfman and Schultz, 1984; Warner, 1988, 1990), and there is now strong evidence for social learning and information transfer across a wide variety of contexts, and in many different fishes (Magurran and Higham, 1988; Suboski and Templeton, 1989; Krause, 1993; Oliveira et al., 1998; Reebs, 2000; see Brown and Laland, in press, for a review). In the light of these findings, it would be valuable to establish which individuals are most likely to generate novel learned behaviour. Second, guppies exhibit significant variation in behavioural dimensions such as boldness and shyness, which may reflect variation in a general response to novel situations, and thereby influence rates of innovation. For instance, the tendency to inspect unfamiliar predators varies considerably between individuals (Magurran et al., 1993), as does latency to emerge from refuge after exposure to a predator (Krause et al., 1998). Third, guppies have a number of practical advantages. They are easy and inexpensive to keep in small laboratory populations owing to their small size and simple feeding requirements. This means that large numbers of experimental populations can be established, presented with novel foraging tasks, and monitored to determine the characteristics associated with behavioural innovation.

In the following sections we (1) describe experiments that demonstrate that guppies are capable of social learning, the diffusion of novel learned behaviour, and behavioural traditions, (2) outline the findings from a series of studies of foraging innovation in the guppy, and (3) discuss how these findings can help us to interpret innovation in other taxa.

Social learning, diffusion, and traditional behaviour

When a novel behaviour spreads through an animal population through social learning processes, typically a single individual will have invented it. Therefore, the study of innovation is central to the research on animal social learning and culture. While it may be difficult to observe innovations when they occur in natural populations, the processes that underlie innovation can be studied experimentally in the laboratory by presenting animals with novel tasks and investigating which individuals solve the problem first and by what means. The argument that guppies can be used as a model system for the study of innovation, and that this model system potentially generates findings that apply to animals that exhibit cultural behaviour, will only be compelling if guppies themselves are capable of social learning. Hence, we begin our review with a description of social learning, diffusion, and traditional behaviour in the guppy.

Empirical and theoretical findings suggest that social learning does not require advanced cognitive abilities (Galef, 1988; Lefebvre and Palameta, 1988; Whiten and Ham, 1992; Heyes, 1994). In fact, most cases of social learning in animals appear to be mediated by simple processes, such as local or stimulus enhancement, where the behaviour of a conspecific draws an animal's attention to a location or stimulus about which the observer subsequently learns something. These forms of social learning are likely to be facilitated by a tendency on the part of individuals to aggregate and influence each other's movements, which increases the probability that individuals will be exposed to the same locations and stimuli as other group members. Shoaling in fishes represents such a tendency (Pitcher and Parrish, 1993), and below we show how it can mediate the social learning of novel foraging information.

Laland and Williams (1997) provided the first clear demonstration that guppies can learn the route to a hidden food source from knowledgeable conspecifics. Naïve observer fish were allowed to swim with demonstrator fish trained to use one of two equivalent routes to a feeder. After 5 days of swimming with trained conspecifics, the observer fish were tested singly, and preferentially used the route their demonstrators had taken. This experiment therefore provides unequivocal evidence for the social learning of foraging information in fish. Simply by shoaling with knowledgeable conspecifics, guppies can learn the route to a food source. The experiment also indicated that frequency-dependent learning may be important in guppies. Different numbers of demonstrators were used in the first experiment, and the performance of the observers increased significantly with increasing number of demonstrators. It would seem that single demonstrators may be less reliable than multiple demonstrators, or larger numbers of demonstrators may provide more visible or salient demonstrations for naïve individuals.

Guppies do not respond well to being tested in isolation. Consequently, Laland and Williams (1997) conducted a second experiment using a transmission chain, which is a more suitable design for social species. Small founder populations of four individuals were trained to take one of two routes to a novel food source, after which founder members were gradually replaced by naïve individuals. Three days after all the original founder members had been removed, the group still showed a strong preference for the route that their founders had originally been trained to use. Laland and Williams (1998) went on to show that preferences for long and circuitous routes were socially transmitted even if a shorter and more energy efficient route was available. These studies demonstrate that arbitrary and even maladaptive (i.e. energetically costly) traditions can be maintained in guppy populations under laboratory conditions. We suggest that equivalent processes may underlie the traditional use of mating sites, schooling sites, and migration routes characteristic of some fish species in nature (Helfman and Schultz, 1984; Warner, 1988, 1990).

Reader and Laland (2000) investigated whether a maze-learning problem could diffuse through experimental populations of the guppy. In these diffusions, knowledge of a route to a feeder could spread through the group by subjects learning from others, discovering the route for themselves, or, most likely, by some combination of these social learning and asocial learning processes. Mixed-sex populations of guppies were presented with three novel foraging tasks, and time to complete the task was recorded for each individual over 15 trials. In one experiment, the populations were made up of equal numbers of food-deprived

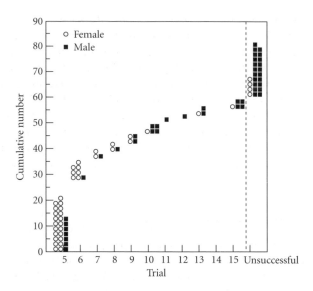

Figure 7.1 Cumulative number of adult males and females reaching a learning criterion (completion of the task in three out of four consecutive trials in significantly less time than the group mean for trial 1). Subjects that did not reach the learning criterion within 15 trials are shown on the right-hand side of the dashed line; $N = 86$. Data from Reader and Laland (2000), reprinted with permission from Elsevier.

and non-food-deprived individuals, whereas a second experiment compared small, young fish with large, older fish.[1] The study revealed a striking sex difference, with novel foraging information spreading at a significantly faster rate through subgroups of females than of males (Figure 7.1). This finding is most likely due to motivational differences between the sexes (discussed below). In addition, food-deprived fish were faster than non-food-deprived conspecifics at learning the tasks and, although there was no overall effect of body size, there was a significant interaction between sex and size. Adult females completed the tasks much faster than adult males, but no sex difference was found in juveniles. Thus, learned information may diffuse in a non-random or 'directed' (Coussi-Korbel and Fragaszy, 1995) manner through guppy populations.

These studies have demonstrated that guppies (and likely other shoaling fish species) are able to learn socially, that traditions can be maintained in guppy populations, and that novel behaviour patterns can diffuse through populations. There is some indication that the processes involved in social learning in fish are simple local and stimulus enhancement mechanisms. These and other experiments have also given us some insight into which features of demonstrators have important influences on social learning. Social learning appears to be frequency-dependent or conformist, as fish learn better with more demonstrators present (Laland and Williams, 1997; Lachlan *et al.*, 1998; Day *et al.*, 2001). Fish prefer to

---

[1] Mass is correlated with age in guppies, so Reader and Laland did not attempt to distinguish between these two factors.

shoal with, and learn better from, familiar individuals (Lachlan *et al.*, 1998; Swaney *et al.*, 2001). We have repeatedly found that naïve guppies prefer to follow conspecifics trained to find food or expect feeding rather than other naïve fish, which raises the possibility that they can select successful demonstrators (Lachlan *et al.*, 1998; Swaney *et al.*, 2001). Laboratory populations of guppies have been reported to learn predator evasion tactics (Brown and Laland, 2002) and copy mate choices (Dugatkin and Godin, 1992, 1993) from conspecifics, while social learning of foraging and predator evasion behaviour has also been found in natural populations of the Trinidadian guppy (Reader *et al.*, in press). Collectively, these studies demonstrate that guppies are capable of social learning and traditional behaviour, and hence there are grounds to regard the guppy as a suitable subject for investigations of innovation.

Experimental studies of innovation

Observations of natural populations suggest that particular classes of individual may be more prone to innovation than others. For instance, juveniles and females are often considered to be the innovators in primate populations (Reader and Laland, 2001), perhaps due to a number of high-profile case studies in the literature, such as Imo the famous sweet-potato-washing macaque (Kawai, 1965), and because researchers have expressed expectations as to which classes of individual should innovate (Kummer and Goodall, 1985; Chapter 10). However, these are assumptions and they are not all supported by empirical evidence (Reader and Laland, 2001). Controlled empirical studies on animal innovation are required. It is not clear whether innovators are primarily particularly creative, intelligent or non-conformist individuals, individuals exposed to particular environmental factors, or individuals in certain motivational states. Below, we review several laboratory studies on innovation in guppies that investigated the features of innovators and the spread of innovations through guppy populations.

Necessity is the mother of (guppy) invention

Laland and Reader (1999a) exposed small populations of guppies, composed of individuals varying in sex, hunger level and body size, to novel foraging tasks. The tasks involved swimming a simple maze to locate a food source, which consisted of an unfamiliar but desirable food accessible from an unfamiliar feeder. Each population was presented with a foraging task by inserting the maze apparatus into the experimental tank, and the category of the first fish to complete the task was recorded. Strictly speaking, the first individual to solve a maze problem cannot be described as an innovator as there is no indication that the individual concerned has yet learned the maze task, and by definition (see Chapter 1) innovation requires learning. However, in their investigation of the diffusion of novel learned behaviour, Reader and Laland (2000) found a significant correlation between latency to complete a maze the first time and number of trials to learn the maze according to the trials-to-criterion measure. This finding legitimizes the use of the first individual in a population to swim the maze as a reliable proxy measure of innovation.

In a first experiment, Laland and Reader (1999a) investigated whether the likelihood that an individual guppy will innovate to locate and exploit a novel food source is influenced by its sex or hunger level. Thirty-six experimental populations of 16 fish were established, each containing four food-deprived and four non-food-deprived fish of each sex. Female guppies are typically larger than males, so in order to rule out size as a confounding variable small females and large males were selected, such that at test there was no significant weight difference between the sexes. In 27 of the 36 populations the first fish to complete the task was a female (Figure 7.2(a)), while in 25 of the 36 populations it was food-deprived (Figure 7.2(b)). No interaction was found between sex and hunger and no effect of task, tank, or task location was found.

A second experiment investigated whether the likelihood that an individual guppy will innovate to locate and exploit a novel food source is influenced by its body size. Using the same procedure, 35 experimental populations of 16 fish were established, each containing equal numbers of males and females, and equal numbers of small and large fish. In contrast to experiment 1, here the sexes differed in mean weight, males being lighter than females, as is the situation in the wild (Magurran *et al.*, 1995). In 25 of the 35 populations the first fish to complete the task was a female, while in 24 of the 35 populations it was small (Figure 7.2(c)). No interaction was found between sex and size and no effect of task, tank or task location was found.

Thus, the second experiment provides clear evidence that females and smaller guppies are more likely to innovate to locate a novel food source than males and larger fish, respectively. Since males were substantially smaller than females, the sex difference found in this experiment cannot be explained as resulting from mass differences since small fish are more, not less, likely to locate the goal zone than large fish. Experiment 1 examined the unnatural situation where males and females were of similar size, but the results of experiments 1 and 2 combined suggest that sex differences in foraging innovation may exist in natural guppy populations. Moreover, in both experiments the male and female fish were of similar ages, which suggest that an age difference is unlikely to account for the difference between the sexes in experiment 1. In the above two experiments, innovators were neither the most active fish (male guppies are typically more active than females; Griffiths and Magurran, 1998), nor those with the fastest swimming speed (large fish are typically faster swimmers than small fish; Pitcher and Parrish, 1993), and statistical analyses have ruled out pseudo-replication. Pilot studies investigating odour cues established that fish are unable to solve these mazes by swimming up odour gradients. Moreover, these procedures have been repeated with no food in the mazes and the observed sex, hunger level, and size differences disappear (Laland, unpublished data). Below we describe how hunger differences can account for all the above results.

To investigate further how motivational state affects innovation, Laland and Reader (1999b) explored the relationship between competitive ability and foraging innovation. They predicted that those fish least successful at scramble competition would be the individuals most likely to innovate when presented with novel foraging tasks. They tested the hypothesis directly by (1) monitoring over a 2-week period the change in weight of individuals in two mixed-sex populations of guppies, (2) recording the success of each individual at scramble competition, and (3) introducing into each population novel maze tasks and recording each fish's time to complete the task. These tasks required subjects to swim a series of mazes to locate hidden food sources. They predicted that there would be

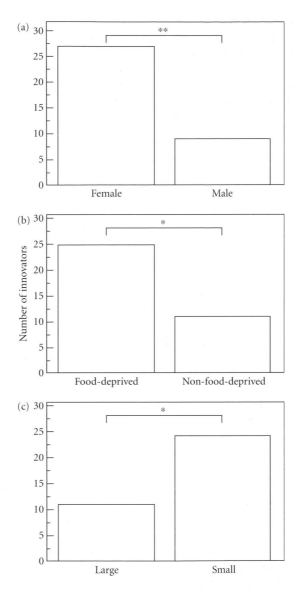

Figure 7.2 The number of innovators that were (a) female or male, (b) food-deprived or non-food-deprived, or (c) large or small. We defined an innovator as the first individual in a population to complete a novel foraging task. $*p < 0.05$; $**p < 0.01$. Data from Laland and Reader (1999a), reprinted with permission from Elsevier.

a positive relationship between weight gain and latency to innovate, as measured by the time taken to complete the novel foraging tasks. They also predicted that there would be a positive relationship between number of food items previously consumed in scramble competition and latency to innovate.

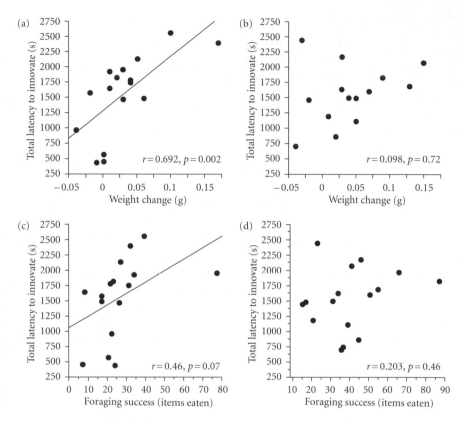

Figure 7.3 A strong correlation is found between weight change and total time to innovate in (a) males, (b) but no relationship is found in females. Results are similar for the relationship between past foraging success and total time to innovate in (c) males and (d) females. Data from Laland and Reader (1999b), reprinted with permission from Oxford University Press.

Males showed a strongly significant correlation between total latency to complete the tasks (henceforth 'total latency to innovate') and weight change (Figure 7.3(a)) and a weaker, but notable, relationship between total latency to innovate and foraging success (Figure 7.3(c)). Females, in contrast, exhibited no significant relationship between total latency to innovate and weight change (Figure 7.3(b)) and total latency to innovate and foraging success (Figure 7.3(d)). Females appeared more motivated to solve the foraging tasks than males, regardless of how they had fared during the scramble competition.

How can we explain the above-mentioned individual differences in guppy foraging innovation? First, consider the observed difference in innovative behaviour between hungry and satiated fish. It is perhaps not surprising that food-deprived fish are more likely to innovate to locate food, since they are more active and exploratory in their behaviour, better motivated to find food, and more likely to take risks to locate food (Godin and Smith, 1988; Milinski, 1993; Krause *et al.*, 1998). Swimming these mazes should be regarded as involving a degree of risk-taking on the part of these fish. Guppies appear to be reluctant to swim through small holes and into compartments, probably because in the wild small cavities and enclosed spaces often contain predators (such as cichlid piscivores and *Macrobrachium* prawns). They also express clear preferences for large over small shoal sizes (Lachlan *et al.*, 1998), and are reluctant to leave their shoal, or at least to lose visual contact with their shoalmates (Day *et al.*, 2001).

Next consider the sex differences. Laland and Reader (1999a, b) observed that these sex differences are well accounted for by considering parental investment patterns. In many vertebrate species in which female parental investment exceeds that of males, male repro-ductive success is most effectively maximized by prioritizing mating and is limited by access to receptive females (Trivers, 1972). In contrast, female reproductive success is limited by access to food resources, particularly in guppies, a species in which the females can store sperm (Reznick and Yang, 1993). Because guppies are viviparous, female parental investment is much greater than that of males (Houde, 1997). Females also have indeter-minate growth, and there is a direct correlation between energy intake and fecundity in females, whereas males stop growing at sexual maturity (Houde, 1997). Consequently, finding high-quality food has greater marginal fitness value for females than for males and this may explain why females should be more investigative than males and are constantly searching for new food sources. Males on the other hand devote much more of their time than females to pursuing the opposite sex (Magurran and Seghers, 1994), and only switch to prioritising foraging when they are really hungry, as evidenced by the correlations in the above study. Exploratory behaviour is often both energetically costly and carries increased predation risk (Hart, 1993; Milinski, 1993), yet adult female guppies are apparently pre-pared to accept these costs in order to find food. In fishes, female fecundity increases with accelerating returns with increasing body length, while a male's ability to obtain matings probably increases linearly or with diminishing returns with body length (Sargent and Gross, 1993), which means that a conservative foraging strategy is less likely to be adaptive in females than in males. Presumably, the proximate cue that drives elevated rates of foraging in females compared with males is hunger, which means that the same causal explanation can be given as above.

Finally, consider the reported size differences in innovation (Laland and Reader, 1999a). Prior to this experiment, the fish were housed in populations of mixed size. So it is likely that the larger fish would have out-competed the smaller fish in scramble competition, leaving the small fish hungrier than the large fish at the outset. It is well established that large fish have lower relative metabolic rates than smaller fish (Krause *et al.*, 1998, 1999), which means that small fish usually get hungrier sooner than larger fish. Thus, the most parsimonious explanation for all of the observed individual differences in problem solving reported above is that innovators are driven to find novel solutions by hunger, or by the metabolic costs of growth or pregnancy. This suggests that foraging innovation in the guppy can be regarded as a state-dependent phenomenon, and fits with the adage 'necessity is the mother of invention'.

Are there innovator fish?

The finding that state-dependent factors can explain foraging innovation tendencies in the guppy did not rule out the possibility that over and above these sources of variance there

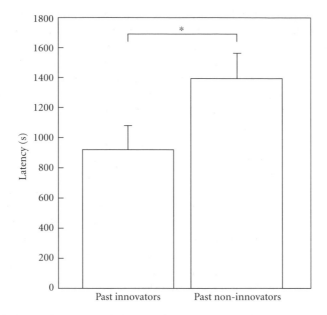

Figure 7.4 The latency (mean ± standard error) of past innovators and past non-innovators to complete a novel foraging task. Past innovators had undergone two rounds of selection, where we chose those guppies quickest to complete two different novel foraging tasks. Past non-innovators were chosen as those completing these tasks the slowest. $*p < 0.05$. Data from Laland and Reader (1999a), reprinted with permission from Elsevier.

are 'personality' differences that affect an individual's propensity to innovate. To investigate this, Laland and Reader (1999a) reasoned that if such trait or personality differences exist, then guppies identified as proficient learners in one novel foraging task would exhibit a similar level of performance if presented with a second foraging task.

However, as guppies are shoaling fish, some fish identified as 'innovators' might be fish that followed other fish through the maze rather than being 'innovators' themselves. Were this to occur, such fish would dilute any differences in innovative tendency between the fish identified as 'past innovators' and 'past non-innovators'. To address this problem, Laland and Reader (1999a) presented guppies with three foraging tasks, the first two identifying fish that had innovated (and failed to innovate) twice, and the third testing whether 'twice-past innovator' fish were faster to solve the novel foraging task than 'twice-past non-innovators'.[2] Past double innovators completed the task significantly faster than past double non-innovators (see Figure 7.4). There were no significant differences in mass

---

[2] Laland and Reader (1999a) used the terms 'innovators' and 'non-innovators' to refer to the fastest and slowest 50 per cent of each population in completing the maze. Strictly speaking, the term innovation should not be used in this context, and 'fast' and 'slow solvers' might be more appropriate terms. We persist with 'innovators' and 'non-innovators' here in order to be consistent with the original study.

between past double innovators and past double non-innovators, thus eliminating mass differences as an alternative explanation. Among fish taking part in the final test, there was a significant positive correlation between their performance on the first and final maze task. Although there was almost certainly some following taking place, this would serve only to weaken any differences between the fish categorized as past double innovators and past double non-innovators. The equal numbers of each sex, and the procedure adopted by Laland and Reader of feeding to excess between tests, render both sex and hunger as implausible explanations for the observed behavioural difference between past innovators and non-innovators. The simplest explanation for the findings of this experiment is that guppies express personality differences in their natural proclivity towards foraging innovation. It is not clear at this time whether such differences reflect variation in mental abilities (e.g. intelligence, creativity), sociality (e.g. a tendency to stay with or leave the group), boldness (e.g. a tendency to approach unfamiliar objects), exploratory behaviour (e.g. a tendency to investigate unfamiliar spaces), some other factor, or some combination of these factors. Some readers may find it surprising that we find evidence for innovative individuals in a species not particularly renowned for its intelligence or problem-solving capabilities. However, evidence is beginning to emerge that the cognitive abilities of fish have been severely underestimated (Bshary et al., 2002).

Conformity and social release

Another study conducted in our laboratory (Lachlan et al., 1998) suggested that the observed tendency exhibited by guppies to shoal with the largest number of fish might generate positive frequency-dependent social learning, or 'conformity'. Conformity is likely to affect both the likelihood of innovation and the probability that innovations will spread. The most common behaviour in a shoal would be adopted more rapidly if individuals adopt the behaviour of the majority, which would act against the transmission of novel alternative behaviour patterns exhibited by only a few individuals (Lachlan et al., 1998).

Day et al. (2001) found that conformity may both increase and reduce opportunities for social learning, depending on environmental conditions. They investigated how shoal size affects foraging efficiency in guppies. In a first experiment, fish in varying shoal sizes were required to locate food, and large shoals were found to locate food faster than small shoals, a finding that has been reported for other species of fish (Pitcher et al., 1982; Morgan and Colgan, 1987; Pitcher and House, 1987; Morgan, 1988; Ryer and Olla, 1991, 1992). However, in a second experiment, the fish had to swim through an opaque maze partition to reach a food source, and in this situation the exact opposite result was found, namely that smaller shoals located food faster than larger shoals (Figure 7.5(a)). The third experiment was designed to clarify the contradictory results of the first two experiments. The second experiment was repeated, using a transparent rather than an opaque maze partition. Fish in larger shoals were once again found to locate the food source faster in these conditions (Figure 7.5(b)).

The apparent contradiction in these findings can be explained in terms of individuals tending to adopt the most common behaviour pattern in the group and preferring to shoal in larger groups rather than breaking visual contact with the group. In the first experiment, conformity accelerated the rate at which large shoals located food relative to small shoals,

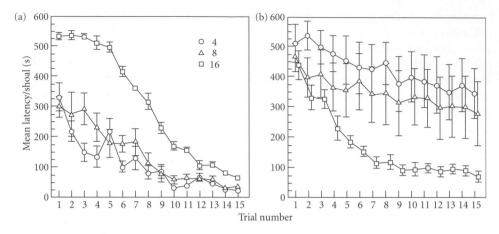

Figure 7.5 Mean (±standard error) shoal latencies for fish in each shoal size to (a) pass through the opaque partition and (b) pass through the transparent partition to enter the goal zone. Data from Day *et al.* (2001), reprinted with permission from Elsevier.

as large aggregations of fish approaching or at a food site attracted conspecifics to the food more rapidly than smaller aggregations. However, where fish are required to break visual contact with their shoal mates to locate food, as was the case in the second experiment, the presence of a larger shoal on the opposite side of the partition to the feeder deters individuals from leaving the shoal to locate food, and deters observers from following them if they do leave the shoal. Hence, in the second experiment, at least early on in the trials, conformity compelled fish not to swim through the partition or follow others that did so, but to remain with the shoal, and the strength of this conformity increased with shoal size.

The conformity hypothesis was supported by the findings of the third experiment, where larger shoals found food faster than smaller shoals despite being in a structurally identical apparatus to fish in the second experiment. The only possible cause of the opposite findings of the second and third experiments was the transparency of the partition, which allowed visual contact to be maintained between fish on either side of the partition in the third experiment. Thus, innovative individuals that solved the maze task, and those that followed them, would not perceive themselves to be leaving the shoal and would not be impeded by conformity. Conformity will cause the most common behaviour pattern to become adopted more rapidly than might otherwise be expected, while new behaviours may be slow to spread (Boyd and Richerson, 1985; Lachlan *et al.*, 1998). In the third experiment, those factors that benefit larger shoals in open-water foraging tasks, such as in the first experiment, come into play.

Frequency-dependent social learning has been previously demonstrated in guppies (Laland and Williams, 1997; Lachlan *et al.*, 1998), and also reported in rats, *Rattus norvegicus* (Beck and Galef, 1989) and pigeons, *Columba livia* (Lefebvre and Giraldeau, 1994), and may be a common feature of animal social learning. The results of Day *et al.*'s experiments raise the possibility that novel behavioural innovations, particularly those that require individuals to break contact with the group, may be more likely to spread in smaller groups of animals than larger groups.

At first sight the hypothesis that conformity is a common feature of animal social learning that inhibits the spread of novel behavioural variants appears to conflict with the observation that there are many reports of behaviour spreading through animal populations (Lefebvre and Palameta, 1988). To explain this contradiction, Brown and Laland (2002) proposed a 'social release' hypothesis, which suggests that animals will be released from conformity in the absence of clear demonstration.

Brown and Laland investigated the transmission of information about an escape route from an artificial predator from trained demonstrator fish to naïve observers. The artificial predator was a trawl apparatus with two differently coloured holes, which the fish had to swim through to avoid being trapped in a small gap between the trawl and the end of the tank. The fish appear to be highly motivated to escape from the trawl net, but are not harmed in any way by the procedure, making this avoidance task a useful simulation of a situation in which individuals can learn an adaptive escape response from more knowledgeable conspecifics. Five shoals of fish incorporating untrained (sham) demonstrators were assigned as controls. Of the six experimental populations, three shoals had demonstrators trained to take a red-outlined hole in the trawl and three had demonstrators trained to take a blue-outlined hole in the trawl. Over 15 trials exposing each shoal to the trawl apparatus, the escape latency and colour of escape hole taken was recorded for each fish in each of these shoals. The demonstrators were then removed and the remaining observers were again exposed to the trawl for 15 trials, recording escape latency and colour of escape hole taken. In the presence of demonstrators, naïve observer fish exhibited a strong preference for the escape route that their demonstrators had been trained to take. They also escaped more quickly than fish in control populations. Once the demonstrators had been removed, however, the observer fish no longer exhibited a preference for the demonstrator route, being equally likely to escape via either escape route.

One interpretation of these results is that the observer fish had not learned to escape when demonstrators were present, but simply shoaled with their escaping demonstrators. However, two findings refute this. First, after their demonstrators had been removed, the observer fish in the experimental shoals still escaped more quickly and frequently than control fish. Second, there was no deterioration in performance in either the latency or the frequency of escape among observers in experimental shoals during the trials when demonstrators were absent. There is clear evidence, then, that observers did improve their escape responses by interacting with demonstrators during the trials when demonstrators were present. The interesting conclusion of this study is that while arbitrary features of the escape response, such as the colour of the escape route chosen, were rapidly lost in the absence of demonstrators, functional aspects of the response, such as the reduced latency to escape, were conserved.

Conformity can help to ensure that animals in a social group exhibit similar behaviour, conferring benefits such as appropriate anti-predator responses, and can facilitate social learning so that animals acquire locally adaptive behaviour. However, in some conditions, conformity may in fact hinder the spread of adaptive behaviour by blocking the diffusion of innovations. The studies outlined in this section also indicate that under certain circumstances animals may be released from conformist social learning, which is how innovations may spread through animal populations.

Implications for the study of innovation in other species

We have repeatedly referred to the solving of the maze tasks as innovation. It might be objected that the swimming of a maze has little in common with, for example, the kind of innovation shown by Imo, the sweet potato-washing macaque, or Sultan, one of the chimpanzees studied by Köhler (1925) thought to have solved complex problems through 'insight', such as using a stick to knock fruit to the ground. Such objectors might prefer to reserve the term 'innovation' for qualitatively new motor patterns, goal-directed behaviour, or cognitively demanding tasks. However, there is no reason to believe that Köhler's chimps moved their bodies in ways they had never moved before, while there is a suspicion that food washing is both natural and common in macaques (Visalberghi and Fragaszy, 1990; Galef, 1992; Lefebvre, 1995). The initial impression that an animal has invented a new type of solution can be misleading, and to restrict the use of the term innovation to such cases is likely to be counterproductive. There is also little evidence that Imo or Sultan solved their problems for the first time in a more goal-directed manner than our innovator fish, and interpretations of their behaviour in terms of insight, intelligence, or other cognitively sophisticated processes have been subject to controversy (Beck, 1980; Epstein *et al.*, 1984). In the field of animal behaviour there is a long history of judgements of an animal's intelligence being biased by the preconceptions of the researcher, usually in a manner that favours those species closely related to humans (Boakes, 1984). For these reasons, we feel it is important that the purported innovations of animals from diverse taxa are evaluated on a level playing field. The key characteristic of our definition of innovation is the introduction of a novel-learned pattern of behaviour into a population, and the kind of innovation expressed by guppies satisfies this definition.

The legitimacy of this stance is supported by the observation that the same explanations that we have described to account for differences in guppy foraging innovation may equally apply to primate innovation. For instance, reasoning based on a consideration of state-dependent and motivational factors, such as hunger, has proved to be the most compelling means of deriving explanations for guppy innovation, and there is reason to believe that the same may be true in primates. Studies in primates indicate that innovators are frequently low in social rank, or poor competitors on the outskirts of social groups (Sigg, 1980; Kummer and Goodall, 1985; Fragaszy and Visalberghi, 1990). In an exhaustive survey of primate innovation, Reader and Laland (2001) found that the incidence of innovation across a wide range of contexts was three times higher in low-ranking than high-ranking chimpanzees. Hence, in guppies and primates alike, innovators may be individuals who are poor competitors, driven to search for alternative behaviour patterns when traditional strategies prove unproductive.

Reader and Laland (2001) also found that innovation was disproportionately frequent in male compared with female non-human primates. At first sight this finding appears to conflict with the patterns of innovation in guppies, in which females were the more innovatory sex. Nevertheless, we believe the same basic explanation applies in both cases.

To account for sex differences in guppy foraging innovation, Laland and Reader (1999a) noted that, in guppies, females are the larger sex and have unconstrained growth, whereas males stop growing at sexual maturity. Females therefore have greater metabolic costs of growth and pregnancy. Yet, as female fecundity is directly related to body size, there are accelerating returns to foraging success. Because of size-related differences in fecundity, there can be considerable variability in female reproductive success. Consequently, there may be greater fitness benefits associated with female than male foraging innovation.

This explanation may equally apply to foraging innovations in other species where maternal investment is greater than paternal investment, that is, to most mammals, including primates. In primates, however, the sex roles are reversed compared with the guppy. Here, it is the males that are usually the larger sex, with in many cases (e.g. in the orang-utan) male growth continuing well beyond sexual maturity, and the males exhibit greater variability in mating success than the females (Trivers, 1972; Daly and Wilson, 1978). In such circumstances it may pay males, particularly those of low status, to take risks and try out novel solutions that either allow access to the foods they require to fuel the growth they need to be an effective competitor, or constitute alternative strategies to reproductive success. Support for this line of reasoning comes from the observation that the sex difference in primate innovation is only significant in dimorphic species, where sexual selection is deemed to have been more active (van Bergen, unpublished data).

It would seem that innovation in guppies and primates may have more in common than many would suspect. The studies outlined in this chapter indicate that the adage 'necessity is the mother of invention' applies to foraging innovations in the guppy. This may in fact be a common feature of animal innovation, since proximate mechanisms such as hunger can also explain much behavioural innovation in primates. Making premature distinctions jeopardizes the ability to see genuine relationships between different kinds of novel behaviour in diverse taxa.

The experiments we have described illustrate that it is possible to study behavioural innovation in animals under controlled experimental conditions. We value field reports of innovation, but in addition would like to encourage researchers to adopt an experimental approach with their study species when they address this topic. We agree with Kummer and Goodall (p. 213, 1985; Chapter 10) that 'systematic experimentation (such as the introduction of a variety of carefully designed ecological and technical "problems" [...]) both in free living and captive groups' is the way forward in innovation research.

Summary

We have described a series of laboratory experiments that investigated the nature and characteristics of behavioural innovation in guppies, and spelled out the significance of these findings for the study of innovation in other species. These experiments have clearly demonstrated that guppies are capable of social learning, that novel-learned foraging behaviour can diffuse through a population, and that social learning processes can mediate behavioural traditions. Variation in novel problem solving is best accounted for by state-dependent factors, such as sex, size, competitive ability, and hunger level. Females, small individuals, poor competitors, and hungry individuals were found to be more likely to be the first in the population to solve a novel foraging task than males, large individuals, good competitors, and satiated individuals, respectively. These and other studies suggest that the adage 'necessity is the mother of invention' is an apt description of foraging innovation in guppies, and may apply to other species. In addition to state-dependent factors, further experimentation revealed stable 'personality' differences in problem-solving ability among guppies. Empirical data also suggest that a positive frequency-dependent form of learning, or conformity, acts to prevent the diffusion of novel-learned behaviour through fish populations, but innovations may spread through a 'social release' process. Explanatory hypotheses derived to account for individual differences in innovation in the guppy also explain variation in innovative behaviour in primates and other species. Experimental approaches to innovation, such as those described in this chapter, are a fruitful avenue for future research.

Acknowledgements

Research supported in part by a Royal Society University Research Fellowship to KNL and a BBSRC studentship to YVB.

References

Beck, B. B. (1980). *Animal tool behaviour: The use and manufacture of tools by animals.* New York: Garland STPM Press.

Beck, M. and Galef, B. G. Jr. (1989). Social influences on the selection of a protein-sufficient diet by Norway rats (*Rattus norvegicus*). *Journal of Comparative Psychology*, **103**, 132–9.

Boakes, R. (1984). *From Darwin to behaviourism: Psychology and the minds of animals.* Cambridge: Cambridge University Press.

Boyd, R. and Richerson, P. J. (1985). *Culture and the evolutionary process.* Chicago, IL: The University of Chicago Press.

Brown, C. and Laland, K. N. (2002). Social learning of a novel avoidance task in the guppy: Conformity and social release. *Animal Behaviour*, **64**, 41–7.

Brown, C. and Laland, K. N. (in press). Social learning in fishes: A review. In *Learning in fishes: why they are smarter than you think. Special issue of fish and fisheries* (ed. C. Brown, K. N. Laland, and J. Krause).

Bshary, R., Wickler, W., and Fricke, H. (2002). Fish cognition: A primate's eye view. *Animal Cognition*, **5**, 1–13.

Coussi-Korbel, S. and Fragaszy, D. M. (1995). On the relation between social dynamics and social learning. *Animal Behaviour*, **50**, 1441–53.

Daly, M. and Wilson, M. I. (1978). *Sex, evolution and behavior: Adaptations for reproduction.* North Scituate, MA: Duxbury Press.

Day, R. L., Macdonald, T., Brown, C., Laland, K. N., and Reader, S. M. (2001). Interactions between shoal size and conformity in guppy social foraging. *Animal Behaviour*, **62**, 917–25.

Dugatkin, L. A. and Godin, J.-G. J. (1992). Reversal of mate choice by copying in the guppy (*Poecilia reticulata*). *Proceedings of the Royal Society of London, Series B*, **249**, 179–84.

Dugatkin, L. A. and Godin, J.-G. J. (1993). Female mate copying in the guppy (*Poecilia reticulata*): Age-dependent effects. *Behavioral Ecology*, **4**, 289–92.

Epstein, R., Kirshnit, C. E., Lanza, R. P., and Rubin, L. C. (1984). 'Insight' in the pigeon: Antecedents and determinants of an intelligent performance. *Nature*, **308**, 61–2.

Fragaszy, D. M. and Visalberghi, E. (1990). Social processes affecting the appearance of innovative behaviours in Capuchin monkeys. *Folia Primatologica*, **54**, 155–65.

Galef, B. G. Jr. (1988). Imitation in animals: History, definitions and interpretation of the data from the psychological laboratory. In *Social learning: Psychological and biological perspectives* (ed. T. Zentall and B. G. Galef, Jr.), pp. 3–28. Hillsdale, NJ: Erlbaum.

Galef, B. G. Jr. (1992). The question of animal culture. *Human Nature*, **3**, 157–78.

Godin, J.-G. J. and Smith, S. A. (1988). A fitness cost of foraging in the guppy. *Nature*, **333**, 69–71.

Griffiths, S. W. and Magurran, A. E. (1998). Sex and schooling behaviour in the Trinidadian guppy. *Animal Behaviour*, **56**, 689–93.

Hart, P. J. B. (1993). Teleost foraging: Facts and theories. In *Behaviour of teleost fishes* (ed. T. J. Pitcher), 2nd edn, pp. 253–84. London: Chapman & Hall.

Helfman, G. S. and Schultz, E. T. (1984). Social tradition of behavioural traditions in a coral reef fish. *Animal Behaviour*, **32**, 379–84.

Heyes, C. M. (1994). Social learning in animals: Categories and mechanisms. *Biological Reviews*, **69**, 207–31.

Houde, A. (1997). *Sexual selection and mate choice in guppies*. Princeton, NJ: Princeton University Press.

Hutt, S. J. (1973). Constraints upon learning: Some developmental considerations. In *Constraints on learning: Limitations and predispositions* (ed. R. A. Hinde and J. Stevenson-Hinde), pp. 457–67. London: Academic Press.

Kawai, M. (1965). Newly-acquired pre-cultural behaviour of the natural troop of Japanese monkeys on Koshima islet. *Primates*, **6**, 1–30.

Köhler, W. (1925). *The mentality of apes*. New York: Harcourt and Brace.

Krause, J. (1993). Transmission of fright reaction between different species of fish. *Behaviour*, **127**, 37–48.

Krause, J., Loader, S. P., McDermott, J., and Ruxton, G. D. (1998). Refuge use by fish as a function of body length-related metabolic expenditure and predation risks. *Proceedings of the Royal Society of London, Series B*, **265**, 2373–9.

Krause, J., Loader, S. P., Kirkman, E., and Ruxton, G. D. (1999). Refuge use by fish as a function of body weight changes. *Acta Ethologica*, **2**, 29–34.

Kummer, H. and Goodall, J. (1985). Conditions of innovative behaviour in primates. *Philosophical Transactions of the Royal Society of London, Series B*, **308**, 203–14.

Lachlan, R. F., Crooks, L., and Laland, K. N. (1998). Who follows whom? Shoaling preferences and social learning of foraging information in guppies. *Animal Behaviour*, **56**, 181–90.

Laland, K. N. and Reader, S. M. (1999a). Foraging innovation in the guppy. *Animal Behaviour*, **57**, 331–40.

Laland, K. N. and Reader, S. M. (1999b). Foraging innovation is inversely related to competitive ability in male but not in female guppies. *Behavioral Ecology*, **10**, 270–4.

Laland, K. N. and Williams, K. (1997). Shoaling generates social learning of foraging information in guppies. *Animal Behaviour*, **53**, 1161–9.

Laland, K. N. and Williams, K. (1998). Social transmission of maladaptive information in the guppy. *Behavioral Ecology*, **9**, 493–9.

Lefebvre, L. (1995). Culturally-transmitted feeding behaviour in primates: Evidence for accelerating learning rates. *Primates*, **36**, 227–39.

Lefebvre, L. and **Giraldeau, L.-A.** (1994). Cultural transmission in pigeons is affected by the number of tutors and bystanders present. *Animal Behaviour*, **47**, 331–7.

Lefebvre, L. and **Palameta, B.** (1988). Mechanisms, ecology and population diffusion of socially learned, food-finding behaviour in feral pigeons. In *Social learning: Psychological and biological perspectives* (ed. T. Zentall and B. G. Galef, Jr.), pp. 141–63. New Jersey: Erlbaum.

Magurran, A. E. and **Higham, A.** (1988). Information transfer across fish shoals under predation threat. *Ethology*, **78**, 153–8.

Magurran, A. E. and **Seghers, B. H.** (1994). A cost of sexual harassment in the guppy, *Poecilia reticulata*. *Proceedings of the Royal Society of London, Series B*, **258**, 89–92.

Magurran, A. E., Seghers, B. H., Carvalho, G. R., and **Shaw, P. W.** (1993). Evolution of adaptive variation in antipredator behaviour. *Marine Behaviour and Physiology*, **23**, 29–44.

Magurran, A. E., Seghers, B. H., Shaw, P. W., and **Carvalho, G. R.** (1995). The behavioural diversity and evolution of the guppy, *Poecilia reticulata*, populations in Trinidad. *Advances in the Study of Behaviour*, **24**, 155–202.

McGrew, W. C. (2002). Twenty lessons learned from labours in cultural primatology. In *How the chimpanzee stole culture: Culture and meanings in apes, ancient humans and modern humans* (ed. F. Joulian). Kluwer/Plenum Press.

Milinski, M. (1993). Predation risk and feeding behaviour. In *Behaviour of teleost fishes* (ed. T. J. Pitcher), 2nd edn, pp. 285–303. London: Chapman & Hall.

Morgan, M. J. (1988). The influence of hunger, shoal size and predator presence on foraging in bluntnose minnows. *Animal Behaviour*, **36**, 1317–22.

Morgan, M. J. and **Colgan, P. W.** (1987). The effects of predator presence and shoal size on foraging in bluntnose minnows. *Environmental Biology of Fishes*, **20**, 105–11.

Oliveira, R. F., McGregor, P. K., and **Latruffe, C.** (1998). Know thine enemy: Fighting fish gather information from observing conspecific interactions. *Proceedings of The Royal Society of London, Series B, Biological Sciences*, **265**, 1045–9.

Pitcher, T. J. and **House, A. C.** (1987). Foraging rules for group feeders: Area copying depends upon food density in shoaling goldfish. *Ethology*, **76**, 161–7.

Pitcher, T. J. and **Parrish, J. K.** (1993). Function of shoaling behaviour in teleosts. In *behaviour of teleost fishes* (ed. T. J. Pitcher), 2nd edn, pp. 363–439. London: Chapman & Hall.

Pitcher, T. J., Magurran, A. E., and **Winfield, I. J.** (1982). Fish in larger shoals find food faster. *Behavioral Ecology and Sociobiology*, **10**, 149–51.

Reader, S. M. and **Laland, K. N.** (2000). Diffusion of foraging innovations in the guppy. *Animal Behaviour*, **60**, 175–80.

Reader, S. M. and **Laland, K. N.** (2001). Primate innovation: Sex, age and social rank differences. *International Journal of Primatology*, **22**, 787–805.

Reader, S. M., Kendal, J. R., and **Laland, K. N.** (in press). Social learning of foraging sites and escape routes in wild Trinidadian guppies. *Animal Behaviour*.

Reebs, S. G. (2000). Can a minority of informed leaders determine the foraging movements of a fish shoal? *Animal Behaviour*, **59**, 403–9.

Reznick, D. and **Yang, A. P.** (1993). The influence of fluctuating resources on life-history patterns of allocation and plasticity in female guppies. *Ecology*, **74**, 2011–19.

Ryer, C. H. and **Olla, B. L.** (1991). Information transfer and the facilitation and inhibition of feeding in a shoaling fish. *Environmental Biology of Fishes*, **30**, 317–23.

Ryer, C. H. and **Olla, B. L.** (1992). Social mechanisms facilitating exploitation of spatially variable ephemeral food patches in a pelagic marine fish. *Animal Behaviour*, **44**, 69–74.

Sargent, R. C. and **Gross, M. R.** (1993). Williams' principle: An explanation of parental care in teleost fishes. In *Behaviour of teleost fishes*. (ed. T. J. Pitcher), 2nd edn, pp. 333–61. London: Chapman & Hall.

Sigg, H. (1980). Differentiation of female positions in hamadryas one-male units. *Zeitschrift fur Tierpsychologie*, **53**, 265–302.

Suboski, M. D. and **Templeton, J. J.** (1989). Life skills training for hatchery fish: Social learning and survival. *Fisheries Research*, **7**, 343–52.

Sugita, Y. (1980). Imitative choice behaviour in guppies. *Japanese Psychological Research*, **22**, 7–12.

Swaney, W., Kendal, J. R., Capon, H., Brown, C., and **Laland, K. N.** (2001). Familiarity facilitates social learning of foraging behaviour in the guppy. *Animal Behaviour*, **62**, 591–8.

Trivers, R. L. (1972). Parental investment and sexual selection. In *Sexual selection and the descent of man, 1871–1971* (ed. B. Campbell), pp. 136–79. Chicago, IL: Aldine.

Visalberghi, E. and **Fragaszy, D. M.** (1990). Do monkeys ape? In *Language and intelligence in monkeys and apes* (ed. S. T. Parker and K. R. Gibson), pp. 247–73. Cambridge: Cambridge University Press.

Warner, R. R. (1988). Traditionality of mating-site preferences in a coral reef fish. *Nature*, **335**, 719–21.

Warner, R. R. (1990). Male versus female influences on mating-site determination in a coral-reef fish. *Animal Behaviour*, **39**, 540–8.

Whiten, A. and **Ham, R.** (1992). On the nature and evolution of imitation in the animal kingdom: Reappraisal of a century of research. *Advances in the Study of Behaviour*, **21**, 239–83.

CHAPTER 8

THE ROLE OF NEOPHOBIA AND NEOPHILIA IN THE DEVELOPMENT OF INNOVATIVE BEHAVIOUR OF BIRDS

RUSSELL GREENBERG

Why study ecological innovation?

Ecological innovation is the adoption of behaviours that allow individuals in a population to exploit newly available, previously unused, or familiar resources in a new way. Innovative behaviours have often been considered the stuff of anecdotes and short communications in natural history oriented journals. Still, in recent decades, innovative behaviour has attracted the attention of researchers investigating very different questions aimed at diverse levels of biological organization. Psychologists and ethologists have been fascinated by the origins of innovation and creativity that might result from play behaviours (Bekoff and Byers, 1998); behavioural ecologists have been documenting the way that innovative behaviours might arise and spread within social groups and through populations (Fisher and Hinde, 1949; Kummer and Goodall, 1985; Fragaszy and Visalberghi, 1990); and ecologists have been investigating how different levels of ecological flexibility develop and the implications this has for different life history strategies of related species (Morse, 1980) and their ability to colonize new places (Sol *et al.*, 2002). However, innovative behaviours have arguably gained greatest theoretical importance when related to the role they might play in rapid macro-evolutionary changes. It has been suggested that behavioural innovations that spread through populations and change the way that animals interact with their environment can eventually lead to new morphological and physiological adaptations (Mayr, 1963).

The idea that the development and spread of innovative behaviour may play a critical role in the propensity towards macroevolutionary shifts is an old one. Over 100 years ago, Lloyd Morgan (1886) argued that behavioural plasticity paved the way to major genetically based adaptations to new environments. The importance of ecological innovation in driving morphological evolution gained centre stage upon the publication of a paper by Wyles *et al.* (1983). Wyles *et al.* hypothesized that the rapid morphological evolution of birds and mammals was facilitated by the spread of new behaviours. By this hypothesis, the propensity to engage in new behaviours and the speed at which they are socially transmitted are key factors leading to morphological change in vertebrates. This line of thinking leads to the concept of innovation-prone taxa. These taxa are bestowed with qualities that either increase the occurrence of innovative behaviour in individuals, lead individuals to replicate these behaviours within their own repertoire, or speed up or render more reliable the transmission of these behaviours between individuals and groups of individuals.

How do we study innovation?

Although the importance of innovative behaviour has long been recognized, the means to subject innovation to scientific study has proven elusive. Innovative behaviours have proven to be difficult to study within the traditional experimental paradigm of comparative psychology. Successful innovations, almost by definition, are rare events that may not occur within the confines of a particular controlled environment (see also Lefebvre and Bolhuis, Chapter 2 and Byrne, Chapter 11). Until recently, field studies of innovative behaviour have focused primarily on longitudinal studies of primate populations (Kummer and Goodall, 1985) and some prominent changes in behaviour in birds, such as the opening of milk bottles by British tits (Fisher and Hinde, 1949). For most vertebrates, it has been difficult to amass a systematic data set that would lend itself to broad comparative analysis. In the past decade, Lefebvre and his coworkers have developed an analytical technique that relies upon the publication of notes describing unusual behaviours of birds in the ornithological literature (Lefebvre *et al.*, 1997, 2001; Nicolakakis and Lefebvre, 2000; Lefebvre and Bolhuis, Chapter 2). Techniques such as these may allow us to examine the evolutionary correlates of innovative tendencies across taxa.

An alternative approach to the direct study of the occurrence of innovative behaviours is a more reductionistic focus on the behavioural syndromes that favour the acquisition and spread of innovative behaviours in populations. In this chapter I will examine the possible link between novelty responses and the probability that an innovation will arise in a particular species. The concept of innovation is inexorably wed to that of newness. In fact, the first dictionary definition of the word 'innovate' is 'the introduction of something new' (*Webster's New College Dictionary*, 1995). Therefore, it should not be surprising that the thesis of this chapter is that the response to novelty in animals plays a pivotal role in the probability that innovative behaviours will develop and spread.

This chapter will address the possible role of novelty responses in shaping innovations related to foraging in birds. However, novelty responses are widespread in vertebrates and hopefully this discussion will illuminate broader issues as well.

Can we recognize innovative behaviours?

US Justice John Paul Stevens once noted that while he could not define pornography, he knew it when he saw it. This approach also seems to describe attempts to define innovative behaviour.

Definitions of innovative behaviour are those behaviours that form a qualitative break with species- or population-typical behaviour. Although innovation can involve behaviours associated with any aspect of an animal's life from its foraging behaviour to its social interactions, in this chapter I will focus on innovations that comprise new techniques for exploiting resources or the use of new types of resources. The use of these new resources offers a new opportunity for individuals that are colonizing new habitats or are facing competition for resources in a species-typical habitat.

Operationally, innovative behaviour has been difficult to rigorously define. By including all published accounts of unusual behaviours, Lefebvre *et al.* (1997) relied upon the judgement of fieldworkers and editors of scientific journals to determine what comprises innovation. This attempt to define innovation has broken a logjam in attempts to analyse its occurrence, but not without some conceptual problems (see also Lefebvre and Bolhuis, Chapter 2). These conceptual problems centre around the *ad hoc* and potentially anthropocentric nature of a classification based on the judgement of fieldworkers as to what constitutes a meaningful qualitative break from species-typical behaviour.

Preconditions for innovative foraging behaviour in birds

Aspects of innovation

The search for innovation prone taxa forces us to establish a framework of hypothetical preconditions for the successful development and spread of innovative behaviour. I briefly present such a framework here, emphasising where novelty responses are important.

As stated above, ecological innovative behaviours can involve both the selection of a new food resource as well as the employment of a specialized behaviour. Attempts to quantify innovative behaviour have not clearly distinguished between innovative behaviours that involve an unusual motor pattern (e.g. dropping shells from the sky) from those that involve the use of distinctly different food types (e.g. consuming blood from freshly moulted feathers of other birds), or those that involve both an unusual motor pattern applied to a new food type. These different classes of innovation may have different underlying control mechanisms and distinct implications for the further evolution of morphology and behaviour in a population or species.

Ecological plasticity and stereotypy

The phenomenon of innovation is integrally related to the concepts of ecological plasticity and stereotypy (Morse, 1980). In defining plasticity, Morse focused attention away from the ecologically static notion of generalist vs specialist, which describes the range of resources used or behaviours employed under a particular set of conditions, towards the rigidity or flexibility of behaviour in the face of changing conditions. In Morse's view, ecological generalization and specialization is a character theoretically independent of plasticity and stereotypy. For example, chickadees and certain species of tropical antbirds can be observed to be foraging primarily by searching dead curled leaves hanging in the forest understorey. During the course of observations, one could classify both as specialists. However, in the case of chickadees it is a short-term preference based on a particular pattern of prey distribution, whereas the antbirds may search dead leaves day in and day out regardless of how resource distribution might change. Although it is valuable to separate the concepts, in nature the two are probably related (i.e. plastic species tend to be generalists as well). Whereas the specialist–generalist classification can be based on comparisons of descriptive data, plasticity vs stereotypy should be teased apart by experimental perturbation of an animals environment.

Earlier, Klopfer (1967) had made a critical distinction between perceptual and motor aspects of stereotypy and plasticity—a difference we will briefly explore here. To briefly quote Klopfer 'an animal that responds only to narrow bands of wave lengths, for instance, would be considered perceptually more stereotyped than one responding to a wider band. On the motor side, stereotypy refers to the availability of only a small variety of movements by means of which an animal can accomplish a given act'. The distinction that Klopfer made is simple, but fundamental to our understanding of the development of innovation. I will only amend it here to emphasize that stereotyped animals may respond to only limited stimuli with a small repertoire of movements because they will not, not because they cannot. That is, the restriction may be regulated by something other than physical or perceptual abilities.

Motor vs consumer plasticity

The probability of pairing a new manoeuvre with a new resource is likely to be dependent upon the inherent plasticity of motor patterns associated with foraging, which I will refer to as *motor plasticity*. Beyond the variety of motor patterns employed, simply the quantity of complex manipulation may be important as well. Primatologists, in particular, have noted a relationship between the development of innovative feeding behaviours (including tool use), the identification of problem-solving abilities, and the quantity of physical manipulation while foraging (Clarke and Boinski, 1995; Boinski *et al.*, 2000; Day *et al.*, 2003).

As I emphasized above, ornithologists invoke the term ecological or behavioural plasticity to cover dissimilar concepts. An individual can be quite restricted in the motor patterns expressed while foraging and yet be quite catholic in its choice of what to feed upon, approach, or search. In fact, I have found that in the wood warblers that I studied, those species exhibiting the most restricted range of foraging manoeuvres tended to be those with the greatest plasticity in where they foraged and on what they fed. In fact, I would argue that consumer plasticity is a survival strategy for animals that lack a behavioural specialization to exploit particular resources. To clearly distinguish the plasticity of movement patterns from the plasticity of choice of foraging site or dietary item, I will refer to the plasticity of choice to approach foods, objects, or places as *consumer plasticity*. It is in the realm of consumer plasticity that novelty responses operate.

Solving new problems: cognition vs emotion

In most discussions, innovation-prone species are characterized by their cognitive and problem-solving abilities, as well as their tendencies to learn from conspecifics (Lefebvre and Giraldeau, 1996; Reader and Laland, 2002). As a short cut for comparisons of cognition and social learning across many taxa, relative brain size (or other more specific metrics, such as forebrain size) is often used as a surrogate for cognitive ability (Lefebvre *et al.*, 1997; Timmermans *et al.*, 2000). However, the ability to solve *new* problems, involving unfamiliar stimuli, not simply any problem, is at the core of the development of ecological innovations. Although it has received much less attention than cognitive abilities, an animal's emotional responses to the situation surrounding a problem (apparatus in lab; microhabitat in the field) may play a decisive role in its problem-solving ability. In particular, ethologists have long recognized that emotional responses to novel situations may greatly influence an animal's apparent cognitive abilities (Scott and Fuller, 1965). The tendency to approach or try novel objects or foods will not foster the development of foraging innovations alone; it is best seen as a necessary but not sufficient condition for the development of innovations. After the initial approach to the novel situation, then trial and error learning kicks in. The animal must then be able to learn to associate the new behaviours with the particular novelty and repeat the innovative performance, incorporating the new motor pattern and consumer preference into its foraging repertoire.

The social dimension

In order for the innovation to spread between individuals and through a population, the fear of novelty must be abated and the attraction to the novelty be enhanced through social transmission. Attending to the discoveries of other, perhaps more neophilic, members of a group may reduce the onus on an individual to be exploratory. Once again, much more attention has been paid to the mode of transmission of learned behaviours than the effects of sociality on novelty responses (Reader and Lefebvre, 2001). However, by influencing the propensity to approach novel situations, social responses to novelty become a critical feature in the development of innovative behaviour.

Responses to novelty

Types of responses

The classification of responses to novelty (neotic responses: Corey, 1977) has been rather simple and initially appears straightforward. We shall see that the relationship between neophobia and neophilia is more complex than the simple definitions might imply. Neophilia is the spontaneous attraction of an animal to a food item, object, or place because it is novel. Neophobia is the aversion that an animal displays towards approaching a food item, object, or place simply because it is novel. The hallmark of both responses is that although various intrinsic features of objects seem to influence the intensity of the neotic response, the response is differentially directed at new objects or stimulus and wanes with repeated exposure to the object. Among the features that contribute to the variation in the initial neotic response are stimulus complexity and the degree of discontinuity from the familiar background objects (Thorpe, 1956).

Neophobia

Fear of novelty or preference for the familiar?

A common response of birds to new foods or objects is to display signs of fear, for instance pileal erection, displacement behaviours, and 'jumping jacks', where approaches are punctuated by short backwards hops (Coppinger, 1969; Greenberg, 1983; Heinrich, 1988;

Beissinger *et al.*, 1994; Raudensush and Frank, 1999). However, the term neophobia is often used interchangeably to refer to (1) the mere preference for feeding on or visiting familiar foods, objects, or places and (2) the aversion or fear demonstrated to the same novel stimuli. The distinction has ramifications both for the way in which neophobia is detected and the role that it might play in the development of innovative behaviour. Lack of attention to novel stimuli suggests that novel foods or sites may be avoided because they are not recognized. An active aversion, on the other hand, indicates that the object is engendering the attention of an animal, but that habituation of the fear response is required for further exploration and learning to occur. Experimentally, neophobia is often inferred in choice tests between familiar and novel foods or sites where the animal prefers the familiar entity. However, such a preference may not reflect aversion or any other emotional response to the novel object. So an assessment of the response of an animal to a novel object presented singly, without a familiar choice, is probably more likely to elucidate an emotional response to the object. The relationship between aversion and preference may be more complicated than has generally been thought. For example, it has been suggested that after the initial neophobia is disengaged, that birds may continue to show a systematic preference for familiar foods in a process that Marples and Kelley (1999) have named 'dietary conservatism'. Furthermore, as I will discuss below, neophobia and neophilia can be displayed simultaneously and their joint intensity may signal the degree of overall attentiveness to novelty.

Testing for neophobia

In many adult birds, the response to a novel object is seemingly passive. The bird remains at a distance from the object with no obvious behavioural response. Occasionally birds make short exploratory forays to the object. However, without detailed physiological monitoring (heart rate, corticosteroids, etc.) it would be difficult to infer that the bird is showing any emotional response to novelty. One experimental strategy that has been employed for captive birds to detect the strength of neophobia is to place the object next to a familiar and preferred food and measure the increased latency to feed in the presence of the object. These types of experiments ('go/no go' experiments) have been employed for a number of taxa (ducks, tits, warblers, parrots, etc.); these studies will be discussed subsequently.

Although a neophobic response can be measured using these approaches, they only allow an assessment of the relative expression of neophobia (i.e. the latency is greater or lesser for different objects). A rarely asked question is what is the absolute effect of neophobia on the ecology of a species in the wild? In other words, how does a difference in seconds or minutes in latency translate into the probability that a novel object will be visited in the wild? The other question is the degree to which initial neophobia explains preferences for familiar foods or objects. In fact, in the go/no go experiments on neophobia, once a novel object is visited for feeding the subsequent latency to feed is low. The object is treated as if it is familiar. As mentioned before, when offered a choice between familiar and novel foods, biases against novelty may persist beyond the period of initial neophobia (Marples and Kelley, 1999). This has led to the suggestion that initial neophobia itself does not have a meaningful long-term impact on consumer choice in birds. However, feeding experiments, even those conducted in the field, are generally designed to essentially force an interaction between bird and novelty that may not occur in nature if the initial response of a bird to novelty is passive avoidance. When birds are making hundreds or thousands of foraging decisions in a day, familiarization to novelty may not occur unless there is a behavioural catalyst that is equivalent to confining a bird to a closed space with novel objects and food.

Intrinsic vs functional neophobia

Tests for neophobia in birds have often been conducted in the wild or on wild-caught birds. The individuals in these experiments may have had distinctly different prior experiences. Although relatively uncontrolled for prior experience, these tests provide information on how individuals vary in their approach to novelty under a natural regime of behavioural development. I refer to this aspect of novelty responses as *functional neophobia*. Experiments with captive-raised individuals can allow for more careful control of the objects experienced in the life of the subject prior to the actual neophobia trials. By raising individuals of different species in similar conditions and then testing for neophobia we can assess differences in *intrinsic neophobia*. The use of the term intrinsic is a shorthand expression for neophobia measured under controlled rearing conditions. It is not meant to imply that all potential environmental or maternal effects can be eliminated. For example, these experiments generally simplify as well as control the rearing environment. Adaptive differences in neophobia may result from complex ontogenetic interactions with the environment that are eliminated in the captive environment. Therefore, we need to develop ways of integrating highly controlled laboratory and field approaches to understand the mechanisms that underlie differences in neophobia—a topic I will return to below.

Functions of neophobia

Neophobia appears to be generally present in all species of mammals and birds, and perhaps many other vertebrates as well (Corey, 1977). This suggests that neophobic responses must have a very general function that serves a wide range of animal taxa. Far less attention has been paid to variation in neophobia, yet any hypothesis that attempts to explain the role of neophobia in innovation-prone taxa would have to make robust predictions about what factors affect the different levels of its expression.

In general, a differential response to novelty is not based on specific cost and benefits ascertained by an animal, but rather the uncertainty of either the costs or benefits of approaching a novel object. Two theories have been proposed to explain the underlying adaptive advantage for neophobia and the variation in the expression of this trait: the neophobia threshold hypothesis (NTH) and the dangerous niche hypothesis (DNH). I will discuss the hypotheses and how they relate to the innovation-prone behaviour patterns.

The neophobia threshold hypothesis

The NTH provides the clearest link between variation in neophobia and ecological plasticity. The hypothesis posits that the degree of aversion to novelty plays a key role in the probability that a new resource will be investigated and hence incorporated into the niche of an adult bird. More specialized birds will remain so because of a higher level of neophobia, and reduced neophobia should be the hallmark of generalists. By this hypothesis, intrinsic levels of neophobia are largely genetically determined and subject to natural selection. Therefore, neophobia can readily evolve in different populations or strains of species or between different species. It is important to note that this hypothesis addresses the proximate regulation of ecological plasticity. By this hypothesis, the ultimate factors that select for neophobia are those that select for specialization. Although ecologically plastic (non-neophobic) birds are free to explore and discover new resources, such exploration might be wasteful in a bird that is specialized on productive and predictable resources. The hypothesis posits that a period of juvenile exploration determines the familiar world for an individual bird and that in later life neophobia determines the probability that an individual will incorporate new objects or food, in a similar manner to the way that preferences develop in filial imprinting (Bateson, 1971). Stronger adult neophobia would act to protect preferences acquired during the juvenile period. There is good empirical support for aspects of the hypothesis. Intrinsic differences in neophobia have been established between various closely related species and strains of species, particularly domesticated vs wild forms (Barnett, 1958; DesForges and Wood-Gush, 1975; Barnett and Cowan, 1976; Mitchell, 1976; Jones, 1986). Particularly instructive are the experiments of Drent *et al.* (2003) and Dingemanse *et al.* (2002) where selection experiments led to two distinct groups of Great Tits that differed in their exploratory tendencies in a few generations. Tests that quantified exploratory tendencies included the rate of exploration of an unfamiliar room and the approach to a novel object in a familiar room, thereby encompassing novelty responses at two scales.

Comparing neophobia in ecologically plastic vs stereotypic species

The prediction that ecologically plastic species are less neophobic than related specialists has been tested only a few times. In a series of experiments on warblers in the genus *Dendroica*, Greenberg (1983, 1984a) determined that the more ecologically plastic Bay-breasted Warbler (*Dendroica castanea*) was consistently less neophobic than its more specialised congener the Chestnut-sided Warbler (*Dendroica pensylvanica*). Relevant to the topic of innovations, Greenberg (1984b) provides a number of examples of unusual foraging behaviour in the former, but not the latter species. For example, Bay-breasted Warblers appear to learn that lights (particularly ultra-violet lights) attract large number of insects, and they were observed concentrating their foraging around lights. The neophobia experiments were conducted on young ones of the year, collected during their first migration, but by no means naïve. Although long-term experiments with hand-raised Chestnut-sided Warblers showed that what was explored during the juvenile exploratory period was treated as familiar in 3- and 6-month-old birds (which would normally be on their tropical wintering grounds; Greenberg, 1984a), no such experiments were conducted on Bay-breasted Warblers, so the comparative level of neophobia in birds reared under controlled and similar conditions has not been tested.

Webster and Lefebvre (2000) tested feeding neophobia of a foraging specialist, the Bananaquit (*Coereba flaveola*), and the more generalized Lesser Antillean Bullfinch (*Loxigilla noctis*) by placing novel objects near a feeder in the wild. They found that the dietary generalist showed a lower aversion to approaching the feeder with novel objects than did the dietary specialist. Mettke-Hoffman *et al.* (2002) conducted feeding neophobia tests on various species of captive parrots and found that the latency to feed was related to several aspects of the ecology of the species, including whether the species has an insular or continental distribution (island forms have lower neophobia) and whether the species lives in complex vs simple habitats (the latter are more neophobic). Specifically, Mettke-Hoffman *et al.* found that parrots from savannah environments are slow to approach novel objects compared to species that occupied several habitats or forest edge. They reasoned that the savannah environment was simple and predictable and hence selection to explore new resources was less intense.

In another series of captive and field experiments, Greenberg (1989, 1990a, b, 1992) tested the NTH as it might apply to two species of North American sparrows in the genus *Melospiza*. The prediction of the studies was that the Song Sparrow (*Melospiza melodia*), a widely distributed species that occupies a broad range of habitats and is a good colonist of islands, would be less neophobic than the Swamp Sparrow (*Melospiza georgiana*), a specialist of shrubby wetlands. In experiments using feeders in the field and presentation of novel objects at food cups in captivity, Song Sparrows were, indeed, far less neophobic, supporting the predictions of the NTH.

Intrinsic and functional neophobia revisited

In the sparrow study, further experiments were done on naïve individuals of both species reared under similar conditions. These experiments produced results opposite to those with wild-raised individuals. The intrinsic level of neophobia was much higher in the Song Sparrow. These surprising results suggest that something about the experience of young Song Sparrows in the wild results in their lower neophobia. One possible explanation is that immatures of the more ecologically plastic species have explored a more unpredictable and dangerous array of habitats. Reduced neophobia found in wild-reared birds is the result of an interaction between juvenile exploration and the environment.

Neophobia experiments were also conducted for Mallards (*Anas platyrhyncos*) and American Black Ducks (*Anas rubripes*) (Bolen *et al.*, unpublished data), with the prediction that the widely distributed, generalist Mallard would be more neophilic than its more specialized congener. Feeding neophobia experiments on groups reared with and without hens showed a clear difference in neophobia, but with Mallards being the more neophobic species. Furthermore, ducks reared in an enriched environment showed reduced neophobia to those reared in a depauperate environment.

These experiments also suggest that the intrinsic level of neophobia is greater in the more ecologically plastic species, perhaps as an adaptation to greater potential dangers in the life of a young Mallard. They further show that an enriched early experience does reduce later life neophobia. Along with the work on *Melospiza* sparrows, these experiments demonstrate that intrinsic fear of novelty measured under controlled rearing conditions may actually be greater in the ecologically more plastic species.

Although the studies are few, they begin to suggest that the more ecologically plastic species may actually have lower levels of neophobia when wild-caught birds are tested, yet show higher intrinsic levels of neophobia. Much more comparative work on neophobia in both wild-reared and naïve captive-reared birds needs to be conducted. However, the results contrary to the NTH suggest that the alternative and more traditional view of the adaptive value of neophobia, the DNH, needs more scrutiny.

The dangerous niche hypothesis

The DNH follows from a more general view that the primary function of neophobia is to protect animals from the unknown potential dangers of new things rather than to maintain foraging specialization. A bird that encounters more novelty may require the protection of heightened caution when exploring new objects and new foods, particularly if toxic foods or a high level of threat from predators characterize its environments. This has been the classic explanation for why some foraging opportunists, such as House Sparrows, rats and ravens, show such high levels of neophobia (Barnett, 1957; Heinrich, 1988; Rana, 1989; Brunton et al., 1993) or why such seemingly adaptable animals are often so difficult to trap and poison. These species explore novel choices, but do so cautiously. However, to further add to the paradox of neophobia in an ecologically plastic species, there is no evidence that rats and ravens in the wild show reduced neophobia based on their exposure and exploration of diverse habitats and niches. Our ignorance of the long-term dynamics of exploration and habituation to novelty in generalist species stems from the lack of study of these processes under natural (non-laboratory) conditions over long periods of time. The DNH would place the emphasis of the control of ecological plasticity in generalist species on some other behavioural mechanism. One possibility is that the innate templates for ecological decision-making are themselves less plastic. Increased neophilia would be another candidate. High neophilia in species displaying high protective neophobia would seem contradictory, unless neophilia and neophobia are not necessarily the ends of a single behavioural continuum, a point I will develop further below.

Neophilia

Uncurious adults

Neophilia is the attraction that an animal shows towards an object simply because it is unfamiliar. Studies of neophilia in birds are based on the speed and frequency of tendency to spontaneously approach and manipulate new objects rather than the increase in latency to approach a food source associated with a novel object. Ornithologists have reported few instances of neophilia in adult birds, particularly in wild settings. There is evidence of object exploration that is not directly related to foraging in some taxa (Ortega and Beckoff, 1987; Mettke-Hoffmann, 1999). However, the phenomenon appears to be either generally uncommon in adults of most bird species or it is masked and hard to detect.

Neophilia and neophobia: opposites or partners?

A masking of neophilia could occur because of the potentially complex interaction of neophobia and neophilia at a novel object. I began the discussion by contrasting neophobia and neophilia as opposite reactions along a potential gradient of reactions to novelty.

It would be fair to say that most ecologically based studies of novelty responses equate neophilia with the lack of neophobia and vice versa. However, the actual relationship between neophilia and neophobia is a continuing area of mystery. The commonly observed ambivalent approach–withdrawal responses can be explained by a two-factor model, which holds that neophobia and neophilia are best considered as two independent responses to a novel stimulus (Russell, 1973). A neophilic response simultaneous to an initially strong neophobic response will contribute to any habituation to a novel food or object. If the time function for habituation of neophilia and neophobia are different then a period of neophobia will be followed by a period of exploration, allowing the bird to habituate. Greenberg and Mettke-Hoffman (2001) presented a two-by two table of the possible combinations of degree of neophobia and neophilia (Table 8.1). They noted that some of the most innovative taxa (corvids and psittacines) show strong tendencies towards both neophobia and neophilia, leading to a complex but intense reaction to novelty. The hypothesis is that these species depend upon exploring new situations to survive, but they do so with a high level of fear and arousal, thus protecting themselves in the face of the potential, unknown dangers that are associated with novelty. Furthermore, Greenberg and Mettke-Hoffman hypothesized that neophilia and neophobia are shaped by different selective factors: neophilia is related to the potential benefits of exploring for new resources and neophobia is a response to the inherent dangers from predators etc. The resulting behaviour is a dynamic balance between the two response functions associated with these behaviours. In addition to the specific nature of the response, the overall attention to novelty may be the hallmark of an innovator. It may be necessary to focus more on the intensity of novelty responses, rather than the 'sign' of the responses (e.g. attraction vs repulsion).

The young and neophilic

Although not well studied, neophilia and object exploration are well developed in fledgling and juvenile birds, as least in passerines; this period of intense exploration and neophilia could be the primary creative period in the development of foraging repertoires as stated

Table 8.1 Simple two-by-two matrix demonstrating the hypothetical relationship between neophilia and neophobia in response to general environmental variables. Based on Greenberg and Mettke-Hoffman (2001), adapted with permission.

	More complex, variable, and cryptic resources →	
More dangerous foraging or greater competition ↓	Low neophobia Low neophilia (e.g. pigeons)	Low neophobia High neophilia (e.g. island colonists)
	High neophobia Low neophilia (e.g. tropical forest specialists)	High neophobia High neophilia (e.g. corvids and some parrots)

in the NTH. For example, Heinrich (1995) reported on the intensity with which young ravens manipulated and explored novel objects. Heinrich found that the apparent attraction to particular stimulus features (in this case, shiny objects) was more readily explainable as an attraction to novel objects. Further, the attraction was greatest to those objects that showed the greatest stimulus discontinuity with the background environment, of which shininess is a prime example. The process of maturation in ravens involved a winnowing of attractive responses to a narrower range of objects that were increasingly similar to real food objects. Heinrich interpreted neophilia as an exploratory phase of a learning process. On the other hand, the same patterns can be explained by a simple diminution of neophilia along with a separate process of trial and error learning. One would like to experimentally examine the possibility that ravens can be trained to approach particular objects as adults, objects dissimilar to what is typically preferred in wild birds, but that they preferred to explore when they were young. In other words, is neophilia a phase that young birds pass through that leads to a change of behaviour (or development of innovation) in adult birds?

The interaction between the positive attraction to novelty and its inhibition through neophobia appears to be played out along a developmental time line in young birds. Object play appears to reach the apex of importance during the late stages of dependence between young and parent (Pellis, 1981) and this is the period when the transition between fledgling and adult foraging patterns occur (Davies and Green, 1976). The period of high reactivity with foreign objects was found to increase through the juvenile period, peaking at 12–15 weeks of age in Great Tits *Parus major* (Vince, 1960). It should be noted that the reactivity and lack of inhibition referred to in Vince's work was not restricted to novel objects. Reactivity to the same object (e.g. pulling a string) occurred day after day in juvenile songbirds. However, novel objects elicit the greater attention and more vigorous response of juveniles during object play. This internal inhibition (=neophobia?) increases slowly while the spontaneous attraction to novel objects declines rapidly. It is therefore during this period of high exploratory and manipulative behaviour of objects and motor plasticity that the possibility of coupling new motor patterns with a new feeding resource is the greatest. Juvenile neophilia might be considered a prime mover in the evolution of innovation-prone taxa and we might look to differences in the long-term retention of this plasticity and the resulting behaviours for behavioural bases for innovative behaviour in birds. Alternatively, this behavioural plasticity may have other intrinsic causes or functions without contributing to the development of long-term innovative behaviour in adults.

Critical to the long term significance of object play in juvenile birds is the degree to which consistent preferences are expressed in what is explored and how these relate to adult foraging preferences. I addressed this question in a study of juvenile object exploration in a specialized and generalized insectivorous songbird (Greenberg, 1987). Hand-raised nestlings of the Worm-eating Warbler (*Helmitheros vermivorus*), a tropical migrant species that specializes in probing dead curled leaves draped on understorey plants, and Carolina Chickadees (*Parus carolinensis*), a resident species with highly generalized foraging microhabitat preferences, were presented with a range of objects, both natural and artificial, to explore during the post-fledgling period. The birds were allowed to explore and manipulate the objects, which they did actively between feeding periods. The activity was most intense 3–4 weeks after fledging. The warblers showed a high degree of consistent individual preferences for exploring the different objects, even those warblers that were reared separately. In particular, the dead curled leaves were substantially more preferred than other objects. In general the rank correlation of preference (as determined by relative number of exploratory visits) was approximately 0.9 for all of the pairwise comparisons among warblers, even those reared apart. This suggests that the birds have a rather rigid template for what attracts their attention during exploration, which is not perceptively influenced by the activity of conspecifics. In contrast, the chickadees showed no consistent preference for any objects and individuals reared apart had a particularly low correlation of preference. Generalist species may develop idiosyncratic preferences in what they explore during juvenile object play and also allows room for social influences.

Neophobia and cognition

It seems intuitively obvious that the intensity of neophobia will influence the problem-solving ability of animals and, hence, the development of innovative behaviour. Animals that shy away from unfamiliar situations are unlikely to explore the situation enough to assess the costs and benefits of a potential foraging site or dietary item. The effect of any emotional response to novelty is generally reduced in most learning experiments through the selection of study animals that show reduced neophobia or through pre-training on experimental apparatus. Seferta *et al.* (2001) addressed this issue directly in experiments where the ability to solve a learning problem in doves and pigeons was (among other things) inversely related to the neophobia displayed in the test birds. Similarly, Webster and Lefebvre (2001) found that in a comparison of simple problem solving in five species of birds in Barbados, the neophobia of individuals was strongly negatively correlated with their overall performance on the task. Furthermore, species showing the greatest innovation rate (based on the aforementioned Lefebvre *et al.* literature review) performed the best on the cognitive task.

The importance of emotionality in general, and fear of novelty in particular, in the performance of problem-solving tasks should not be of any great surprise to students of animal learning. Consider the classic experiments of artificial selection of maze-solving ability of rats (Tyron, 1940). The initial experiments demonstrated that rats could be bred for an ability to solve a maze with few errors (maze bright) and many errors (maze dull rats) in a few generations. Subsequent testing, however, showed that the strains differed in a number of behavioural traits that indicated differences in emotionality (fear in new situations) and that these differences might have accounted for much of the apparent variation in cognitive abilities (Searle, 1949). Further research indicated that the difference in maze-learning ability disappeared for rats reared in enriched environments (Cooper and Zubeck, 1958). Taken together, these results suggest that experience and reaction to novel spaces and other aspects of novelty may have been the primary factor determining maze-solving abilities.

The brain structures important in processing novelty and engendering emotional neotic responses are poorly known in birds, but such responses are processed within the Hypothalmic-limbic system in mammals (Corey, 1977). Cognitive abilities are probably controlled in the avian forebrain, with the hyperstriatum ventrale playing an important role (Bayley, 1984; Timmermans *et al.*, 2000). The degree to which cognition and novelty responses covary reflects independent behavioural responses that originate in different neural centres acting in consort. Therefore, the degree of neophobia can be seen to be independent of, yet possibly correlated with, problem solving and innovation. Certainly the correlation between brain areas important for cognition and for neotic responses would be an important area for further research.

Social transmission of novelty responses

If neophobia is an important intervening variable in the development of problem solving and innovative behaviour, as suggested by the Webster and Lefebvre study, then social transmission of neotic responses should be a central issue in the spread of innovations through populations. The role of group living in neotic responses is complex and has received relatively little attention. Outside of the null hypothesis that sociality has no effect, we can propose a number of ways that group living can act on novelty responses. First, simply being in a group can influence the response of individuals making them more or less neophobic (Coleman and Mellgren, 1994). Second, we could imagine different individuals might show different responses to novelty that can be transmitted to the rest of the group. For example, dominance rank may influence the propensity to incur risk while foraging and risk-prone individuals may explore novelty, providing cues to which more risk averse individuals can respond.

With respect to the effect of simply being in a group, Coleman and Mellgren (1994) examined the group effect in captive Zebra Finches (*Taeniopygia guttata*) concluding that the average latency to feed at a novel feeder was reduced in a small group compared with solitary birds because the response of slower (more neophobic) individuals was improved. This suggests that more neophobic individuals attend to the behaviour of the less neophobic individuals. Working on capuchin monkeys Visalberghi *et al.* (1998) found that social facilitation speeds up the familiarization process necessary to consume new foods.

It has been hypothesized that less dominant individuals are more prone to take risks to uncover new food sources, because dominant individuals are able to displace subordinates from any resource that proves both safe and profitable (Wilson *et al.*, 1994). This has been observed for a variety of taxa (Hegner, 1985; Stahl *et al.*, 2001). The results are somewhat mixed for the few studies of the role of social dominance in shaping novelty responses. Work on corvids (Katzir, 1982, 1983; Heinrich *et al.*, 1995) showed that individuals of low to mid dominance rank were likely to initiate the approach to novel foods, spaces, or objects and were then joined by other flock members. Mettke-Hoffman *et al.* (2002), however, found no such effect in their studies of neophobia in parrots. One cautionary noise must be uttered: I know of no study where neophobia was compared for individuals in and out of a social group, thus assessing the direct effect of dominant–subordinate behavioural interactions on a particular decision to approach or avoid novelty.

In one of the few studies of interspecific interactions, Greenberg (1987b) showed that highly neophobic Chestnut-sided Warblers showed no reduction in their aversion to feed in the presence of novel objects when in the presence of Bay-breasted Warblers, their dominant and less neophobic congeners.

While this overview does not cover all the studies of sociality and novelty responses, it shows that the role of novelty is complex and critical if we are to understand the role of sociality in the spread of innovative behaviours that are derived from the approach of novel objects or situations.

Novelty responses in major vertebrate groups

The propensity towards innovative behaviour has been proposed to vary between different taxonomic groups. Wyles *et al.* (1983) argued that the rapid rate of morphological evolution in birds and mammals relates to the frequent development and rapid social dissemination of behavioural innovations in these groups. In developing their hypothesis, they emphasized higher cognitive abilities and a greater importance of social learning in mammals and birds than other vertebrates. Research on birds has suggested that variation innovation and cognition (or the development of portions of the brain related to cognition) are associated with particular taxa (Lefebvre *et al.*, 1997; Sol and Lefebvre, 2000), with groups like corvids and parrots showing the greatest tendency to display innovative behaviours. Finally, some comparative analyses suggest that even among or within closely-related species differences in the frequency of innovation can be found (Sol *et al.*, 2002).

Attention and response to novelty of one sort or another is widespread throughout the vertebrates. However, the intensity of expression of neophilia and neophobia are known to be highly variable between closely-related species and within species and hence is a highly labile trait. Therefore, if intrinsic levels of neophilia or neophobia are associated with the development of innovation, it neither appears to be a difficult trait to evolve, nor, once it does evolve, would it necessarily lead to the evolution of innovative tendencies in a large evolutionary clade. The tendency to be attracted to and explore novel objects has been found in some reptiles (Burghardt, 1998) including turtles, varanid lizards, and crocodiles (Glickman and Sroges, 1966). However, comparative details on the quantitative nature of novelty responses for different vertebrates are generally lacking. For example, I know of no studies that examine comparative neophobia in reptiles using the latency to feed in the presence of novel objects. Do non-avian or non-mammalian vertebrates respond to novelty when it affects the context of foraging and not the potential prey itself in a manner similar to birds and mammals? Do any other vertebrates approach birds and mammals in the importance that novel object exploration has in the activity budget of juveniles?

Perhaps what varies in an evolutionarily important way is the intensity and persistence of the attention to novelty as reflected both in fear and attraction. I know of no reports of the intense period of novel object exploration found in reptiles similar to that found in many juvenile birds and mammals. The intensity of activity and length of this period may vary in important ways between major groups of birds, but has not been explored systematically. A hint of this kind of variation was provided in the broad comparison (of primarily mammals) presented in the work of Glickman and Sroges (1966). These researchers used a simple assay of 'curiosity' (neophilic object exploration) that consisted of placing a diverse set of novel objects (one at a time) in the cage of over 200 species of mammals and reptiles in-housed in zoos. Based on this they were able to develop a broadly comparative picture of the pattern of object exploration within and between major vertebrate taxa. They found little evidence of object exploration in reptiles compared with mammals, significant variation between and within major mammalian orders, and some correlation with foraging behaviour. Similar standardized assays could be done, both in the field and in captivity, for neophobia and juvenile neophilia in birds.

Novelty responses at different ecological scales

The recent flourishing of work on ecological innovation and the ecological basis of novelty responses have progressed under a general assumption that these are general attributes of particular individuals and species that cut across different spatial scales of decision-making, from the selection of diet, to the approach of specific objects and microhabitats, to the choice of habitat. However, such consistency of responses between different aspects of decision-making in animals has been questioned both for cognitive ability and risk-taking behaviour and associated socially-related temperament (Wilson et al., 1994). That there really are innovation-prone taxa characterized by greater ecological plasticity is an assumption that sits well with the intuition of naturalists, but clearly needs more rigorous empirical testing. Furthermore, the degree to which attentiveness, attraction, and aversion to novelty are correlated between decisions regarding space use, object manipulation, and diet choice has rarely been addressed. The research on behavioural syndromes in Great Tits (Verbeek et al., 1994, 1996), where space and object exploration appear to be correlated with each other and to social dominance, provides tantalising evidence that individuals may have a general temperament with respect to various aspects of foraging behaviour. Clearly more work along these lines on a diversity of species needs to be undertaken.

Future research

Most of the small amount of attention that has been paid to novelty responses in the development of ecological plasticity and innovation has been focused on neophobia, the dominant response of adult birds to novelty. Neophobia has the potential of providing a brake that slows down the engagement of animals with the type of new resource that generate innovative behaviour. The NTH provides a clear framework for relating differences in the intensity of neophobic responses to ecological plasticity. The hypothesis that neophobia varies adaptively is attractive because this character has been shown repeatedly to differ within and between closely related species and thus appears to be evolutionarily labile. However, the empirical evidence for relating neophobia to plasticity and, by inference to innovation, is weak and even at times contradictory. We lack a meaningful understanding of how neophobia operates in birds outside the laboratory. We also have a poor understanding of how intrinsic levels of neophobia, those that would be measured with naïve birds in a controlled environment, relates to the functional neophobia of wild birds that have experienced variable environments. Finally, few studies have looked at neophobia in different behavioural contexts. For example, are birds that are neophobic to objects similarly neophobic to new habitats or habitat patches on one hand, or new potential dietary items on another? Finally, recent experiments have shown that intrinsic levels of neophobia may actually be much higher in seemingly more adaptable species. The adaptive significance of different levels of intrinsic neophobia could be explained by the danger (DNH) or unpredictability of the environment. Clearly, much more phenomenological and comparative work needs to be done to determine the relationship of neophobia to the development of innovative behaviour.

Second, the dynamic interaction of neophilia and neophobia in adult birds needs more research. The model that neophobia is the dominant adaptive response to novelty in wild adult birds needs to be reconsidered in light of the potential importance that neophilia and exploration have in ameliorating the effects of neophobia. In the end, what may prove to be important in relating neotic responses to innovation is the overall intensity of the response and the degree of attention that is paid to novelty, whether the initial response is purely aversive or some combination of aversion and attraction. Attention may be better studied looking at patterns of neural activity in the brain or physiological measures of emotional arousal.

An area of research that is perhaps the most promising in terms of examining the behavioural basis of innovative behaviour is the investigation of pattern and function in juvenile neophilia. This is the time period when individuals show the greatest motor plasticity associated with object exploration and play. It is also the period of greatest consumer plasticity in the form of positive attraction to novel stimuli and active object exploration. The nexus of these behavioural patterns is intriguing, but the long-term effect of behaviours expressed and objects experienced during this period is poorly known. The degree to which behaviours developed during this period can be spread to older birds in a population is also poorly known. However, on the surface it appears to be a period of potential expansion of behaviours and preferences. It would be fascinating and challenging to attempt to induce innovative behaviour by presenting unusual resources during the juvenile exploratory period to captive birds and tracking the stability of such behaviour in adults and its subsequent spread through a captive population. It would be fruitful to quantify the variation in the intensity of exploration and neophilia in this period, making carefully designed phylogenetic and ecological comparisons.

This brief review has presented reasons to believe that novelty responses probably play a key role in the expression of cognitive abilities of birds in the wild and in determining the probability that ecological innovations will develop and spread. I have been unable to provide definitive statements on what that role is, but hopefully signposts have been set out to help inspire future research in this area. Although considerable research has been conducted on neophobia and neophilia, the study of these phenomena remain at the margins of research on animal cognition in the wild. In preparing this chapter, I perused a number of recent books in the newly flowering field of cognitive ecology and failed to find any meaningful discussion of the role of novelty responses. At the very least, I can hope that novelty responses will be viewed as a meaningful piece to the understanding of the development of innovative behaviour in the future, rather than an annoying intervening variable, obscuring underlying truths.

Summary

Innovative behaviour is closely associated with the way that animals explore or avoid novel foods, objects, or places. Whereas most studies of innovative behaviour have focused on cognition and the social transmission of learned behaviours, emotionally based novelty responses are the first line of attack or defence in the response to the novelty that is often associated with innovation.

For foraging adult birds, neophobia is the most apparent response to novelty. Although neophobia is a widespread, if not universal response, of adult birds, the intensity of expression varies considerably between individuals within a species and between closely-related species. Two hypotheses account for adaptive variation in neophobia: the NTH and the DTH. The NTH focuses on the role of neophobia in regulating ecological plasticity and the latter concentrates on the protective nature of neophobia holding that neophobia is more intense in more dangerous environments. Experiments that test these hypotheses are few. Those that examine 'functional' neophobia (e.g. the neophobia found in naturally reared birds) have found support for the NTH. However, the very few experiments on intrinsic neophobia (neophobia found in birds reared in similar and controlled environments) seem to support the DNH.

Uninhibited neophilia (aka, exploration, curiosity) is commonly expressed in juvenile birds, particularly passerines. The nexus of neophilia, object play, and a high degree of motor plasticity in juveniles make this life history stage an important one to examine for the origins of innovative behaviour. It is likely that neophilia and neophobia can function simultaneously in adult birds and that initial neophobia masks any attraction to novelty. The dynamics of the two responses can generate strongly ambivalent responses characteristic of cautious generalists (such as corvids). If this is true, then perhaps it is the intensity of the attention and emotionality in the face of novelty that is the hallmark of the innovation-prone species rather than the positive or negative nature of the response. I suggest that future research on innovative behaviour should focus, in part, on the neural substrates, long-term development, and the role of social behaviour in novelty responses in more and less innovative taxa.

Acknowledgements

I would like to thank the following people for discussions of ideas surrounding novelty responses over the years: Eugene Morton, Ginger Bolen, Claudia Mettke-Hoffman, and Louis Lefebvre. Simon Reader, Kevin Laland and two anonymous reviewers provided valuable comments on an early draft. The Smithsonian Institution's Scholarly Studies Program provided funding for research on ducks and warblers. NSF grant supported work on sparrows.

References

Barnett, S. A. (1958). Experiments in 'neophobia' in wild laboratory rats. *British Journal of Psychology*, **49**, 195–201.

Bateson, P. G. (1971). Imprinting. In *Ontogeny of vertebrate behaviour* (ed. H. Moltz), pp. 369–87. New York: Academic Press.

Bayley, M. B. (1984). Intelligence and the ecological niche. In *Animal intelligence* (ed. R. J. Hoage), pp. 34–56. Washington, DC: Smithsonian Institution Press.

Beissinger, S. R., Donnay, T. J., and Walton, R. (1994). Experimental analysis of diet specialisation in the Snail Kite: the role of behavioural conservatism. *Oecologia*, **100**, 54–65.

Bekoff, M. and Byers, J. A. (1998). *Animal play: Evolutionary, comparative, and ecological perspectives.* Cambridge: Cambridge University Press.

Boinski, S. R., Quatrone, R., and Swarts, H. (2000). Substrate and tool use by brown capuchins in Suriname: Ecological contexts and cognitive bases. *American Anthropologist*, **102**, 741–61.

Brunton, C. F. A., MacDonald, D. W., and Buckle, A. P. (1993). Behavioural resistance towards poison baits in brown rats, *Rattus norvegicus*. *Applied Animal Behaviour Science*, **38**, 159–74.

Burghardt, G. M. (1998). The evolutionary origins of play revisited: Lessons from turtles. In *Animal play: Evolutionary, comparative, and ecological perspectives* (ed. M. Bekoff and J. A. Myers), pp. 1–26. Cambridge: Cambridge University Press.

Clarke, A. S. and Boinski, S. (1995). Temperament in nonhuman primates. *American Journal of Primatology*, **37**, 103–26.

Coleman, S. L. and Mellgren, R. L. (1994). Neophobia when feeding alone and in flocks in Zebra Finches *Taeniopygia guttata*. *Animal Behaviour*, **48**, 903–7.

Cooper, R. M. and Zubeck, J. D. (1958). Effects of enriched and restricted early environments on learning abilities of bright and dull rats. *Canadian Journal Psychology*, **12**, 159–64.

Coppinger, R. P. (1969). The effect of experience and novelty on avian feeding behaviour with reference to the evolution of warning colouration in butterflies. Part I. Reactions of wild caught blue jays to novel insects. *Behaviour*, **35**, 45–60.

Corey, D. T. (1977). The determinants of exploration and neophobia. *Neuroscience and Biobehavior Review*, **2**, 235–53.

Cowan, P. E. (1977). Neophobia and neophilia: New-object and new-place reactions in three *Rattus* species. *Journal of Comparative and Physiological Psychology*, **90**, 190–7.

Davies, N. B. and Green, R. E. (1976). The development and ecological significance of feeding techniques in the Reed Warbler (*Acrocephalus scirpaceus*). *Animal Behaviour*, **24**, 213–29.

Day, R. L., Coe, R. L., Kendal, J. R., and Laland, K. N. (2003). Neophilia, innovation and social learning: A study of intergeneric differences in callitrichid monkeys. *Animal Behaviour*, **65**, 559–71.

DesForges, M. F. and Wood-Gush, D. G. M. (1975). A behavioural comparison of domestic and Mallard ducks. *Animal Behaviour*, **23**, 692–7.

Dingemanse, N. J., Both, C., Drent, P. J., Van Oers, K., and Van Noordwijk, A. J. (2002). Repeatability and heritability of exploratory behaviour in great tits from the wild. *Animal Behaviour*, **64**, 929–38.

Drent, P., Van Oers, K., and Van Noordwijk, A. J. (2003). Realised heritability of personalities in the great tit (*Parus major*). *Proceedings of the Royal Society Series B*, **270**, 45–51.

Fisher, J. and Hinde, R. A. (1949). The opening of milk bottles by birds. *British Birds*, **42**, 347–57.

Fragaszy, D. M. and Visalberghi, E. (1990). Social processes affecting the appearance of innovative behaviours in capuchin monkeys. *Folia Primatologica*, **54**, 155–65.

Glickman, S. E. and Sroges, R. W. (1966). Curiosity in zoo animals. *Behaviour*, **24**, 151–88.

Greenberg, R. (1979). Body size, breeding habitat, and winter exploitation systems in *Dendroica*. *Auk*, **96**, 756–66.

Greenberg, R. (1983). The role of neophobia in foraging specialisation of some migrant Warblers. *American Naturalist*, **122**, 444–53.

Greenberg, R. (1984a). Differences in feeding neophobia between two species of tropical migrant warblers (*Dendroica castanea* and *D. pensylvanica*). *Journal of Comparative Psychology*, **98**, 131–6.

Greenberg, R. (1984b). The winter exploitation systems of Bay-breasted and Chestnut-sided warblers in Panama. *University of California Publications in Zoology*, **117**, 1–124.

Greenberg, R. (1984c). Neophobia in the foraging site selection of a Neotropical migrant bird: An experimental study. *Proceedings of the National Academy of Sciences (USA)*, **81**, 3778–80.

Greenberg, R. (1987a). The development of dead leaf foraging in a Neotropical migrant Warbler. *Ecology*, **68**, 130–41.

Greenberg, R. (1987b). Social facilitation does not reduce neophobia in Chestnut-sided Warblers (Parulinae: Dendroica). *Journal of Ethology*, **5**, 7–11.

Greenberg, R. (1989). Neophobia, aversion to open space, and ecological plasticity in song and swamp sparrows. *Canadian Journal of Zoology*, **67**, 1194–9.

Greenberg, R. (1990a). Feeding neophobia and ecological plasticity: A test of the hypothesis with captive sparrows. *Animal Behaviour*, **39**, 375–9.

Greenberg, R. (1990b). Ecological plasticity, neophobia, and resource use in birds. *Studies in Avian Biology*, **13**, 431–7.

Greenberg, R. (1992). Differences in neophobia between naïve song and swamp sparrows. *Ethology*, **91**, 17–24.

Greenberg, R. and Mettke-Hoffmann, C. (2001). Ecological aspects of neophobia and neophilia in birds. *Current Ornithology*, **16**, 119–69.

Hegner, R. E. (1985). Dominance and anti-predatory behaviour in Blue tits (*Parus caeruleus*). *Animal Behaviour*, **33**, 762–8.

Heinrich, B. (1988). Why do ravens fear their food? *Condor*, **90**, 950–2.

Heinrich, B. (1995). Neophilia and exploration in juvenile Common Ravens Corvus Corax. *Animal Behaviour*, **50**, 675–704.

Heinrich, B., Marzluff, J. M., and Adams, W. (1995). Fear and food recognition in naïve Common Ravens. *Auk*, **112**, 499–503.

Jones, R. B. (1986). Responses of domestic chicks to novel food as a function of sex, strain and previous experience. *Behavioral Processes*, **12**, 261–71.

Katzir, G. (1982). The relationship between social structure and the response to novel space. *Behaviour*, **81**, 231–63.

Katzir, G. (1983). Relationship between social structure and the response to novelty. *Behaviour*, **87**, 183–208.

Klopfer, P. H. (1967). Behavioural stereotypy in birds. *Wilson Bulletin*, **79**, 290–300.

Kummer, H. and Goodall, J. (1985). Conditions of innovative behaviours in primates. *Philosophical Transactions of the Royal Society of London*, **308**, 203–14.

Lefebvre, L. and Giraldeau, L. A. (1996). Is social learning an adaptive specialisation? In *Social learning in animals: The roots of culture* (ed. C. M. Heyes and B. G. Galef, Jr.), pp. 107–28. New York: Academic Press.

Lefebvre, L., Whittle, P., Lascaris, E., and Finkelstein, A. (1997). Feeding innovations and forebrain size in birds. *Animal Behaviour*, **53**, 549–60.

Marchetti, K. and Price, T. (1989). Differences in the foraging of juvenile and adult birds: The importance of developmental constraints. *Biological Review of the Cambridge Philosophical Society*, **64**, 51–70.

Marples, N. M. and **Kelley, D. J.** (1999). Neophobia and dietary conservatism: Two distinct processes? *Evolutionary Ecology*, **13**, 641–53.

Mayr, E. (1963). *Animal species and evolution.* Cambridge, MA: Belknap.

Mettke, C. (1995). Explorationsverhalten von Papageien—Adaptation an die umwelt? *Journal fur Ornithologie*, **136**, 468–71.

Mettke-Hofmann, C. (1999). Different reactions to environmental enrichment of nomadic and resident parrot species *Psittacidae. International Zoo Yearbook*, 37.

Mettke-Hoffman, C., Winkler, H., and **Leisler, B.** (2002). The significance of ecological factors for exploration and neophobia in parrots. *Ethology*, **108**, 249–72.

Mitchell, D. (1976). Experiments on neophobia in wild and laboratory rats: A revaluation. *Journal of Comparative and Physiological. Psychology*, **90**, 190–197.

Morgan, L. (1886). *Habit and instinct.* London: Arnold.

Morse, D. H. (1980). *Behavioural mechanisms in ecology.* Cambridge, MA: Harvard University Press.

Nicolakakis, N. and **Lefebvre, L.** (2000). Forebrain size and innovation rate in European birds: Feeding, nesting and confounding variables. *Behaviour*, **137**, 1415–29.

Nicolakakis, N., Sol, D., and **Lefebvre, L.** (2003). Behavioural flexibility predicts species richness in birds, but not extinction risk. *Animal Behaviour*, **65**, 445–52.

Ortega, J. C. and **Bekoff, M.** (1987). Avian play: Evolutionary and developmental trends. *Auk*, **104**, 338–41.

Pellis, S. M. (1981). Exploration and play in the behavioural development of the Australian Magpie *Gymnorhina tibicen. Bird Behavior*, **3**, 37–49.

Rana, B. D. (1989). Some observations on neophobic behaviour among House Sparrows, *Passer domesticus. Pavo*, **27**, 35–8.

Raudensush, B. and **Frank, R. A.** (1999). Assessing food neophobia: The role of stimulus familiarity. *Appetite*, **32**, 261–71.

Reader, S. M. and **Laland, K. N.** (2002). Social intelligence, innovation and enhanced brain size in primates. *Proceedings of the National Academy of Science (USA)*, **99**, 4436–41.

Reader, S. M. and **Lefebvre, L.** (2001). Social learning and sociality. *Behavioural and Brain Sciences*, **24**, 353–5.

Russell, P. A. (1973). Relationship between exploratory behaviour and fear: A review. *British Journal of Psychology*, **64**, 417–33.

Scott, J. P. and **Fuller, J. L.** (1965). *Genetics and the social behaviour of the dog.* Chicago, IL: University of Chicago Press.

Searle, L. W. (1949). The organization of heredity of maze-brightness and maze-dullness. *Genetic Psychology Monographs*, **39**, 279–335.

Seferta, A., Guay, P.-J., Marzinotto, E., and **Lefebvre, L.** (2001). Learning differences between feral pigeons and zenaida doves: The role of neophobia and human proximity. *Ethology*, **107**, 281–93.

Smith, S. (1973). Factors directing prey-attack behaviour by the young of 3 passerine species. *Living Bird*, **12**, 57–67.

Sol, D. and **Lefebvre, L.** (2000). Forebrain size and foraging innovations predict invasion success in birds introduced to New Zealand. *Oikos*, **90**, 599–605.

Sol, D., Lefebvre, L., and **Timmermans, S.** (2002). Behavioural flexibility and invasion success in birds. *Animal Behaviour*, **63**, 495–502.

Stahl, J., Tolsma, P. H., Loonen, M. J. J. E., and **Drent, R. H.** (2001). Subordinates explore but dominants profit: Resource competition in high Arctic barnacle goose flocks. *Animal Behaviour*, **61**, 257–264.

Thorpe, W. H. (1956). *Learning and instinct in animals.* London: Methuen and Co.

Timmermans, S., Lefebvre, L., Boire, D., and Basu, P. (2000). Relative size of the hyperstriatum ventrale is the best predictor of feeding innovation rate in birds. *Brain, Behavior and Evolution*, **56**, 196–203.

Tyron, R. C. (1940). Genetic differences in maze-learning ability in rats. *Yearbook of the National Society for the Study of Education.*

Verbeek, M. E. M., Boon, A., and Drent, P. J. (1996). Exploration, aggressive behaviour and dominance in pair-wise confrontations of juvenile male great tits. *Behaviour*, **133**, 945–63.

Verbeek, M. E. M., Drent, P. J., and Wiepkema, P. R. (1994). Consistent individual differences in early exploratory behaviour of male great tits. *Animal Behaviour*, **48**, 1113–21.

Verbeek, M. E. M., DeGoede, P., Drent, P. J., and Wiepkema, P. R. (1999). Individual behavioural characteristics and dominance in aviary groups of great tits. *Behaviour*, **136**, 23–48.

Vince, M. A. (1960). Developmental changes in responsiveness in the Great Tit *Parus major. Behaviour*, **15**, 219–42.

Visalberghi, E., Valente, M., and Fragaszy, D. (1998). Social context and consumption of unfamiliar foods by Capuchin monkeys (*Cebus apella*) over repeated encounters. *American Journal of Primatology*, **45**, 367–80.

Webster, S. and Lefebvre, L. (2000). Neophobia in the Lesser-Antillean bullfinch, a foraging generalist, and the bananaquit, a nectar specialist. *Wilson Bulletin*, **112**, 424–7.

Webster, S. and Lefebvre, L. (2001). Problem solving and neophobia in a Passeriformes-Columbiformes assemblage in Barbados. *Animal Behaviour*, **62**, 23–32.

Wilson, D. S., Clark, A. B., Colman, K., and Dearstyne, T. (1994). Shyness and boldness in humans and other animals. *Trends in Ecology and Evolution*, **9**, 442–6.

Wyles, S. J., Kunkel, J. G., and Wilson, A. C. (1983). Birds, behaviour, and anatomical evolution. *Proceedings of the National Academy of Science (USA)*, **80**, 4394–7.

CHAPTER 9

CHARACTERISTICS AND PROPENSITIES OF MARMOSETS AND TAMARINS: IMPLICATIONS FOR STUDIES OF INNOVATION

HILARY O. BOX

Introduction

Innovative behaviours result from interactions between ecological opportunities and a suite of characteristics and propensities that vary among individuals of different species within which the influences of age, sex, and social context are of critical interest.

The aim of this chapter is to discuss a number of specific issues within these contexts with reference to the marmosets and tamarins—Callitrichidae (Rylands *et al.*, 2000)—that occur in a diversity of habitats from Panama to North Eastern Paraguay and South Eastern Brazil. Marmosets and tamarins include the pygmy marmosets *Cebuella* (the smallest species of monkey), weighing around 85–140 g, to the lion tamarins (*Leontopithecus* spp.), weighing around 500–550 g. Callitrichids live in small groups from 3 to 15 individuals depending upon genus and species; there is normally one or sometimes two breeding females.

There has been a substantive increase in information on callitrichid biology in recent years, presenting a good opportunity to open up discussions of innovation in marmosets and tamarins, despite the lack of a well established literature directly concerned with innovation. Moreover, the group has characteristics and propensities, and experiences social and ecological influences, that are unique and unusual among the primates (e.g. social behaviour, morphological specialisations, reproductive behaviour), and so present interesting challenges to the study of animal innovation.

There are two main aims in presenting this chapter. First, to draw attention to information about marmosets and tamarins that is relevant to comparative perspectives in animal innovation. Callitrichids are still a little-known group compared with many other primate taxa. Second, to consider more specifically experiments to discuss differences among individuals of different sex, age, breeding status, species, and genera in their behavioural responses to environmental challenges.

Characteristics and propensities of marmosets and tamarins

Marmosets and tamarins all eat the same primary types of food—insects, fruits, flowers and nectar, and exudates of plants. The differences among genera and species in the frequencies with which they utilize these foods, point up features that involve morphological, behavioural, and mental specializations. For example, there is a developing interest among researchers in the association between mental abilities and diet in callitrichids (Garber, 1989). Recently, Platt and his coworkers (Platt, 1994; Platt *et al.*, 1996) used a set of visuospatial memory tasks to compare two species with considerable overlap in their diets, but which have morphological and behavioural specializations with regard to their principal foods. The marmoset *Callithrix kuhli*, like all marmosets, has a dental specialization unique to primates and an enlarged caecum to gouge and help digest gums. Gums are renewable relatively rapidly, and are abundant. Moreover, a small core area with few gum trees is used frequently. By contrast, golden lion tamarins (*Leontopithecus rosalia*) use widely separated insect and ripe fruit foraging sites that are renewed relatively slowly. The need of golden lion tamarins to remember information (e.g. location and quality) about food for longer is reflected for example, in their ability to perform better on spatial memory tasks, whereas the marmosets performed better on tasks with short retention intervals. Lion tamarins also remember information about the colour of food for longer periods than the marmosets. In all, these experiments indicate that there is a number of visuospatial memory systems 'that are specifically adapted for tracking the spatial and temporal distributions in their primary foods' (Platt *et al.*, 1996, p. 392).

Lion tamarins are also specialized for manipulative foraging of insects in holes in trees, and especially in the leaf axils of bromeliads; they take relatively large insects and other relatively large prey. A few species of Saguinus tamarins are also specialized for manipulative foraging (but to a lesser extent, and not in bromeliads) and take relatively large prey. Further, studies of Saguinus tamarins have shown a number of distinct insect foraging patterns (Garber, 1993); there are differences in techniques for hunting that involve the substrates that are used, the vertical stratification within the habitat and modes of positional behaviour. Saddle-backs (*Saguinus fuscicollis*) and some others show the most distinctive patterns; these species take relatively large and cryptic animals. There have been long-standing interests in morphological (as with elongated cheiridia: Coimbra-Filho, 1970; Hershtrovitz, 1977) and ecological (as in the relative availability of insect prey: Rylands, 1989) differences in all these regards (Singer and Schwibbe, 1999). Recently, with allometric analyses of museum specimens of all the genera, Bicca-Marques (1999) has shown that the shape of the hands (length and thickness) is a good predictor of non-manipulative and manipulative foraging techniques. The hands of the lion tamarins are relatively longer and thinner than those of other genera. The hand shape of some other manipulative *Saguinus* tamarins is also significantly different from that of all other non-manipulative tamarins. Interestingly, the analyses also showed significant differences in hand morphology between species of tamarins that form mixed species associations, and these differences are reasonably associated with resource partitioning and coexistence among closely related species. Further, Garber and Leigh (2001) report that differences in positional behaviour of saddle-back and red-chested tamarins (*Saguinus labiatus*), together with Goeldi's monkey

(*Callimico goeldii*) living in a mixed species troop, correlated with species differences in limb proportions and locomotor anatomy that provide a framework for understanding niche partitioning in these animals.

It is clearly important to emphasize morphological characteristics among species that influence positional and foraging behaviour; some potential innovative behaviour in feeding will be influenced by these differences. Hence, some species, as with golden lion tamarins, have potential access to new foods that will not be available to many other species of callitrichids; they are able to forage in microhabitats that influence their ecological opportunities. Moreover, in a more general context, there is theoretical interest in the mental abilities that extractive foraging requires (Parker and Gibson, 1977), and by the cognitive demands imposed by different kinds and distribution of food resources eaten by species of different genera (Milton, 1988). On both counts we might consider that lion tamarins will have comparative advantages over other callitrichids to respond to new ecological opportunities.

The fact that a number of species live in mixed species associations, and demonstrate closely coordinated movements and activities is also important for our understanding of the responsiveness of marmosets and tamarins in their natural environments (Pook and Pook, 1982; Terborgh, 1983; Yoneda, 1984; Garber, 1988; Heymann, 1990; Norconk, 1990; Peres, 1992; Buchanan-Smith, 1999). Moreover, experimental studies of single and mixed species of *Saguinus* tamarin groups have increased our appreciation of the adaptive advantages that mixed groups may provide. One of these advantages is to gain information about food. For example, work by Prescott and Buchanan-Smith (1999) has shown that adult male–female pairs of congeneric species in captivity may be as effective as conspecifics as demonstrators in a novel foraging task. Further, experimental field studies by Bicca-Marques (2000) indicate that dominant emperor tamarins (*Saguinus imperator*) may improve their foraging efficiency by using information from the foraging behaviour of saddle-backs. Comparisons of the two species both in and out of mixed association also showed that the saddle-backs foraged more efficiently out of association, and that both species incurred costs (decreased feeding time and food intake) when they formed groups of mixed species. Hence, there are experimental paradigms to investigate innovative behaviour. Again, the natural vertical stratification that has been reported for all mixed species of tamarins, as with saddle-backs lower than congeners, raises various possibilities. For instance, Hardie (1995) has shown that red-chested tamarins (*S. labiatus*) were positively facilitated by the responses of saddle-backs to approach unfamiliar objects that were lower in their shared enclosure, but that these objects were not responded to without such facilitation. Hardie (1995) also found that groups of both saddle-backs and red-chested tamarins rapidly responded differently to objects associated or not associated with food; that the information was retained for at least several (seven) weeks, and that information could be transferred between the two species. Such information will certainly encourage further empirical work in different conceptual and contextual domains. However, it is also important as a guiding principle to note that, for example, foraging by marmosets and tamarins under natural conditions involves the integration of spatial, temporal, social, and ecological information (Garber, 2000).

From a different perspective, marmosets and tamarins are cooperative breeders and in that context, they show delayed dispersal, alloparental care and delayed breeding by subordinates (Tardif, 1997). With reference to studies of innovation, questions involve whether breeding females and/or males behave differently in new situations than subordinates, and in what contexts. Some of these concerns are discussed in later sections of the chapter.

We may also consider what differences might be expected more generally with the development and transmission of new behaviour among members of singular cooperative breeders compared with those that are competitive breeders. Interestingly, in this context, Snowdon (2001) has argued that the unusual social dynamics of a cooperative breeding system might lead to more efficient cognitive skills including social learning, information donation and imitation, than would be found among species with different social systems.

Further, a driving force behind much interest in marmosets and tamarins is in their unique mode of anthropoid primate reproduction—as in that females have multiple births usually twins, and a postpartum oestrous that allows them to combine pregnancy with lactation. There is a high reproductive output (potentially two sets of twins a year in some species) but with very high energetic costs, although infants are cared for communally by all the members of a social unit. Moreover, as a group, these monkeys are highly cooperative. *Saguinus* tamarins, for example, the largest and most diverse group of the callitrichids, are characterized by high levels of cooperation, tolerance, and adaptability (Caine, 1993). Aggression is rarely observed in nature, and differences among individuals in social status are not detectable by conventional methods (Caine, 1993). Groups move as cohesive units when foraging and travelling and, in some species at least, individuals take turns to act as sentinels (Zullo and Caine, 1988). Caine (1993) has emphasized that the number of tasks, such as vigilance and infant care, that are shared by all members of social groups indicate the unusual nature of social life of these animals among the primates, and suggests that this comparative perspective is given additional interest because such cooperation can 'take place outside the context of kin selection or perhaps even without the benefit of long-term associations' (p. 211).

Cooperation and tolerance are also important influences as social processes that facilitate new behaviours and their transmission (Coussi-Korbel and Fragaszy, 1995; Box, 1999; van Schaik et al., 1999; van Schaik, in press). Individual red-chested tamarins have been observed to be tolerant at a newly discovered source of food (Mayer et al., 1992), and give food calls that are likely to attract other monkeys, as when the food is particularly palatable, and even when it is in small quantities (Addington et al., 1991). Moreover, among the cooperative activities of marmosets and tamarins that include infant care, vigilance, mobbing and food calls, food sharing both by spontaneous offering and in response to begging, is a conspicuous part of infant life (Feistner and McGrew, 1989). As cooperative breeders, both parents and natal adults may provision immature callitrichids, and although food sharing in response to begging (passive food sharing) is observed in various mammals, marmosets and tamarins, and especially some tamarins (Tardif et al., 1993) show active food sharing that is unique among mammals; individuals with food initiate food sharing with specific behaviours (Wallander, 1998). Differences in food-sharing behaviour among both species and individuals, however, must be recognized with regard to the development of explanatory hypotheses (Brown and Almond, personal communication; Price and Feistner, 2001).

Food sharing has been found to continue well after weaning among Saguinus species (Feistner and Chamove, 1986) and lion tamarins (Feistner and Price, 2000). It is commonly found in contexts where food items are rare or difficult to obtain or process (Ruiz *et al.*, 1999). Among lion tamarins for instance, there is a significant dependence upon animal prey and fruit food items that are difficult to deal with, and food sharing, particularly by offering, is of critical importance. Hence, food sharing is interesting among these primates as a means of providing opportunities to develop and transmit information about new foods. For example, Roush and Snowdon (2001) have shown that the food calls that accompany food transfers to young cotton-top tamarins (*Sanguinus oedipus*) may facilitate learning about both appropriate foods and the vocalizations associated with feeding in adults; indeed, that it may provide a form of information donation by adults.

Once again, evidence for the donation of information is rare among the primates, and is of interest among callitrichids in that context (see King, 1999). Moreover, experiments by Rapaport (1999) with captive golden lion tamarins have shown that, compared with obtaining foods independently, immature animals were more likely to accept new foods acquired from other members of the social group. Further, adults transferred to immatures food that was known to adults, but new to immatures, and foods that were new to all, was transferred more frequently than foods that were familiar to both adults and immatures. Hence, she suggests that adults change their behaviour in ways that result in the facilitation of learning about food.

At this point we may emphasize implications for innovation from the information that has been discussed so far. For example, differences among genera and species within genera in hand morphology suggest more innovation in extractive manipulative tasks among the lion tamarins, and to a lesser extent some *Saguinus* tamarins, than among the marmoset genera. Differences in spatial memory among species of different genera may create further new ecological opportunities.

Again, although there is natural vertical stratification between species living in mixed species associations that present ecological challenges and opportunities of niche separation, there is also experimental evidence that innovative behaviour may be transferred across species. Hence, an advantage of mixed species associations is in the potential for the development and spread of innovation between species that is not necessarily limited to the characteristics and propensities of the individual species. More studies are required here on the contexts in which this may occur.

Further, the cooperative breeding system and highly tolerant social dynamics of callitrichids suggest that there may be comparatively efficient diffusion of innovative behaviours compared with primate taxa of different social dynamics. We may also expect there to be substantive differences among genera, and some species within genera of callitrichids with reference to degrees of social tolerance and feeding ecology. Moreover, evidence for food sharing among callitrichids provides an excellent comparative perspective within which to study experimentally the development and specific social routes of transmission for the acquisition of new behaviours. So far there is evidence that new information about food is directed from adults to infants. This makes sense with regard to experience of the food repertoire, but other social possibilities should be considered in different contexts of potential innovation as with exploration and play within young twin pairs (Box, 1975; Ingram, 1977).

The second main aim of this chapter is to consider differences among individuals of different sex, age, breeding status, species, and genera in their behaviour to unfamiliar environmental challenges. The main interest, however, is in the study of sex differences.

Sex differences

Male and female marmosets and tamarins are very similar morphologically and behave similarly in many respects. However, there is a growing body of information describing differences between males and females, including sex differences in response to unfamiliar conspecifics (e.g. French and Snowdon, 1981; French and Inglett, 1991), vigilance behaviour (e.g. Savage et al., 1996; Koenig, 1998), scent marking (e.g. Epple, 1994; Heymann, 1998), vocalizations (Benz et al., 1990; Norcross et al., 1999), infant care (Yamamoto et al., 2000), and responses to food tasks (Box et al., 1995). There are hypotheses for callitrichids as a group as well as for sex differences in particular species. Sex differences in primates as a whole are also of interest: for example, the review by Reader and Laland (2001) reported elevated rates of innovation in male primates compared with females. However, further analyses that break the data down into dimorphic and monomorphic species suggest a more subtle pattern (Laland, van Bergen and Reader, personal communication). Their finding in this regard is that adult males of dimorphic species are more innovative than females, whereas among monomorphic species there is a tendency for females to be more innovatory; that the former is related to sexual selection whereas the latter (at least with regard to foraging) 'is perhaps most clearly related to the costs of reproduction' (Laland, personal communication).

Within this general framework, the aim of this section is to discuss the responses of marmosets and tamarins to sources of food, and to unfamiliar tasks that involve food. Subsequently, comparisons are made with experiments with challenges that do not involve food.

Female priority to food

One area of interest is that adult females among species of marmosets and tamarins have been shown to demonstrate a priority of access to food that both males and females want, in conditions that involve the spatial and temporal restriction of food. These effects refer only to responses to food; it is not social dominance—social dominance refers to aggressive outcomes in a variety of circumstances (Kappeler, 1990). Such priority of access is, once again, unusual among the primates (Hrdy, 1981; Jolly, 1984; Richard, 1987). Priority of foraging access deserves theoretical and practical consideration, and has implications for studies of innovation as when differences between males and females may lead to new feeding opportunities.

There are examples of female priority of access to food among callitrichids in a variety of social and ecological conditions (Box, 1997). For instance, studies with different species have shown that reproductive females living in family groups demonstrate priority of access to restricted food. Petto and Devin (1988) found that adult female common marmosets (*Callithrix jacchus*) showed priority over supplementary food as recorded by the time they spent feeding, their access to preferred foods, and the variety of foods they ate.

Again, Tardif and Richter (1981) showed that adult females were the highest consumers in family groups of both common marmosets and cotton-top tamarins when fruit was given as an addition to the normal diet. It is also relevant, for example, that in a study with three family groups of marmosets (two of common marmosets and one of the pencilled *Callithrix penicillata*) in which there were fortuitously an equal number of males and females at all ages, all the females maintained significantly greater proximity to an additional preferred food given daily in conditions of limited access than the males. There was no aggressive behaviour, but baseline observations throughout the experiment showed that the females also became more responsive to the food than the males, by maintaining more proximity to it over the duration of the study (Box and Smith, 1995). There are also a number of relevant observations (e.g. Maier *et al.*, 1982; Garber, 1993) on different species in nature. For example, a study by Garber and Kitron (1997) on the feeding ecology and group structure of moustached (*Saguinus mystax*) and Panamanian tamarins (*Saguinus geoffroyi*) found that the dominant breeding females had selected and swallowed larger seeds than the males. These seeds had larger amounts of pulp to process per unit time, and were of greater nutritional value. The selection of these seeds demonstrated priority of access at feeding sites. Further, these females had priority of access to baited traps (that allowed room for one animal at a time) in which the seeds were expelled and collected.

Behavioural strategies of males and females

In order to generate hypotheses about the development of innovative behaviour, it is relevant to consider the behavioural strategies of males and females of different species in response to food. Among common marmosets, for instance, breeding females may be aggressively assertive within their family groups: there are observations in nature and in captivity. Tardif and Richter (1981) found that the females often prevented the adult males from obtaining food. For cotton-tops, however, and although the females also competed for food within their families, they did not compete with the males. In fact, Tardif and Richter (1981) suggested that the males may have inhibited their own feeding. There are also differences, for example among common marmoset families, in responses to environmental events. These differences reflect a complex mix of social dynamics, demography and individual differences in temperament (Box, 1982), but it is useful to generate hypotheses that take account of the fact that differences between species of these two genera are generally consistent with differences in their natural feeding ecology and social organization (Box, 1997). Hence, compared with *Saguinus* species, the dental specialization of marmosets (*Callithrix*) to obtain defensible food resources, together with living in smaller home ranges, and in mating systems that tend more towards monogamy (Ferrari and Lopes Ferrari, 1989) fit well with their observed behaviour in which adult females obtain priority of access to important static food sources by competitive strategies, and in which males are relatively secure in their paternity (Box, 1997). The food of *Saguinus* species is more widely scattered; their home ranges are also larger, and their social units tend to be relatively unstable with more adults per group. Moreover, there are notable differences among species in the flexibility of their mating systems (Goldizen, 1988; Epple, 1990).

Further, although there is little information as yet, there may be also significant differences in behavioural strategies of males and females of species of genera other than *Callithrix* and *Saguinus*. For example, Addington (1999) studied pygmy marmosets, a highly gumnivorous species with relatively small home ranges. Addington demonstrated experimentally that although aggressive behaviour did increase under conditions in which food was monopolizable, aggression was much less common than tolerant feeding behaviour; females were no more aggressive to males than vice versa. Hence, there were indications that interactions during feeding were egalitarian and not hierarchical.

Again, differences in behavioural strategies between male and female lion tamarins compared to other genera are suggested by discussions of their feeding ecology, social demography (Rylands, 1993), and the behavioural control of potentially reproductive animals within a social unit (e.g. French and Inglett, 1991). These issues deserve specific attention (Box, 1997). For example, the presence of a number of adults of both sexes in natural social units, the largest home ranges of all callitrichids, and a frugivorous–insectivorous diet would suggest that female lion tamarins will not demonstrate priority of access to food as shown among *Callithrix* marmosets, but that they will behave more like *Saguinus* tamarins in response to new food challenges.

Further, because relatively few species of any genus have been studied, and given the wide ecological variation and marked diversity of the group as a whole, we should also pay attention to differences among species in the more familiar genera of *Callithrix* and *Saguinus*. This is in keeping with a growing appreciation of diversity among callitrichids. Moreover, there are indications of species differences in response to food albeit in conditions that did not involve spatial restriction among individuals. For instance, Bicca-Marques (2000) studied the responses of emperor and saddle-back tamarins to food on food platforms large enough to accommodate whole groups at an experimental field site. Saddle-backs differed significantly from emperor tamarins in that emperor tamarins demonstrated a consistent order in visiting the platforms (immatures, breeding females, adult males, and non-breeding adult females) but the saddle-backs were not consistent; there was no clear order by age or sex. In fact, all individual saddle-backs had relatively equal access, including a lactating female. It is interesting to note that saddle-backs are the most behaviourally flexible of tamarin species (Garber, 1993); saddle-backs are ecologically very successful, and notably flexible in mating system.

Taken together this information shows differences among males and females of different species in their strategies with regard to female priority of access to food that are at least consistent with differences among genera in their social organization and feeding ecology, but that there may be also alternative and/or additional influences (as yet unspecified) upon the behavioural strategies of adult males and females of different genera, and even of different species within the genera in which females demonstrate priority of access to food.

Hypotheses for female priority to food

We may consider that the behavioural strategies of adult males of different species and genera will demonstrate a range of functional possibilities in feeding situations, including vigilance, mate guarding and, as in some lemurs (Hrdy, 1981; Richard, 1987) deference to adult females that are the mothers of their offspring in protection of their genetic invest-ment (Box, 1997). For adult females the high energetic requirements of reproduction (combining pregnancy and lactation) among marmosets and tamarins compared with other monkeys and apes (Tardif et al., 1993; Nievergelt and Martin, 1999) are reasonably associated with their responsiveness to food in different species and genera, as with prior-ity of access to food, and particularly in situations where highly nutritious and/or preferred foods are spatially restricted. Inevitably, however, general hypotheses to associate aspects of feeding ecology and social organization with sex differences in feeding behaviour, will be refined as more information on the biology of these animals becomes available. Energy requirements certainly provide compelling hypotheses, but it is also important to recognize complexities, and to refine hypotheses within this general perspective. For instance, Nievergelt and Martin (1999) investigated changes in energy intake and body weight dur-ing pregnancy and lactation in mated pairs of captive common marmosets. Females did not increase their energy intake during pregnancy, but did so up to 100 per cent during lac-tation, and also gradually lost weight. We might expect, for example, that reproductive females will demonstrate more evidence of priority of access to food at these times. In fact, in some of our own preliminary work with common marmosets (Box et al., 1999) this hypothesis was supported. Breeding females were the most assertive over food of their families, and were more assertively responsive to food sources after a birth than before it. Breeding females were also more assertive than the non-breeding females in their families. It was interesting to find, however, that the mated pairs of all the families increased their assertiveness compared with their offspring when food was present. In fact, the whole domain of 'energy requirements' is shot with complexities. For example, natural populations of animals exhibit mechanisms of behavioural compensation (such as reducing activity) at times of high energy expenditure (Nievergelt and Martin, 1999).

Further, we are dealing with communal rearing systems (Tardif, 1997), and therefore need to consider energetic costs spread among individuals at different stages of breeding investment, and the differing contributions to rearing of helper monkeys after a birth. For instance, the costs of infant carrying among captive cotton-top tamarins were non-invasively evaluated by Sanchez et al. (1999) by measuring body weight among members of family groups. They showed that breeding males, and male helpers, incurred major costs by their carrying activities; they lost weight and decreased their feeding time and energy intake. By contrast, breeding females could reduce their carrying time and increase their energy intake during the pre-ovulatory time when the infants were totally dependent.

In a replication and extension of the Sanchez et al. (1999) study, Achenbach and Snowdon (2002) showed that the weight loss of adult male cotton-tops is inversely proportional to the number of helpers present. For example, both studies showed a weight loss of around 10–11 per cent for males with no helpers. Moreover, the potentially considerable costs that may be incurred by males lead us to ask why there should be a sex bias in competition for food (Snowdon, personal communication). Snowdon notes that similar questions are raised from work in which male common marmosets increase in weight during the pregnancy of their mates, and that male cotton-tops show weight gain during the mid-point of their mates pregnancy, although these increases in weight occur after the females have stopped nursing (Ziegler, unpublished data). Snowdon poses the question that perhaps both sexes are equally competitive, and require energy, but demonstrate relevant behaviour at different times in the reproductive cycle (Snowdon, personal communication).

In a different domain, recent work on visual capacity among callitrichids as by Caine and Mundy (2001) has far reaching implications for interpretations of sex differences in foraging success. The point is that, in keeping with the majority of other New World monkeys, males and females differ in abilities to perceive colour. All males are dichromatic, whereas the majority of females are trichromatic. These differences have long been associated with foraging. Interestingly, Caine and Mundy (2001) have now demonstrated, for example, that the trichromatic females in groups of marmosets (*Callithrix geoffroyi*) were more successful at finding orange pieces of food in the grass of their large enclosures, than were the dichromatic animals. Importantly, this work suggests that differences between males and females in foraging success may be related to differences in visual capacity and not to sex differences *per se*. Hence, as Caine points out, if trichromatic females happen to be used in a foraging study, they may have an advantage over males in that they see colour cues that help them to find food. Again, when females are not more successful, they may be dichromatic females, or perhaps that the cues associated with food are camouflaged—the trichromats have a disadvantage in such cases. Importantly also, trichromatic females may be differently motivated to search for food in different conditions. Females may spend more time searching for food far away, for instance, because they can discriminate food that is far away, whereas dichromatic individuals may look closer to hand. 'To the extent that there are trichromats in a group, sex differences may appear; to the extent that there are no trichromats in a group, sexual dimorphism in foraging behaviour may be minimized' (Caine, personal communication, 2002).

Sex differences in response to foraging tasks

In light of the discussion in the preceding section, it is interesting to consider the behaviour of non-reproductive adult males and females in foraging tasks that do not involve visual cues that may favour one sex or the other. For example, 14 non-reproductive male–female adult pairs of three species of *Saguinus* tamarins (red-chested, cotton-tops, and saddle-backs) were given limited access to a series of unfamiliar food tasks in their home cages, in which they learnt to reach in and take pieces of a preferred food, pieces of apple, with no colour discrimination involved (Box *et al.*, 1995). Although there were no significant differences between males and females in the amount of time they spent investigating the tasks generally, the females were more successful in solving the tasks; they responded significantly more quickly, more persistently and obtained more food than the males. There was very little aggressive behaviour, and no retrospective evidence that the females were pregnant during the study. Moreover, it is relevant to consider the lack of significant differences between paired male and female common marmosets in their energy intake during the ovarian cycle (Nievergelt and Martin, 1999; Nievergelt, personal communication).

Again, a study by Michels (1998) with adult male–female pairs of common marmosets, and in which food presented in the home cage was either concentrated or dispersed, and either difficult or easy to obtain, showed that the females obtained more food in all the conditions because they were more aggressive, and spent more time searching for food.

In these experiments then, there is evidence that females are more successful and respons-ive in unfamiliar food tasks without the direct influence of reproductive status, and poten-tially increased energy requirements. Once again, there were also differences between males and females of different species in their strategies of behaviour, as with aggression and per-sistence (Box et al., 1995; Box, 1997; Michels, 1998), that are consistent with observations discussed in the previous section. As in studies mentioned earlier with female priority of access to food, the food was both temporarily and spatially restricted. In these cases it is also relevant that female 'priority' was observed in a different social context because differences in social and ecological conditions may lead to differences in the expression of sex differences in behaviour (Goldfoot and Neff, 1985). Hence, at this point it is also of interest to consider sex differences in new situations that do not involve food.

Sex differences in response to new areas

Behaviour in unfamiliar areas is interesting in our present context because innovative responses to new areas have ecological validity, perhaps following dispersal for example (see also Sol, Chapter 3). Hence, there has been a number of projects in which groups of callitrichids have been released from a cage into a free ranging environment (as in zoos) in order to build up information about behavioural competence, and sometimes as part of a pre-release strategy for reintroduction into nature.

Studies by Price et al. (1991) and Price (1992) found that when cotton-top tamarins were moved into a wooded area, females of family groups were the most responsive. In Price et al. (1991) the females were usually the first to explore new areas, and find new travel routes, whereas the males stayed in the familiar areas and were more vigilant. In the Price (1992) paper, females were again reported as more responsive, for example, in the use of new trees. The breeding female was infrequently the first tamarin to enter a new tree, but she did use almost as many as her daughters. There is a possibility here that the behaviour of the females is at least partly due to differences in their visual capacities compared to males. However, McGrew and McLuckie's (1986) study of groups of cotton-tops in which nuclear families were given access to outdoor exercise areas, showed that subadult females were the first to leave the home area; that they stayed away for longer periods, and traversed the greatest distances. Subadult males were the next most responsive, with breeding pairs the least responsive. By some contrast, albeit again in a very small sample, Chamove and Rohrhuber (1989) observed the adult males of two cotton-top families to be the first to move into a large unfamiliar area. The immatures, however, did spend more time outside.

The responses of young non-breeding cotton-top females to unfamiliar areas is worth fol-lowing up in the context of potential development of innovative behaviours. It may indicate a behavioural propensity, but there is no evidence that females are the sex most likely to dis-perse for example (Savage et al., 1996). It is generally important that immatures are persis-tently regarded as the most responsive to a variety of new situations, across many taxa; that age is important with regard to the willingness or ability of individuals to acquire new behaviour (Kappeler, 1987). This is regarded as functionally significant in that responses to new situations may provide opportunities to learn adaptive skills, and juveniles represent an important developmental stage in this regard (see also Greenberg, Chapter 8). It is also important to consider that the recent across-primate review by Reader and Laland (2001) found more reports of innovation in adults, and less in subadults, than expected by chance (see also Reader and Laland, Chapter 1).

Experiments with adult male–female pairs of callitrichids raise additional perspectives. Responses to new spatial opportunities may contrast with sex differences in response to a new feeding opportunity, as mentioned earlier, for pairs of tamarins (Box et al., 1995). No sex differences were observed when these same pairs were subsequently given daily trials in a large outside area to which their home cages were tunnelled (Box and Rohrhuber, unpublished data). Again, although non-reproductive adult male–female pairs of common marmosets behaved very similarly to new areas tunnelled to their home areas, as in the time that they took to enter, and the time that they spent in the new areas, the total number of activities, which involved marking, sniffing, and gnawing in the new areas, however, was significantly greater for the females. Moreover, when unfamiliar objects were added to the new area, the females touched and manipulated them significantly more than did the males (Box, 1988).

Responses to unfamiliar objects

As with responses to all classes of unfamiliar stimuli, theoretical interest in behaviour to new and innocuous objects centres on the dual and complex interactive influences of neophobia and neophilia. A discussion with specific reference to innovation is given by Greenberg in Chapter 8. Observations on primates, however, lags far behind that of other taxa such as birds. There are few studies of the responses of wild primates to unfamiliar objects; relevant studies include those of Menzel (1969) on rhesus monkeys (*Macaca mulatta*), Menzel (1966) with Japanese macaques (*Macaca fuscata*) and Visalberghi et al. (in press) with capuchins (*Cebus apella*), in which individuals were found to be notably cautious. Again, the few studies of first responses of different species to new foods in field conditions have also shown neophobic responses, although interestingly of course, new foods have been included in the diet of a variety of species including macaques, capuchins, and common chimpanzees, and with juveniles less neophobic as among capuchins (Fragaszy et al., 1997).

One important issue regarding the neophobia of monkeys to new objects in the field, for example, concerns the contrast with capuchins in captivity, where new objects typically elicit interest, manipulation and play (e.g. Visalberghi, 1988; Fragaszy and Adams-Curtis, 1991; Visalberghi and Anderson, 1999). New objects in nature may be potentially but not specifically associated with food, and may be associated with potential danger (Visalberghi et al., in press), but the contrast with responses in captivity may pose serious questions for functional hypotheses.

Sex differences in response to new objects in captivity have been reported in a few cases (e.g. Joubert and Vauclair, 1986; Visalberghi, 1988; Drea, 1998). For example, Visalberghi (1988) found that male capuchins within two social groups interacted more with plain wooden blocks than did females. The blocks elicited high levels of response throughout the study. Responses to new objects in captivity among callitrichids (at least in family groups) have not included sex differences. For example, an experiment by Menzel and Menzel (1979) with saddle-back tamarins, and by Millar et al. (1988) with cotton-tops and common marmosets, in which junk objects were introduced into the home areas, found that juveniles were the most overtly responsive of their families; they approached and investigated, by visual and manual inspection, and were apparently spontaneously attracted to the new objects; their responses may be described in terms of neophilia. The parents were the least responsive, and there were no sex differences overall.

Caine (1986) reported no sex differences in the visual scanning of unfamiliar objects by group-living red-chested tamarins. By contrast, sex differences to threatening stimuli, such as a potential predator, have been reported for both red-chested tamarins (e.g. Buchanan-Smith, 1999) and common marmosets (e.g. König, 1988); males are more vigilant, and appear to play a special role in vigilance. There are also field observations (e.g. Savage *et al.*, 1997 on cotton-tops), that males are more likely than females to act as 'sentries'. In fact, marmosets and tamarins spend much of their time monitoring their environment, and although much of this may be associated with finding food, Caine (1993) has strongly emphasized the influence of considerable predation pressure. We clearly need to develop profiles of responses for both males and females in different contexts of environmental challenge.

Concluding remarks

The marmosets and tamarins are potentially a stimulating group for studies of innovation. Their close phylogenetic relationships, a number of characteristics that are unique and unusual to the primates, together with differences among taxa in body size, dental morphology, diet, propensity to form mixed species associations, mating systems, and social stability, make them good subjects with regard to examining hypotheses for the ultimate and proximate, ecological and social organizational influences upon the development and social transmission of new behaviour. Hence, although callitrichids share a common range of morphological, physiological, and behavioural characteristics, they are a less homogeneous group than they may at first appear to be (Ferrari, 1993). Moreover, an appreciation of differences in response to new situations by species within and between genera is important for setting up and interpreting studies of innovation. For example, observations in captivity may allow examination of the spontaneous development of innovations (see Kummer and Goodall, 1985). In addition, studies that seek to improve captive conditions such as with environmental enrichment (e.g. Poole, 1990; Box and Rohrhuber, 1993; Buchanan-Smith, 1994) present opportunities to study innovative behaviour.

It is important to note that different species of callitrichids cope with conditions of captivity differently. For example, Wormell *et al.* (1996) found that pied tamarins (*Saguinus bicolor bicolor*) are more sensitive to environmental disturbances than the other tamarins studied. Species also vary in their health and breeding in captivity (Stevenson, 1984). Again, with reference to different species in nature, some species such as common marmosets are geographically widespread and adaptable, whereas others such as the endangered lion tamarins are far less so. Some species are greatly threatened by habitat destruction for example, whereas others survive well in small and degraded areas (Rylands, 1993). Field studies provide the most natural environments and hence the most realistic perspectives to our understanding that adaptive behaviour, including innovation, involves a suite of characteristics with morphological, mental, social, and ecological influences. One important point is that we should not assume that particular characteristics and propensities necessarily confer advantages of behavioural flexibility. For example, we have noted various points about golden lion tamarins, as with their hand morphology and spatial memory, but it is also important that foraging in bromeliads may influence their preference for high elevations in primary and specific forest habitats (Rylands, 1989), and that the use of tree holes as sleeping sites (Coimbra-Filho, 1978) may also influence their preference for relatively undisturbed forests. Hence, compared with other callitrichids, the lion tamarins appear to be ecologically inflexible, as perhaps reflected in their restricted geographical distribution (Ferrari, 1993). By contrast, the smaller saddle-back tamarin is ecologically successful; it is one of the most widespread of the callitrichids, and apparently one of the most adaptive. Their hands show that they can forage for relatively large insects for instance, but a major consideration (as Ferrari, 1993 emphasizes) is their propensity to form mixed species associations, and thereby gain advantages in numerous ways, as in the detection of predators, and the vertical partitioning of resources. We obviously need a great deal more information, but that which we have clearly shows the importance of considering the complexity of mediating interactions with intrinsic and extrinsic environmental influences.

The common mixed-species association lifestyle of many callitrichids opens up possibilities for studies of innovation. For example, experimental work has shown a facilitatory congeneric influence between *Saguinus* tamarin species in solving a novel food task; more studies in different contexts will undoubtedly follow to add knowledge about advantages of mixed species associations. For a general discussion of mixed species associations in callitrichids see Heymann and Buchanan-Smith (2000).

We should be cautious, however, about assuming that the transfer of information between species is always potentially advantageous. We have seen that with reference to foraging different species may take advantage of information from another species, but that costs may be incurred by both species. Moreover, differences in social status between species in mixed associations are likely to influence the relative uptake of information. Bicca-Marques (2000), for example, found that socially dominant emperor tamarins may use information from the foraging behaviour of saddle-backs, but that the saddle-backs foraged more efficiently out of association. Although an experimental construction, the results do provoke additional questions. Further, when analyses were made on the behaviour of individuals of both tamarin species they could be classified as producers, scroungers, and opportunists depending on their investment in searching for food, and these roles were most probably determined by such factors as age and social status. Similar differences within social groups of animals in foraging tasks have been discussed and reported for a variety of taxa (e.g. Barnard and Sibly, 1981; Giraldeau and Lefebvre, 1987). The implications of such individual strategies for the development and transmission of innovations deserve specific attention. One possibility for example, is that when individuals adopt scrounging strategies they may be at a disadvant-age in learning for themselves (e.g. Giraldeau and Lefebvre, 1987; see also Reader and Laland, Chapter 1). Alternatively, it may also be argued that scrounging may potentially enhance social learning (e.g. Fritz and Kotrschal, 1999).

There are a number of issues with reference to individual differences and innovation. Hence, in another context, and although social tolerance potentially facilitates innovation, there are also cases where this is not so. Voelkl and Huber (2000) reported that although a young female common marmoset innovated a new way of opening a film canister to obtain food, no other monkey in her group developed the technique. Interestingly, however, all the marmosets in an adjacent group that could observe the task did demonstrate the new behaviour. Such examples remind us that innovation may be the result of individual propensities, but that the spread of innovations involves additional social and environmental influences (see also Reader and Laland, Chapter 1).

Moreover, the notable social cooperation among callitrichids is one of the special interests that we may have in studies of their innovative behaviour. On the one hand, social cooperation may be regarded in a facilitatory context-as social processes that facilitate new behaviours and their transmission. Alternatively, there may be also costs 'to being so attuned to cues from others in terms of loss of innovation in some contexts' (Snowdon, personal communication) In Snowdon's colony of cotton-tops, for example, individuals may communicate about a noxious change in a familiar edible food without actually sampling it (unpublished data).

Cooperation and social tolerance are combined in callitrichids with a unique anthropoid reproduction mode that includes multiple births. There are a number of issues here relevant to innovation. For example, with regard to the social dynamics of callitrichids, much behavioural interaction in young animals takes place within twin groups (Box, 1975; Ingram, 1977). With such close proximity and coordinated activities as in play and exploration as twins develop, the question arises as to whether this is a social context favouring directed social learning (Coussi-Korbel and Fragaszy, 1995).

Again, potential hypotheses to account for sex differences in response to food challenges include high energetic demands of female reproduction that combine pregnancy with lactation. However, priority of access to food by non-reproductive females, together with a variety of research that demonstrates the complexity of 'energy hypotheses', do not support simple correlational inferences. Further, recent research on differences in visual capacities between all male callitrichids and the majority of females in their perception of colour, may well account for many observed sex differences in feeding situations. It is also of interest, however, that female priority of access to restricted food has also been shown in experimental conditions that do not involve colour discrimination. We might also add that the greater responsiveness of adult females in various feeding situations, irrespective of its immediate motivation, might facilitate the development of innovative behaviour, because they are more likely to interact with a new source of food. Moreover, given that there are robust sex differences in responses to food challenges in different contexts that are at least consistent with natural feeding ecology and social organization in some species, we may argue for example, that such priority is functionally significant in the context of a cooperative breeding system, and is advantageous where there is a small number of breeding partners.

There are also indications of sex differences in response to new spatial opportunities and we might begin to ask to what extent we are dealing with natural behavioural propensities that impact on innovation. We need much more work on the strategies of behaviour in males and females among different species and genera in different environmental challenges in order to build up a profile of the suite of characteristics and propensities involved. Much emphasis in this chapter has been on the behaviour of female callitrichids, due mainly to my interest in the unusual (among primates) finding of female priority to food in some species.

There are also an increasing number of reports that the behaviour of males is significantly different from that of females in a variety of contexts, such as vigilance. Much more attention will be given in future research to the ways in which behavioural strategies (including in feeding situations) impact on innovation.

Callitrichids eat a wide variety of foods, and live in a diversity of habitats, and we might expect that they will develop innovative behaviour patterns. However, we do need specific observations and discussions about the potential advantages and costs of innovative behaviour in this group. In the final analysis, comparative knowledge of callitrichid characteristics and propensities will enable us to assess their status as potential innovators, determining which are innovation-prone species for example (see Greenberg, Chapter 8).

There is obviously a long way to go with all this work. One of the big problems so far is that the sample sizes of animals in experimental and field studies is often very small. However, it is also relevant that that there are, nevertheless, opportunities for the application of formal meta-analyses methods (e.g. Cooper and Hedges, 1994) in examining such data. Another problem is that because we have no established literature on innovation among callitrichids, we are mainly confined to talking about behavioural responses that are not necessarily innovations. At best, however, the behavioural results are predictive for studies of innovation, and in any case, are intended to stimulate ideas in the area.

Finally, the practical applications of studies of innovative behaviour include their use as tools in conservation strategies, as the training of captive-born animals for release and independent life in their natural habitat. A recent study is mentioned to point out some new and cautionary perspectives. Critically, very little of the behaviour in the wide ranging skills necessary for survival in nature is genetically hardwired (Kleiman, 1989; Box, 1991). Hence, it is appropriate to consider training for reintroduction by attempts to design specific tasks for specific skills, as well as to promote environmental conditions that provide a range of physical and social challenges that encourage natural behaviour and behavioural flexibility (Box, 1991).

Moreover, Beck et al. (2002) note that most guidelines on reintroduction procedures recommend structured training and preparation of animals for life in a natural environment. Training is regarded as being necessary for sufficient individuals to survive and reproduce, that is, a successful reintroduction. Training is also regarded as important in increasing the welfare of individuals to be reintroduced, as well as to the cost effectiveness of a project. Interestingly, however, there is very little empirical information on the effect of training on reintroduction (Kleiman, 1996). The observations and discussion by Beck et al. (2002) make an important contribution in this domain. They examined the influence of a number of pre-release environments in captivity with subsequent survivorship of golden lion tamarins in nature in Brazil. A variety of cage conditions with no attempts to specifically stimulate behaviour, was compared with training conditions in cages. These conditions involved specific challenges to stimulate behaviour that is critical in the wild. For instance, tasks were given that encouraged foraging for embedded foods. In other conditions, tamarins lived in wooded areas in zoological gardens, and confronted many different environmental challenges. The results are counter-intuitive and instructive. Despite a variety of caveats, there were no significant differences in subsequent survival for animals from these pre-release conditions. 'Captive born lion tamarins reintroduced directly out of cages, with or without training, and those given the opportunity to range freely on zoo grounds before release are all equally likely to survive up to two years' (Beck *et al.*, 2002, p. 290).

So, with regard to the survival of young tamarins after their release, there is no robust evidence that training confers advantages in either the short or the longer term. Beck *et al.* (2002) note that a number of studies with other animal taxa are in general agreement. A key point, for example, is that training with specific tasks changes the behaviour of individuals, but that these changes do not then positively influence improved survival in nature. This is a very good example of the need to study a variety of mediating influences upon the development of behavioural strategies in new situations.

Summary

There is no established literature in innovation for the marmosets and tamarins. The aim of this chapter is to discuss characteristics and propensities of the group that have implications for innovation. There are morphological and visuospatial differences among genera and species that may affect responses to new ecological opportunities, such as those that involve manipulative tasks. Studies of mixed species associations show that, although costs may be incurred, new behaviour may be facilitated in ways that would not necessarily be developed by species individually. Further, a cooperative breeding system and highly cooperative behaviours suggest cognitive capacities for, and facilitation of, the transfer of new behaviour patterns; food sharing and associated food calling may be an example.

Among individuals, sex differences are discussed in a number of social and ecological contexts. Adult females may demonstrate a priority to food and be more successful in new food tasks. Explanations in terms of energy requirements, however, turn out to be complex, and alternative hypotheses are considered. Moreover, responses to new spatial and object challenges show different patterns of responsiveness among age and sex classes, and different hypotheses deserve attention. Subsequently, a brief discussion of practical applications of these interests as tools in conservation strategies is included.

Acknowledgements

My thanks to Kevin Laland for his invitation to participate in the symposium on Animal Innovation that he convened at the International Ethological Congress in Tuebingen, Germany in 2001. Warm thanks also to both the editors, who together with Chuck Snowdon and two other anonymous referees provided many helpful comments and suggestions on the first draft of this chapter.

References

Achenbach, G. G. and Snowdon, C. T. (2002). Costs of caregiving: Weight loss in captive adult male cotton-top tamarins (*Saguinus oedipus*) following the birth of infants. *International Journal of Primatology*, **23**, 179–89.

Addington, R. G., Caine, N. G., and Schaffner, C. (1991). Factors affecting the food calls of red bellied tamarins. *American Journal of Primatology*, **24**, 85.

Addington, R. L. (1999). Social foraging in captive pygmy marmosets (*Cebuella pygmaea*): Effects of food characteristics and social context on feeding competition and vocal behaviour. Ph.D. Thesis, University of Wisconsin.

Barnard, C. J. and **Sibly, R. M.** (1981). Producers and scroungers: A general model and its application to captive flocks of house sparrows. *Animal Behaviour*, **29**, 543–50.

Beck, B. B., Castro, M. T., Stoinski, T. S., and **Ballou, J. D.** (2002). The effects of pre-release environments and post-release management on survivorship in reintroduced golden lion tamarins. In *Lion tamarins: Biology and conservation* (ed. D. G. Kleiman and A. B. Rylands), pp. 283–300. Washington, DC: Smithsonian Institution Press.

Benz, J. J., French, J. A., and **Leger, D. W.** (1990). Sex differences in vocal structure in a callitrichid primate, *Leontopithecus rosalia*. *American Journal of Primatology*, **21**, 257–64.

Bicca-Marques, J. C. (2000). Cognitive aspects of within-patch foraging decisions in wild diurnal and nocturnal New World monkeys (*Saguinus imperator, Saguinus fuscicollis, Callicebus cupreus, Aotus nigriceps,* Brazil). Ph.D. Thesis, University of Illinois.

Bicca-Marques, J. C. (1999). Hand specialisation, sympatry, and mixed species associations in callitrichines. *Journal of Human Evolution*, **36**, 349–78.

Box, H. O. (1975). Quantitative studies of behaviour within captive groups of marmoset monkeys (*Callithrix jacchus*). *Primates*, **16**, 155–74.

Box, H. O. (1982). Individual and intergroup differences in social behaviour among captive marmosets (*Callithrix jacchus*) and tamarins (*Saguinus mystax*). *Social Biology and Human Affairs*, **47**, 49–68.

Box, H. O. (1984). Behavioural responses to environmental change: Some preliminary observations on common marmosets. *Social Biology and Human Affairs*, **49**, 81–9.

Box, H. O. (1988). Behavioural responses to environmental change: Observations on captive marmosets and tamarins (*Callitrichidae*). *Animal Technology*, **39**, 9–16.

Box, H. O. (1991). Training for life after release. In *Beyond captive breeding: Re-introducing endangered mammals to the wild* (ed. J. H. W. Gipps), pp. 111–23. Oxford: Clarendon Press.

Box, H. O. (1997). Foraging strategies among male and female marmosets and tamarins (*Callitrichidae*): New perspectives in an underexplored area. In *Biology and conservation of New World primates,* (ed. H. O. Box and H. M. Buchanan-Smith) *Folia Primatologica*, **68**, 296–306.

Box, H. O. (1999). Temperament and socially mediated learning among primates. In *Mammalian social learning: Comparative and ecological perspectives* (ed. H. O. Box and K. R. Gibson), pp. 33–56. Cambridge: Cambridge University Press.

Box, H. O. and **Rohrhuber, B.** (1993). Differences in behaviour among adult male, female pairs of cotton-tops (*Saguinus oedipus*) in different conditions of housing. *Animal Technology*, **44**, 19–30.

Box, H. O. and **Smith, P.** (1995). Age and gender differences in response to food enrichment in family groups of captive marmosets (*Callithrix-Callitrichidae*). *Animal Technology*, **46**, 11–18.

Box, H. O., Rohrhuber, B., and **Smith, P.** (1995). Female tamarins (*Saguinus*—Callitrichidae) feed more successfully than males in unfamiliar foraging tasks. *Behavioural Processes*, **34**, 3–12.

Box, H. O., Yamamoto, M. E., and **Lopes, F. A.** (1999). Gender differences in marmosets and tamarins: Responses to food tasks. *International Journal of Comparative Psychology*, **12**, 59–70.

Buchanan-Smith, H. M. (1994). Environmental enrichment in captive marmosets and tamarins. *Humane Innovations and Alternatives*, **8**, 559–64.

Buchanan-Smith, H. M. (1999). Tamarin polyspecific associations: Forest utilisation and stability of mixed-species groups. *Primates*, **40**, 233–47.

Caine, N. G. (1986). Visual monitoring of threatening objects by captive tamarins (*Saguinus labiatus*). *American Journal of Primatology*, **10**, 1–8.

Caine, N. G. (1993). Flexibility and co-operation as unifying themes in *Saguinus* social organisation and behaviour: The role of predation pressures. In *Marmosets and tamarins: Systematics, behaviour and ecology* (ed. A. B. Rylands), pp. 200–19. Oxford: Oxford University Press.

Caine, N. G. (2002). Seeing red: Consequences of individual differences in colour vision in callitrichid primates. In *Eat or be eaten. Predation-sensitive foraging in primates* (ed. L. E. Miller) Cambridge: Cambridge University Press.

Caine, N. G. and Mundy, N. I. (2001). Predator/prey adaptations related to dichromatic and trichromatic colour vision in callitrichids. *Abstracts of the XVIIIth Congress of the International Primatological Society*, p. 79.

Chamove, A. S. and Rohrhuber, B. (1989). Moving callitrichid monkeys from cages to outside areas. *Zoo Biology*, **8**, 151–63.

Coimbra-Filho, A. F. (1970). Consideracoes gerais e situacao atual dos micosleoes escuros, *Leontideus chrysomelas* (Kuhl, 1820) e *Leontideus chrysopyhus* (Mikan, 1823) (Callitricidae, Primates). *Review of Brasilian Biology*, **30**, 249–68.

Cooper, H. and Hedges, L. V. (1994). *The handbook of research synthesis*. New York: Russell Sage Foundation.

Coussi-Korbel, S. and Fragaszy, D. M. (1995). On the relation between social dynamics and social learning. *Animal Behaviour*, **50**, 1441–53.

Drea, C. M. (1998). Social context affects how rhesus monkeys explore their environment. *American Journal of Primatology*, **44**, 205–14.

Epple, G. (1990). Sex differences in partner preference in mated pairs of saddle-back tamarins (*Saguinus fuscicollis*). *Behavioural Ecology and Sociobiology*, **27**, 455–9.

Epple, G. (1994). Gender differences in chemosignalling systems of callitrichid monkeys: A review. *Abstract of the 15th Congress of the International Primatological Society, Bali*, p. 330.

Feistner, A. T. C. and Chamove, A. C. (1986). High motivation toward food increases food-sharing in cotton-top tamarins. *Developmental Psychobiology*, **19**, 437–52.

Feistner, A. T. C. and McGrew, W. C. (1989). Food sharing in primates: A critical review. In *Perspectives in primate biology* (ed. P. K. Seth and S. Seth), Vol. 3, pp. 21–36. New Delhi: Today and Tomorrows Printers and Publishers.

Feistner, A. T. C. and Price, E. C. (2000). Food sharing in black lion tamarins (*Leontopithecus chrysopygus*). *American Journal of Primatology*, **52**, 47–54.

Ferrari, S. F. (1993). Ecological differentiation in the Callithrichidae. In *Marmosets and Tamarins: Systematics, behaviour and ecology* (ed. A. B. Rylands), pp. 314–28. Oxford: Oxford University Press.

Ferrari, S. F. and Lopes Ferrari, M. A. (1989). A re-evaluation of the social organisation of the Callitrichidae with reference to the ecological differences between genera. *Folia Primatologica*, **52**, 132–47.

Fragaszy, D. M. and Adams-Curtis, L. E. (1991). Environmental challenges in groups of capuchins. In *Primate responses to environmental change* (ed. H. O. Box), pp. 239–64. London: Chapman & Hall.

Fragaszy, D. M., Visalberghi, E., and Galloway, A. (1997). Infant tufted capuchin monkeys behaviour with novel foods: Opportunism, not selectivity. *Animal Behaviour*, **53**, 1337–43.

French, J. A. and Inglett, B. J. (1991). Responses to novel social stimuli in callitrichid monkeys: A comparative perspective. In *Primate responses to environmental change* (ed. H. O. Box) London: Chapman & Hall.

French, J. A. and Snowdon, C. T. (1981). Sexual dimorphism in responses to unfamiliar intruders in the tamarin *Saguinus oedipus*. *Animal Behaviour*, **29**, 822–9.

Fritz, J. and Kotrschal, K. (1999). Social learning in common ravens, *Corvus corax*. *Animal Behaviour*, **57**, 785–93.

Garber, P. A. and Kitron, U. (1997). Seed swallowing in tamarins: Evidence of a curative function or enhanced foraging efficiency? *International Journal of Primatology*, **18**, 523–38.

Garber, P. A. and Leigh, S. R. (2001). Patterns of positional behaviour in mixed-species troops of *Callimico goeldii*, *Saguinus labiatus*, and *Saguinus fuscicollis* in northwestern Brazil. *American Journal of Primatology*, **54**, 17–31.

Garber, P. A. (1988). Diet, foraging patterns and resource defence in a mixed species troop of *Saguinus mystax* and *Saguinus fuscicollis* in Amazonian Peru. *Behaviour*, **105**, 18–34.

Garber, P. A. (1989). Role of spatial memory in primate foraging patterns: *Saguinus mystax* and *Saguinus fuscicollis*. *American Journal of Primatology*, **19**, 203–16.

Garber, P. A. (1993). Feeding ecology and behaviour of the genus Saguinus. In *Marmosets and tamarins: Systematics, behaviour and ecology* (ed. A. B. Rylands), pp. 273–95. Oxford: Oxford University Press.

Garber, P. A. (2000). Evidence for the use of spatial, temporal and social information by some primate foragers. In *On the move: How and why animals travel in Groups* (ed. S. Boinski and P. A. Garber), pp. 261–98. Chicago, IL: The University of Chicago Press.

Giraldeau, L.-A. and Lefebvre, L. (1987). Scrounging prevents cultural transmission of food-finding in pigeons. *Animal Behaviour*, **35**, 387–94.

Goldfoot, D. A. and Neff, D. A. (1985). On measuring behavioural sex differences in social contexts. *Handbook of Behavioural Neurobiology*, **7**, 767–83.

Goldizen, A. W. (1988). Tamarin and marmoset mating systems: Unusual flexibility. *Trends in Ecology and Evolution*, **32**, 36–40.

Hardie, S. M. (1995). The behaviour of mixed-species tamarin groups (*Saguinus labiatus* and *Saguinus fuscicollis*). Ph.D. Thesis, University of Stirling, Scotland.

Hershkovitz, P. (1977). *Living New World monkeys part I. (Platyrrhini), with an introduction to primates.* Chicago, IL: Chicago University Press.

Heymann, E. W. (1990). Interspecific relations in a mixed-species troop of moustached tamarins, *Saguinus mystax*, and saddle-back tamarins, *Saguinus fuscicollis* (Primates: Callitrichidae) at the Rio Blanco, Peruvian Amazonia. *American Journal of Primatology*, **21**, 115–27.

Heymann, E. W. (1998). Sex differences in olfactory communication in a primate, the moustached tamarin *Saguinus mystax* (Callitrichidae). *Behavioral Ecology and Sociobiology*, **43**, 37–45.

Heymann, E. W. and Buchanan-Smith, H. M. (2000). The behavioural ecology of mixed troops of callitichine primates. *Biological Reviews of the Cambridge Philosophical Society*, **75**, 169–90.

Hrdy, S. B. (1981). *The woman that never evolved.* Harvard: Harvard University Press.

Ingram, J. C. (1977). Interactions between parents and infants, and the development of independence in the common marmoset (*Callithrix jacchus*). *Animal Behaviour*, **25**, 811–27.

Jolly, A. (1984). The puzzle of female feeding priority. In *Female primates: Studies by women primatologists* (ed. M. F. Small) pp. 197–215. New York: Liss.

Kappeler, P. M. (1987). The acquisition process of a novel behaviour pattern in a group of ring-tailed lemurs (*Lemur catta*). *Primates*, **28**, 225–8.

Joubert, A. and Vauclair, I. (1986). Reaction to novel objects in a troop of Guinea baboons: Approach and manipulation. *Behaviour*, **96**, 92–104.

Kappeler, P. M. (1990). Female dominance in *Lemur catta*: More than just female feeding priority. *Folia Primatologica*, **55**, 92–5.

King, B. J. (1999). New directions in the study of primate learning. In *Mammalian social learning: Comparative and ecological perspectives* (ed. H. O. Box and K. R. Gibson), pp. 17–32. Cambridge: Cambridge University Press.

Kleiman, D. G. (1989). Reintroduction of captive mammals for conservation. Guidelines for reintroducing endangered species into the wild. *Bioscience*, **39**, 152–61.

Kleiman, D. G. (1996). Reintroduction programs. In *Wild mammals in captivity: Principles and techniques* (ed. D. G. Kleiman, M. E. Allen, K. V. Thompson, S. Lumpkin, and H. Harris), pp. 297–305. Chicago, IL: University of Chicago Press.

Koenig, A. (1998). Visual scanning by common marmosets (*Callithrix jacchus*): Functional aspects and the special role of adult males. *Primates*, **39**, 85–90.

Kummer, H. and **Goodall, J.** (1985). Conditions of innovative behaviour in primates. *Philosophical Transactions of the Royal Society of London, Series B*, **308**, 203–14.

Maier, W., Alonso, C., and **Langguth, A.** (1982). Field observations on *Callithrix jacchus*. *Zeitschrift für Säugetierkunde*, **47**, 334–46.

Mayer, K. E., Blum, C. A., and **Caine, N. G.** (1992). Comparative foraging styles of tamarins and squirrel monkeys. *American Journal of Primatology*, **27**, 47.

McGrew, W. C. and **McLuckie, E. C.** (1986). Philopatry and dispersion in the cotton-top tamarin (*Saguinus o. oedipus*): An attempted laboratory simulation. *International Journal of Primatology*, **7**, 401–22.

Menzel, E. W. (1966). Responsiveness to objects in free-ranging Japanese monkeys. *Behaviour*, **27**, 130–50.

Menzel, E. W. (1969). Naturalistic and experimental research on primates. In *Naturalistic viewpoints in psychological research* (ed. E. Willems and H. Raush), pp. 78–132. New York: Holt, Rinehart and Winston.

Menzel, E. W., and **Menzel, D. R.** (1979). Cognitive, developmental and social aspects of responsiveness to novel objects in a family group of marmosets (*Saguinus fuscicollis*). *Behaviour*, **70**, 251–79.

Michels, A. M. (1998). Sex differences in food acquisition and aggression in captive common marmosets (*Callithrix jacchus*). *Primates*, **39**, 549–56.

Millar, S. K., Evans, S., and **Chamove, A. S.** (1988). Older offspring contact novel objects soonest in Callitrichid families. *Biology of Behaviour*, **13**, 82–96.

Milton, K. (1988). Foraging behaviour and the evolution of primate intelligence. In *Machiavellian intelligence: Social expertise and the evolution of intellect in monkeys, apes and humans* (ed. R. W. Byrne and A. Whiten), pp. 285–305. Oxford: Oxford University Press.

Nievergelt, C. M. and **Martin, R. D.** (1999). Energy intake during reproduction in captive common marmosets (*Callithrix jacchus*). *Physiology and Behaviour*, **65**, 849–54.

Norconk, M. A. (1990). Mechanisms promoting stability in mixed *Saguinus mystax* and *S. fuscicollis* troops. *American Journal of Primatology*, **21**, 129–46.

Norcross, J. L., Newman, J. D., and **Cofrancesco, L. M.** (1999). Context and sex differences exist in the acoustic structure of phee calls by newly paired common marmosets (*Callithrix jacchus*). *American Journal of Primatology*, **49**, 165–81.

Parker, S. T. and **Gibson, K. R.** (1977). Object manipulation, tool use, and sensorimotor intelligence as feeding adaptations in early hominids. *Journal of Human Evolution*, **6**, 623–41.

Peres, C. A. (1992). Consequences of joint-territoriality in a mixed species group of tamarin monkeys. *Behaviour*, **123**, 220–46.

Petto, A. J. and Devin, M. (1988). Food choices in captive common marmosets (*Callithrix jacchus*). *Laboratory Primate Newsletter*, 27, 7–9.

Platt, M. L. (1994). Memory and feeding ecology in lion tamarins (*Leontopithecus rosalia*) and marmosets (*Callithrix kuhli*). Ph.D. thesis: University of Pennsylvania.

Platt, M. L., Brannon, E. M., Briese, T. L., and French, J. A. (1996). Differences in feeding ecology predict differences in performance between golden lion tamarins (*Leontopithecus rosalia*) and Wied's marmosets (*Callithrix kuhli*) on spatial and visual memory tasks. *Animal Learning and Behaviour*, 24, 384–93.

Pook, A. G. and Pook, G. (1982). Polyspecific association between *Saguinus fuscicollis, Saguinus labiatus, Callimico goeldii*, and other primates in north-western Bolivia. *Folia Primatologica*, 38, 196–216.

Poole, T. B. (1990). Environmental enrichment for marmosets. *Animal Technology*, 41, 81–6.

Prescott, M. J. and Buchanan-Smith, H. M. (1999). Intra- and inter-specific social learning of a novel food task in two species of tamarin. *International Journal of Comparative Psychology*, 12, 71–92.

Price, E. C. (1992). Adaptation of captive-bred cotton-top tamarins (*Saguinus oedipus*) to a natural environment. *Zoo Biology*, 11, 107–20.

Price, E. C., McGivern, A. M., and Ashmore, L. (1991). Vigilance in a group of free-ranging cotton-top tamarins (*Saguinus oedipus*). *Dodo: Journal of the Jersey Wildlife Preservation Trust*, 27, 41–9.

Price, E. and Feistner, A. T. C. (2001). Food sharing in bare-faced tamarins (*Saguinus bicolour bicolour*). Development and individual differences. *International Journal of Primatology*, 22, 231–41.

Rapaport, L. G. (1999). Provisioning of young in golden lion tamarins (Callitrichidae, *Leontopithecus rosalia*): A test of the information hypothesis. *Ethology*, 105, 619–36.

Reader, S. M. and Laland, K. N. (2001). Primate innovation: Sex, age and social rank differences. *International Journal of Primatology*, 22, 787–806.

Richard, A. F. (1987). Malagasy prosimians: Female dominance. In *Primate societies* (ed. B. B. Smutz, D. L. Cheney, R. M. Seyfarth, R. W. Wrangham, and T. T. Struhsaker), pp. 25–33. Chicago, IL: Chicago University Press.

Roush, R. S. and Snowdon, C. T. (2001). Food transfer and development of feeding behaviour and food associated vocalizations in cotton-top tamarins. *Ethology*, 107, 415–29.

Ruiz, M., Carlos, R., Kleiman, D. G., Dietz, J. M., Moraes, E., Grativol, A. D. *et al.* (1999). Food transfers in wild and reintroduced golden lion tamarins, *Leontopithecus rosalia*. *American Journal of Primatology*, 48, 305–20.

Rylands, A. B. (1989). Sympatric Brazilian callitrichids: The black tufted-ear marmoset, *Callithrix kuhli*, and the golden-headed lion tamarin, *Leontopithecus chrysomelas*. *Journal of Human Evolution*, 18, 679–95.

Rylands, A. B. (1993). The ecology of the lion tamarins, Leontopithecus: Some intrageneric differences and comparisons with other callitrichids. In *Marmosets and tamarins: Systematics, behaviour and ecology* (ed. A. B. Rylands), pp. 296–313. Oxford: Oxford University Press.

Rylands, A. B., Schneider, H., Langguth, A., Mittermeier, R. A., Groves, C. P., and Rodriguez-Luna, E. (2000). An assessment of the diversity of New World primates. *Neotropical Primates*, 8, 61–93.

Sanchez, S., Pelaez, F., Gil-Buermann, C., and Kaumanns, W. (1999). Costs of infant carrying in the cotton-top tamarin (*Saguinus oedipus*). *American Journal of Primatology*, 48, 99–111.

Savage, A., Giraldo, L. H., Soto, L. H., and Snowdon, C. T. (1996). Demography, group composition and dispersal in wild cotton-top tamarin (*Saguinus oedipus*) groups. *American Journal of Primatology*, 38, 85–100.

Savage, A., Snowdon, C. T., Giraldo, H., and Soto, H. (1997). Parental care patterns and vigilance in wild cotton-top tamarins (*Saguinus oedipus*). In *Adaptive radiation of neotropical primates* (ed. M. A. Norconk, A. F. Rosenberger, and P. A. Garber), pp. 187–99. New York: Plenum Press.

Singer, S. S. and Schwibbe, M. H. (1999). Right or left, hand or mouth: Genera-specific preferences in marmosets and tamarins. *Behaviour*, **136**, 119–45.

Snowdon, C. T. (2001). Social processes in communication and cognition in callitrichid monkeys: A review. *Animal Cognition*, **4**, 247–57.

Stevenson, M. F. (1984). The captive breeding of marmosets and tamarins. *Proceedings of Symposium of The Association of British Wild Animal Keepers*, **8**, 49–67.

Tardif, S. D. (1997). The bioenergetics of parental behaviour and the evolution of alloparental care in marmosets and tamarins. In *Co-operative breeding in mammals* (ed. H. G. Solomon and J. A. French), pp. 11–33. Cambridge: Cambridge University Press.

Tardif, S. D., and Richter, C. B. (1981). Competition for a desired food in family groups of the common marmoset (*Callithrix jacchus*) and the cotton top tamarin (*Saguinus oedipus*). *Laboratory Animal Science*, **31**, 52–5.

Tardif, S. D., Harrison, M. L., and Simek, M. A. (1993). Communal infant care in marmosets and tamarins: Relation to energetics, ecology, and social organisation. In *Marmosets and tamarins: Systematics, behaviour and ecology* (ed. A. B. Rylands), pp. 220–34. Oxford: Oxford University Press.

Terborgh, J. (1983). *Five new world primates: A study in comparative ecology*. Princeton, NJ: Princeton University Press.

van Schaik, C. P. (in press). Local traditions in orangutans and chimpanzees: Social learning and social tolerance. In *Towards a biology of traditions: Models and evidence* (ed. D. M. Fragaszy and S. Perry). Cambridge: Cambridge University Press.

van Schaik, C. P., Deaner, R. O., and Merrill, M. Y. (1999). The conditions for tool use in primates: Implications for the evolution of material culture. *Journal of Human Evolution*, **36**, 719–41.

Visalberghi, E. (1988). Responsiveness to objects in two social groups of tufted capuchin monkeys (*Cebus apella*). *American Journal of Primatology*, **15**, 349–60.

Visalberghi, E., Janson, C. H., and Agostini, I. (in press). Response toward novel foods and novel objects in wild tufted capuchins (*Cebus apella*). *International Journal of Primatology*.

Voelkl, B. and Huber, L. (2000). True imitation in marmosets. *Animal Behaviour*, **60**, 195–202.

Wallander, D. L. (1998). Food sharing and the development of feeding behaviour in *Saguinus imperator*, the emperor tamarin (family: Callitrichidae). Ph.D. Thesis: Northwestern University.

Wormell, D., Brayshaw, M., Price, E., and Heron, S. (1996). Pied tamarins *Saguinus bicolour bicolour* at the Jersey Wildlife Preservation Trust: Management, behaviour and reproduction. *Dodo: Journal of the Wildlife Preservation Trust*, **32**, 76–97.

Yamamoto, M. E., Santos, B. G. A. C. L., and Lopes, N. A. (2000). Differential infant care in captive common marmosets (*Callithrix jacchus*). *Abstract of the European Federation of Primatology 2000 Meeting*, p. 29.

Yoneda, M. (1984). Comparative studies on vertical separation, foraging behaviour and travelling mode of saddle-backed tamarins (*Saguinus fuscicollis*) and red-chested moustached tamarins (*Saguinus labiatus*) in Northern Bolivia. *Primates*, **25**, 414–22.

Zullo, J. and Caine, N. G. (1988). The use of sentinels in captive groups of red-bellied tamarins. *American Journal of Primatology*, **14**, 455.

INNOVATION, INTELLIGENCE, AND COGNITION

CHAPTER 10

CONDITIONS OF INNOVATIVE BEHAVIOUR IN PRIMATES

HANS KUMMER AND JANE GOODALL

Innovative behaviour achieved through exploration, learning and insight heavily depends on certain motivational, social, and ecological conditions of short duration. We propose that more attention should be given to what these conditions are and where they are realized in natural groups of non-human primates. Only to the extent that such favourable conditions were frequently realized in a social structure or an extraspecific environment could selective pressures act on innovative abilities. There is hope that research into field conditions of innovative behaviour will help to identify its selectors in evolution.

PART 1 (H. KUMMER)

Introduction

The motive of this paper is that we almost completely lack an ecology of intelligence. No other dimension of behaviour has so systematically *not* been studied in the field.

The major selection pressures that made the order of primates into specialists for mental flexibility have been the subject of some speculation. Chance and Mead (1953) seem to have been the first to develop the argument that the complexity of *social* life was the prime selective agent of primate intelligence. The group member is required to judge the skills and changing inclinations of others and to use them for intelligent predictions for its own benefit. The view was later and independently presented by Alison Jolly (1966) and by Humphrey (1976). The difficulty is that many non-primates live in large groups without having evolved particular intelligence levels. Social complexity is probably the effect as much as the ultimate cause of intelligence.

Katharine Milton (1981) plausibly argued that the necessity of predicting the few productive days of widely dispersed tree patches in tropical forests was a major factor in the evolution of primate mental abilities. Yet, resources dispersed in time and space are hardly a unique feature of primate environments; they are also mastered by predators of highly mobile groups of prey. The complex system of ultimate causes is likely to defy any one-factor explanation. While certain selective pressures may have been particularly formative, it is most probable that a whole *set* of primate characters were mutually preadaptive to each other's evolution, and that they evolved into a syndrome of which mental flexibility is a part. Foremost among these characters is the a priori lack of outstanding competence in any specific skill. Our inventive species probably emerged among the primates rather than from a different order *because* primates are remarkably ill-equipped with innate technologies. The termite is better than the chimp at nest-building by an incredible amount, but the chimp beats the termite by learning to tease him from his mound. Instinct is like a key fitting a single lock. A key is easy to use. The poor a priori competence of primates is comparable to a pick-lock. Its use requires and therefore promotes a kind of skill, which eventually permits the opening of many locks.

If some selectors of primate intelligence were of outstanding importance, it should be possible to locate them empirically. The rationale is this: a behavioural character is tested by selection only in certain situations, those in which the behaviour is used to function. The critical assumption is that the situation that *selected* for a behavioural character in phylogeny also *trains* it in individual life. For example, the genetic basis of flying skill was selected only while the primeval birds were airborne; the learning of flying skill is restricted to the same context. An intelligence or learning disposition selected by the context of being socially subordinate would presumably be most proficiently developed by subordinates. The extent to which this assumption is correct must be ascertained for each behaviour. In general, it should be correct (1) if the behaviour with the higher fitness contribution were also the one with the greater reward—a very frequent case in all animals, (2) if each individual were capable of several behavioural variants so that it could compare the rewards—a frequent case in primates, and (3) if training could substantially improve on the uninstructed inherited ability—a condition fulfilled by definition in learning and intelligence.

From the above one may assume that innovative behaviour was selected primarily in the situations that now most effectively teach it. It would therefore be profitable to look systematically at the everyday situations in which present-day primates learn and invent most proficiently. These are the conditions by and for which their innovative dispositions must have been selected.

Time does not permit even a sketchy classification of the everyday conditions relevant to innovative performance. Instead a few examples from the field and the quasi-natural captive group will show the interest of the topic.

Intellectual training by group life

If the Chance–Jolly–Humphrey hypothesis is correct, that is, if complex social life selected primate intelligence *and* trained it to its greatest proficiency in natural life, we would expect the training effect to be most pronounced in those group members whom group life affects and constrains most, compared with solitaries. Most likely, these are the group members of low dominance rank. Subordinates should have made the most of the genetic disposition for intelligence and learning, since they most depend on the formation of alliances or need to lead their mate to where the dominant male cannot see them, and so forth.

Strayer (1976) and Bunnell *et al.* (1980; Bunnell and Perkins, 1980) tested macaques that normally lived in captive groups on learning tasks. In Strayer's study, a buzzer sounded when a reward could be obtained. The subordinate monkeys made fewer time-out errors, that is, they did not press the lever when the buzzer was silent, whereas dominants pressed even with the buzzer off. Dominant macaques were more dependent on regular reinforcement. In Bunnell's study the subordinate macaques were faster to master a reversal learning task. Dominant animals that fell in rank improved their performance, which speaks against the possibility that subordinates performed better because they were younger. The subordinates in these studies were thus more attentive to the outside conditions of their actions as the test apparatus, and they were better at relearning and at persisting in spite of frustration. This is what we would expect if they had transferred abilities acquired in the highly conditional social life of a low-ranker.

Free time and energy

These are other candidate conditions of innovative behaviour. The reason is that environments are not stationary over time. Animals survive certain periods in which their maximized net energy intake is just sufficient to keep them alive. At other more favourable periods, survival may be so easy that the animal has time and energy to spare. It would be advantageous if it could somehow store them for worse times.

To hoard food in a cache is one form of storage, but the hoard is vulnerable to decay, to parasitism and to plundering by observant conspecifics. It is found in solitary rodents and in insects with hidden or defended nests. Non-human primates do not use hoards, probably because they have no dens and no organization for guarding and sharing. A second form of storage is fatty tissue, which is a well-defendable but burdensome hoard. It is particularly common in marine mammals, where the weight is carried by the water and where the fat is a useful protection against hypothermia. Primates, being arboreal and a prey species, require agility and hardly use it.

By far the most elegant form of storage is new knowledge or skill in communication or ecological techniques. Here, the time and energy spent in exploratory behaviour is transformed into a light-weight unstealable commodity that can save time and energy in a harsher future.

Every zoo colony of primates is a rough experimental test of the hypothesis that free time and energy do indeed promote innovation. Kummer and Kurt (1965) quantitatively compared the behaviour of hamadryas baboons in the Zurich Zoo with the behaviour of wild troops in Ethiopia. Of 68 motor and vocal signals that the zoo baboons used in social communication, nine were not found in any wild troop, suggesting that they were innovations. (On the other hand, all social signals seen in the wild were also found in the zoo colony.) None of them looked pathological. Rather, they were functional elaborations, such as a new gesture that invited a youngster to be carried. In the case of 'protected threat', in which a female directs the attack of the dominant male against her opponent, the zoo version was technically improved and clearly more efficient than the crude hints at the same behaviour in the wild.

The best ecological invention of the zoo baboons was to obtain drinking water from a mound, otherwise unobtainable, by lowering the body, tail first, down a vertical wall and then to suck the water from the tuft (Schönholzer, 1958). The tail-drinking also was an innovation. No wild baboon was seen to approach an incentive backwards.

Within the limits, free time and energy thus seem to further innovation.

Part 2 (Jane Goodall)

Introduction

The study of the chimpanzees at Gombe, in Tanzania, is now in its 24th year. The longitudinal records provide many anecdotes regarding innovative behaviour and the following
discussion concerning conditions favouring (1) the appearance of novel patterns *per se*,
and (2) the transmission of such patterns through the social group, is centred around this
information.

An innovation can be: a solution to a novel problem, or a novel solution to an old one;
a communication signal not observed in other individuals in the group (at least at that
time) or an existing signal used for a new purpose; a new ecological discovery such as a
food item not previously part of the diet of the group. In some species of primates a new
food becomes incorporated into the diet of a troop as a result of a whole series of
independent individual discoveries; other novel behaviours, such as potato-washing, are
discovered by one individual only and acquired by other troop members through observational learning.

Some innovations derive from the ability of the individual to profit from an accidental
happening. Thus when the chimpanzee Mike was afraid to take a banana from my hand he
seized a clump of grasses and swayed them to and fro in the typical threat gesture. As he did
so one of the grasses touched the fruit. Instantly he let go of the grasses, looked round, hastily
broke off a slender, bendy twig, dropped it at once, reached for a thicker stick and hit the
banana to the ground from where he took and ate it. Ten minutes later when I proffered a
second fruit Mike unhesitatingly reached for a stick and repeated the performance. The first
step in such a chance sequence is not always seen, but can sometimes be deduced. At Gombe,
pots and pans are often washed at the edge of the lake. The anubis baboons of one study troop
spend long periods of time digging in the gravel, searching for scraps. One day when the lake
was rough and the successive breaking of large waves made drinking difficult, an adolescent
male, Asparagus, was observed to dig a hole near the high water mark, wait for a wave to fill
it, then quickly drink as the wave retreated. Subsequently he was observed to do the same
thing on another occasion. Six years later 8-year-old Sage, of the same troop, was seen drinking
in the same way on four separate occasions. It seems likely that the two males discovered the
technique independently—probably by first drinking water that filled up holes dug by them
when searching for food and subsequently profiting from that experience, digging the holes
for the specific purpose of drinking. One year after he was seen drinking in this way
Asparagus disappeared (he died or transferred to a new troop). Sage was 3 years old at that
time, and might, therefore, have learnt the behaviour by watching the older male: if so it
seems strange that he was not seen to drink in this way during the next 6 years.

Other innovations result from the ability of the higher primates to use existing behaviour patterns for new purposes. Chimpanzees at Gombe routinely use leaves to wipe their
bodies when soiled: one female twice used handfuls of leaves to brush away stinging insects
(once bees from a hive she was raiding, once ants from the trunk of a tree to which she was
clinging as she raided their nest) and one male used leaves to wipe the inside of a baboon
skull during a meat-eating session: he ate the leaf-brain wad (Wrangham, 1975).

A third kind of innovation is the performance of a completely new pattern, such as the bizarre series of somersaults by which the polio stricken male, McGregor, managed to move from place to place after losing the use of his legs or the odd form of locomotion invented by one infant during lone play (to be described).

Conditions for the appearance of new behaviours

Innovations may be occasioned by sudden change in the environment such as when wild populations are provisioned or during an outbreak of disease (such as the polio at Gombe (Goodall, 1983)). They may also, as has been described, appear during periods of environmental stability, particularly during periods of plenty, which result in an excess of leisure and energy. This is exemplified in some captive situations.

A pre-requisite for the solving of problems in novel ways, be they ecological, technical (involving the use of tools), or social (involving the manipulations of companions) is familiarity with the components of the situation. Chimpanzees have certain manipulative tendencies, which appear to be inborn—at a certain age an infant will reach out, grasp, and brandish a stick (Schiller, 1949). But he must become familiar with sticks during play if he is to use them successfully as tools to solve novel problems (Figure 10.1) (Köhler, 1925; Schiller, 1952). Similarly, chimpanzees have certain inborn communications signals, but must have opportunity to interact with other chimpanzees to use these signals successfully (Menzel, 1964). Social experience and familiarity with the other members of the group are essential if a chimpanzee (or other animal) is to invent novel methods of getting his own way in competitive situations.

Social innovations

Much social skill emerges as a direct result of competition between group members. A subordinate individual A, to attain a particular goal *in spite of* the inhibiting proximity of a social superior B must either form a coalition with a third individual C (so that A and C are superior in strength to B) or follow a more devious route, such as moving out of B's

Figure 10.1 Flint 'clubs' an insect, a behaviour not observed in the adult tool-using repertoire at Gombe.

sight, or persuading B to move away, or distracting B's attention. Some of the solutions to social problems of this sort represent novel ones on the part of the individual concerned. The following are examples of innovation of this sort observed among the chimpanzees at Gombe.

1.   One reproductive strategy of the male chimpanzee is to establish exclusive mating rights over a female by leading her away from the other males to the periphery of the community range (McGinnis, 1973; Tutin, 1979). A male wanting to establish a consortship of this sort faces obvious problems if there are higher ranking males in the group who are interested in the same female. One individual, Satan, was, on several occasions observed to keep very close to a sexually attractive female until she made her nest in the evening, rise before any other chimpanzee the next morning, shake branches at her (a signal for her to follow him), and lead her away before the others left their nests (Tutin, 1979).

2.   The charging display, when a male rushes over the ground or through the branches, swaying, dragging and throwing vegetation, branches and rocks is of major significance to his struggle for a higher position in the dominance hierarchy. Figan not only showed unusual timing and placing of his displays (Bygott, 1974) but repeatedly got up before other chimpanzees in his group and performed wild arboreal displays while his companions were still in their nests. This caused great confusion and was one of the methods he employed during his successful bid for alpha position (Riss and Goodall, 1977). Other males occasionally used this technique, but it seems that only Figan profited from the experience and made it a regular occurrence.

3.   Adolescent males typically challenge older females with bristling, swaggering displays during which branches may be waved with which they may hit the females. Responses to such displays vary depending on the age of the male, rank of his victim in relation to that of his mother and so on, and range from fear and avoidance, through retaliation, to totally ignoring the youngster for as long as possible. Sometimes they hold onto the branches with which they are being flailed. One female, Winkle, twice terminated such displays in unusual ways. Once she firmly removed the branch from a young male even before he had begun to sway it, and once she reached out and vigorously tickled the swaggering youngster so that his aggressive display ended in *laughing*. Of interest was the fact that the same female used two quite different unusual patterns to achieve the same end.

Banana-feeding at Gombe created a new situation. Aggressive competition was greatly increased and the chimpanzees tended to aggregate in large numbers in camp as they waited for us to distribute the fruits. Among the innovations that appeared as a result of the new situation were the following.

1.   Before we had devised a way of distributing the bananas among the various individuals present it was often the case that youngsters got very few or none at all. One day, when the young adolescent male Figan was part of a large group and, in consequence, had not managed to get more than a couple of bananas, he suddenly got up and walked away with a purposeful gait. His mother followed and, as is often the case when one individual sets off as though with a goal in mind, the others followed. Ten minutes later Figan reappeared by himself, and was given bananas. We thought this was a coincidence as indeed it may have been on that first occasion. But when Figan repeated the performance on four subsequent occasions it became clear that he was following a deliberate strategy (Goodall, 1971).

2.   In the early 1960s there were occasionally a few empty four gallon paraffin cans lying about my camp. At one time or another almost all of the 14 adult males had incorporated one of these into a charging display. One of them, Mike, profited from the experience and began to use the cans in almost all displays that he performed in camp. He even learned to keep three ahead of him as he charged towards his superiors. Within a four-month period he had become alpha male, having thoroughly intimidated all his rivals and, so far as we know, without taking part in a single fight (Goodall, 1968, 1971).

Groups of chimpanzees in captivity show, as did the hamadryas, much innovation in the social sphere. With adequate food and drink provided, their health attended to, and most dangers and excitements removed from daily life, chimpanzees not only have more leisure time than their wild counterparts, they are also subjected to a new stress, over and above boredom and confinement. At Gombe, if the social environment becomes tense and hostile, an individual can, and often does, move off quietly by itself or with a small compatible group. In captivity, even in a large enclosure, this is not possible. The novel behaviours, which zoo or laboratory chimpanzees show in their social interactions are undoubtedly, at least in part, a response to this new challenge. And, because they are forced to remain constantly in each other's company, chimpanzees in captivity are probably able to predict more accurately the responses of their companions so that their interactions can become increasingly sophisticated. One adolescent male, Shadow, in Emil Menzel's group, developed a unique courtship display during which he stood bipedally and flipped his upper lip back over his nostrils. Precisely how this strange performance originated is not known: it was clear, however, that the adult females, all dominant to Shadow, responded aggressively to the more usual male courtship (probably because this involves many aggressive elements). With the new display Shadow was able to convey his sexual interest devoid of aggressive overtones (C. Tutin and W. McGrew, personal communication). The chimpanzees of the large Arnhem colony have been carefully studied by de Waal (1982) who observed a great deal of extraordinarily astute social skill. Once, for example, when a subordinate male was surprised during a clandestine courtship by the arrival of the alpha male, he quickly covered his erect penis with his hands, effectively hiding the tell-tale signal. After an aggressive dispute between two of the adult males, a female sometimes showed behaviour, which served to hasten reconciliation between the rivals: she would lead one towards the other, sit between them so that both groomed her, then step quietly away leaving the males grooming each other: harmony was restored. Each of the adult females was seen to act thus at one time or another; such incidents have never been observed in the wild.

Ecological and technical innovations

While social innovations can only appear within the social environment, during interactions with other group members, ecological and technical innovations are more likely to appear during periods when an individual is, to some extent, freed from social distractions. In a hamadryas one-male unit the females have differing spatial positions: one stays close to the unit male; one spends much time at the periphery of a foraging group. While the central female is preoccupied with social interactions, particularly with the male, the peripheral female is alert to danger and is the individual most likely to discover a new food source. Sigg (1980) tested wild family units temporarily kept in field enclosures: he found that there was a marked difference in the abilities of peripheral vs central females in their

ability to learn new ecological tasks. Thus peripheral females readily learned to discriminate between nails painted different colours that were used as markers for food buried in
the ground. They also remembered the location of underground water several hours after
being allowed to watch its burial. Central females seemed unable to learn these tasks, perhaps because their attention was almost totally absorbed in social interactions. In the Tai
Forest of the Ivory Coast chimpanzees crack open nuts with stones or heavy pieces of
wood. Adult females are more efficient in the hammering technique than males on several
measures. This may be due, at least partly, to the fact that males appear to be more easily
distracted by ongoing social events: they look around, away from the task in hand, more
frequently than do hammering females (Boesch and Boesch, 1981, 1983).

Many innovations appear during childhood when a youngster is cared for and protected
by its mother and thus has much time for carefree play and exploration. At Gombe, where
female chimpanzees spend a great deal of time in association with their dependent offspring
only, an infant may be without opportunity for playing with other youngsters for hours or
even days at a stretch. With no social distraction the conditions are ideal for the emergence
of novel behaviours in the ecological and technical sphere. This is particularly true for the
first-born child who does not even have the opportunity to play with elder siblings. One
infant invented a new method of locomotion, swinging her body forward through her arms
(in the typical 'crutching' gait) while keeping one foot firmly tucked into the opposite
groin. Another threw a round fruit into the air and caught it as it fell: he tried to repeat the
performance many times but I did not see him succeed again. A third lay on his back with
his legs in the air and rotated a large round fruit which he held on the soles of his feet. A
fourth used stones and twigs to tickle her clitoris, continuing until she was laughing loudly.
One infant spent many minutes playing with a butterfly, another with a frog. Occasionally
an infant will taste a novel food object: behaviour, which we have never observed in an
adult chimpanzee in the wild.

Once an infant has mastered a newly acquired adult manipulative pattern this may be
practised in a variety of contexts other than that in which it was learned. Thus one infant,
Flint, used the termite-fishing technique on his mother's leg, pushing a blade of grass
between the hairs and then sucking the end. And, on another occasion, Flint used a grass
blade to 'fish' for water caught in the hollow of a tree stump. After sucking off the drops a
few times the blade gradually became more and more crumpled until he had made a
miniature *sponge* (Figure 10.2). This behaviour might have led to the typical leaf sponge
used by the Gombe chimpanzees for soaking water from hollows of this sort (although it
seems more likely that this originated for the removal of dead leaves that had accumulated
in the water). At Gombe the chimpanzees open the large hard-shelled *Strychnos* fruits by
banging them against the tree trunk or a rock. During one *Strychnos* season infant Flint
picked up a rock and smashed an insect on the ground. This pattern might, one day, lead
to the use of a hammer stone for cracking open nuts, as in West African chimpanzee
cultures. On another occasion Flint hit an insect with a 'club' (Figure 10.3).

Innovation and tradition

A comparison of the ecological, technical, and social communication patterns of different
groups of chimpanzees in different geographical localities reveals a wide variety of cultural

Figure 10.2 (a) Flint dips a grass stem into a water bowl, using the termite-fishing technique. (b) After a few such dips the grass becomes crumpled and resembles a miniature *sponge*. (A sponge of crumpled leaves is traditionally used by the Gombe chimpanzees for drinking from water bowls.)

Figure 10.3 (a) Feeding traditions are passed from one generation to the next through mechanisms of social facilitation and observational learning. Here one-and-a-half-year-old Getty watches his grandmother's 6-year-old son (his uncle), Gimble, feeding on leaves. (b) Getty then samples the same food, while Gimble, in turn watches him. This was not a new item in the diet, but the example illustrates the mechanisms for the passing on of information from one individual to another in chimpanzee society.

traditions. Gombe and Mahale, 100 miles (160 km) apart, have many plant species in common: Mahale chimpanzees feed on some that are never used by the Gombe chimpanzees and vice versa, and some foods that *are* eaten at both localities are, nevertheless, prepared differently. Driver and Carpenter ants are present at both places: at Gombe the chimpanzees fish frequently for Driver but not Carpenter ants: at Mahale it is the other way round. Gombe chimpanzees use leaf sponges when drinking from water bowls: at Mahale this has not been observed. At Mahale two chimpanzees grooming often show a unique 'hand-clasp' posture never observed at Gombe (McGrew and Tutin, 1978). Different traditions have been found in hamadryas baboons and various macaques (see, e.g. Kummer, 1971) and it seems certain that groups of almost all higher mammals will be found to show intergroup cultural variation. Traditions obviously began with the innovative performance of particular individuals in the past—performances that were subsequently incorporated into the repertoire of a group through processes of social facilitation and observational learning (Figure 10.3). Only in the Japanese macaques has this process been carefully documented (Kawamura, 1954; Itani, 1965), first in regard to the unwrapping and eating of candies and subsequently for the dramatic inventions of potato-washing and wheat-cleaning initiated by the gifted young female Imo (and which, as mentioned, led to other sea-related activities). It was young monkeys who were the first to perform the new behaviours, their peers and their mothers who were the first to follow their example. Siblings and other closely associating adults were next on the list. As new infants were born, they acquired the behaviours, in the normal learning process, from their mothers. In the case of the unwrapping and eating of candies the reward was immediate and obvious and it is undoubtedly significant that this behaviour was acquired by all members of the troop, whereas the washing and cleaning techniques were never learnt by some of the adult males.

The rhesus monkeys that were transferred from Santiago to a new island (Morrison and Menzel, 1972) rapidly incorporated many new foods into their diet. There was no evidence that these new foods were consistently sampled by any individual monkey or category of monkey: each one tried each new food as he came across it and accepted or rejected it according (presumably) to its taste. When wheat was introduced to one group of Japanese monkeys the whole troop had acquired the new food habit within 4 h (Itani, 1965). Similarly, the baboons (of all ages and both sexes) at Gombe have always unhesitatingly tried new foods offered to them. In chimpanzee society, however, both at Gombe and Mahale, it is only youngsters who experiment with new foods (Nishida et al., 1984). Moreover, the conservative attitude of mothers and elder siblings, who usually snatch or flick away a new food item from an infant, makes it unlikely that the infant will repeat the experiment, and even less likely that the food in question will be incorporated into the diet of the community as a whole.

One new behaviour that did spread through the community at Gombe was the use of sticks as levers to try to open banana boxes. Four and a half months after these boxes had been installed three adolescents began, independently, to use sticks to try to prize open the steel lids. Because a box was sometimes opened when a chimpanzee was working at it, the tool use was occasionally rewarded and, over the next year, the habit spread until almost all members of the community, including adult males, were seen using sticks in this way. That many individuals learnt as a result of watching their companions is suggested by the fact that one female was observed to behave thus on her very first visit to camp: before this she had had ample opportunity to watch what was going on from the surrounding vegetation (Goodall, 1968).

Most of the innovative performances in the social sphere observed at Gombe were seen in single individuals only, and often only once, as when an adolescent male showed a submissive genital display (facing his aggressor with laterally positioned leg, hip, and knee bent and an erection of the penis) exactly comparable to that of the squirrel monkey (Ploog, 1967, plate 10.3). A juvenile female, Fifi, also showed a completely new pattern (at least at Gombe, though it was observed in Washoe (Gardner and Gardner, 1969)) 'wrist-shaking'. This gesture, a rapid to and fro shaking of the hand, was directed at an adult female in an aggressive context. The following week Fifi repeated the gesture and continued to do so, on and off, for the next 10 months, always in the aggressive context. A week after Fifi's first observed wrist-shake the new gesture appeared in a slightly younger female, Gilka, who was Fifi's most frequent playmate. For the next few months Gilka wrist-shook vigorously in many contexts. Then she, too, gradually dropped the gesture from her repertoire (Goodall, 1973). Another new pattern was shown by infant male Flint when he inspected the genital area of a female with a small stick, instead of with his finger as is usual. He continued to stick-inspect occasionally over the next few weeks during which time another infant male imitated the behaviour. After this it was not seen again. Köhler (1925) reported a number of novel behaviours, which became 'fashions' for a few weeks or months and then vanished from the repertoire of his colony.

Behaviours of this sort, which have no obvious practical reward, are unlikely to spread through a group unless they persist long enough, in one female, for her infant to acquire it during normal learning. The female, Madam Bee, developed a unique grooming pattern, vigorously scratching the skin of her partner before intently examining the skin at that place. The appearance of this behaviour followed the polio epidemic during which Madam Bee lost the use of one arm. At that time her elder daughter was about 8 years old, the younger, one and a half years old. Subsequently the younger child was seen using the unique pattern on a number of occasions, but the older daughter was never seen to do so (Goodall, 1973).

Twice the characteristic performances of individual males were incorporated into the repertoire of younger males. When Mike was enhancing his display by the use of empty four gallon cans Figan was an adolescent: on several occasions Figan was observed as he 'practised' with an empty can discarded by Mike in the bush. Figan himself, as described, developed a highly effective arboreal display when his companions were still in their nests. Young Goblin, who had associated closely with Figan from early adolescence onwards, and had watched Figan's rise in the hierarchy, subsequently began to perform early morning displays himself when he, at the age of 14, was making a determined bid for power.

In the social sphere, innovations are often designed to get the better of higher ranking companions. And, at least sometimes, they are quickly dropped when counter-strategies are developed. Thus Emil Menzel (1974) describes how a subordinate female, Belle, who had been shown the whereabouts of hidden food, tried, in various and ever more sophisticated ways to withhold this information from the dominant male, Rock (since, if she led him to the place he invariably took all the reward). But Rock quickly learnt to see through her various subterfuges. If she sat on the food, he learnt to search underneath her. When she began sitting half way towards the food, he learnt to follow the direction of her travel until he found the right place. He even learnt to go in the opposite direction when she tried to lead him directly *away* from the food. And, since she would sometimes wait until he was not looking, Rock learned to feign disinterest, but was ready to race after her once she began to head for the goal. This sequence of events is particularly interesting since it shows how one innovation during competitive interactions in the social sphere can stimulate the rival to make a novel response. Such competition between animals as cognitively advanced as chimpanzees can thus lead to an escalation in the development of new social strategies within a group.

Conclusion

More carefully documented 'anecdotal' reports from field studies should yield a wealth of information on innovative performances in the ecological, technical, and social spheres. Of the many such behaviours observed, only a few will be passed on to other individuals, and seldom will they spread through the whole troop. However, systematic experimentation (such as the introduction of a variety of carefully designed ecological and technical 'problems' and long-term recording of reactions to them) both in free-living and captive groups would provide a new way of studying the phenomena of innovative behaviours and their transmission through and between social groups.

References

Boesch, C. and Boesch, H. (1981). Sex differences in the use of natural hammers by wild chimpanzees: A preliminary report. *Journal of Human Evolution,* **10,** 585–93.

Boesch, C. and Boesch, H. (1983). Optimisation of nut-cracking with natural hammers by wild chimpanzees. *Behaviour,* **34,** 265–86.

Bunnell, B. N., Gore, W. T., and Perkins, M. N. (1980). Performance correlates of social behaviour and organization: Social rank and reversal learning in crab-eating macaques (*M. fascicularis*). *Primates,* **21,** 376–88.

Bunnell, B. N. and Perkins, M. N. (1980). Performance correlates of social behaviour and organization: Social rank and complex problem solving in crab-eating macaques (*M. fascicularis*). *Primates,* **21,** 515–23.

Bygott, J. D. (1974). Agonistic behaviour and dominance in wild chimpanzees. Ph.D. Thesis, University of Cambridge.

Chance, M. R. A. and Mead, A. P. (1953). Social behaviour and primate evolution. *Symposia of the Society for Experimental Biology VII (Evolution),* pp. 395–439.

de Waal, F. (1982). *Chimpanzee politics: Power and sex among apes.* New York, London: Harper Row.

Gardner, R. A. and Gardner, B. T. (1969). Teaching sign language to a chimpanzee. *Science, Washington,* **165,** 664–672.

Goodall, J. van Lawick (1968). The behaviour of free-living chimpanzees of the Gombe Stream Reserve. *Animal Behavior Monographs,* **1,** 161–311.

Goodall, J. van Lawick (1971). *In the shadow of Man.* London: Collins, and Boston, Houghton-Mifflin.

Goodall, J. (1973). Cultural elements in a chimpanzee community. *Symposium of the Fourth International Congress of Primatology* Vol. **1:** *Precultural primate behaviour* (ed. E. Manzel), pp. 144–184. Basel: Karger.

Goodall, J. (1983). Population dynamics during a 15-year period in one community of free-living chimpanzees in the Gombe National Park, Tanzania. *Z. Tierpsychol,* **61,** 1–60.

Humphrey, N. K. (1976). The social function of intellect. In *Growing points in ethology* (ed. P. P. G. Bateson and R. A. Hinde), pp. 303–17. Cambridge: Cambridge University Press.

Itani, J. (1965). On the acquisition and propagation of a new food habit in the troop of Japanese monkeys at Takasakiyama. In *Japanese monkeys: a collection of translations* (ed. K. Imanishi and S. A. Altmann) pp. 52–65. Edmonton: University of Alberta Press.

Jolly, A. (1966). Lemur social behaviour and primate intelligence. *Science, Washington,* **153**, 501–506.

Kawamura, S. (1954). A new type of behaviour of the wild Japanese monkeys—an analysis of an animal culture. *Seibutsu Shinka,* **2**, 1. (Cited in Itani 1965.)

Köhler, W. (1925). *The mentality of apes.* New York: Harcourt & Brace.

Kummer, H. (1968). *Social organization of hamadryas baboons.* Chicago, IL: Aldine.

Kummer, H. (1971). *Primate societies: Group techniques of ecological adaptation.* Arlington Heights: Harlan Davidson.

Kummer, H. and Kurt, F. (1965). A comparison of social behavior in captive and wild hamadryas baboons. In *The baboon in medical research* (ed. H. Vagtborg), pp. 1–46. Austin: University of Texas Press.

McGinnis, P. R. (1973). Patterns of sexual behaviour in a community of free-living chimpanzees. Ph.D. Thesis, University of Cambridge.

McGrew, C. and Tutin, C. E. G. (1978). Evidence for a social custom in wild chimpanzees. *Man* (ns), **13**, 234–51.

Menzel, E. W. (1964). Patterns of responsiveness in chimpanzees reared through infancy under conditions of environmental restriction. *Psychologische Forschung,* **27**, 337–365.

Menzel, E. W. (1974). A group of young chimpanzees in a one-acre field. In *Behaviour of non-human primates* (ed. A. M. Schrier & F. Srollnitz), Vol. 5, pp. 83–153. New York: Academic Press.

Milton, K. (1981). Distribution patterns of tropical plant foods as an evolutionary stimulus to primate mental development. *American Anthropology,* **83**, 534–548.

Morrison, J. A. and Menzel, E. W. (1972). Adaptation of a free-ranging rhesus monkey group to division and transplantation. *Wildlife Monographs,* no. 31.

Nishida, T., Wrangham, R. W., and Goodall, J. (1984). Local differences in plant-feeding habits of chimpanzees between the Mahale Mountains and Gombe National Park, Tanzania (in press).

Ploog, D. W. (1967). The behaviour of squirrel monkeys as revealed by sociometry, cioacoustics, and brain stimulation. In *Social communication among primates* (ed. S. A. Altmann), pp. 149–84. Chicago, IL: University of Chicago Press.

Riss, D. and Goodall, J. (1977). The recent rise to the alpha-rank in a population of free-living chimpanzees. *Folia primatol,* **27**, 134–151.

Sigg, H. (1980). *Z. Tierpsychol.,* **53**, 265–302.

Schiller, P. H. (1949). Innate motor action as a basis of learning. In *Instinctive behaviour* (ed. C. H. Schiller). New York: International Universities Press.

Schiller, P. H. (1952). Innate constituents of complex responses in primates. *Psychological Review,* **59**, 177–191.

Schönholzer, L. (1958). Beobachtungen über das Trinkverhalten von Zootieren. Unpublished doctoral dissertation, University of Zurich.

Strayer, F. F. (1976). Learning and imitation as a function of social status in Macaque Monkeys (*Macaca nemestrina*). *Animal Behaviour,* **24**, 835–48.

Tutin, C. E. G. (1979). Mating patterns and reproductive strategies in a community of wild chimpanzees. *Behavioral Ecology and Sociobiology,* **6**, 29–38.

Wrangham, R. W. (1975). The behavioural ecology of chimpanzees in Gombe National Park, Tanzania. Ph.D. Thesis, University of Cambridge.

CHAPTER 11

NOVELTY IN DECEIT

RICHARD W. BYRNE

Introduction

Novelty is a hard trick to manage. If it were straightforward to explain the knack of creative innovation, there would be no sales for the dozens of courses competing in their claims to promote innovative business solutions. Moreover, for animals living in an environment that is a natural one for their species, most of the time novelty is not even likely to be beneficial. Natural selection has shaped each species' behavioural repertoire and behaviour deployment towards a state that is a local optimum within the niche exploited by the population, so that novelty may be tantamount to error and disruption. Yet the capacity to innovate is of course something that can benefit individuals of many species in the longer term. These conflicting facts present researchers with a problem. Innovative behaviour is a priori likely to be rather rare, and so the more traditional methods of collecting reliable data—for example, focal behaviour recording of natural innovations, or experiments to study innovative behaviour in the laboratory—are not likely to be useful.

One solution that has been adopted in similar situations is to canvass a number of researchers; in particular, those who have been in a position to record the unusual behaviour in question for extended periods of time (Byrne and Whiten, 1985; Whiten and Byrne, 1988b; Caro and Hauser, 1992; Lefebvre et al., 1997). The resulting corpus of records can then be examined as a form of data. This approach relies on the relatively descriptive nature of most ethological work, and the strong tradition within ethology of recording behaviour in a reliable way (Hinde, 1973; Altmann, 1974; Martin and Bateson, 1986). Laboratory researchers underestimate the care that is taken to avoid bias and subjectivity in observational data, and may dismiss such anecdotal records as unreliable when they are not (Byrne, 1997). Of course, a one-off record of behaviour, however suggestive, is of little use as data because of the ambiguity of its interpretation. The point of a survey approach is to identify *patterns* on the basis of replicated records, so that deductions can be made safely. Likewise, there are obvious biases in the behaviour patterns that will be noticed and those that will strike observers as worth recording, so that a survey should ideally be followed up by a more systematic and prospective examination. This chapter aims to begin this process for the non-human primates (hereafter, 'primates').

The social lives of primates are known to be complex, and suspected to present intellectual challenges to the individuals concerned (Jolly, 1966; Humphrey, 1976; Seyfarth et al., 1986). Also, since the 1950s many long-term studies of primates have been started, contributing to a database of intimate details of social behaviour in a large range of species. This conjunction clearly presents an opportunity to examine the distribution of innovative actions across a wide range of species, in an order of mammals known to be specialized in their relatively large brains (Martin, 1990). Sometimes researchers have indeed been so impressed with the unusual nature of primate behaviour that they have *labelled* particular actions as innovations, and the distribution of such reports has been studied (Reader and Laland, 2002).

But perhaps innovation is more widespread? Primate social manoeuvring is undeniably extensive. Thus, primates may invest time in building up affiliative relationships, using grooming in the case of Old World species, and they are sometimes quick to repair the more crucial of those relationships by reconciliation after conflicts; they use an extensive repertoire of gestures and calls to regulate interactions at a distance, and they show remarkable social knowledge in interpretation of these signals (see e.g. Cheney and Seyfarth, 1990; de Waal, 1992; Harcourt, 1992). Yet this kind of social manipulation within primate groups, important though it is in the animals' lives, is of a routine and species-typical character: none of it necessarily requires innovation (Byrne, 1996a).

One skill that may give evidence of innovation, however, is deception. Social manipulation by deception has long been known in some species of primates (Goodall, 1971; Kummer and Goodall, 1985, Chapter 10). Some cases of deceptive communication by animals are found to be species-typical (e.g. stomatopods: Adams and Caldwell, 1990; chickens: Gyger and Marler, 1988), and are considered to be adaptations. But deception in primates is often idiosyncratic, varying from individual to individual within a population, and is therefore much more likely to give clues to the ability of the species to learn novel and flexible responses to challenging circumstances. Note that this does not mean that *when observed*, a particular case of deception in an individual primate is likely to represent the innovative, initial use of a manoeuvre. Far more likely, the animal will have tried the tactic on other victims beforehand, and be using that form of deception somewhat routinely by the time this is noticed by a primatologist. Nevertheless, these idiosyncratic cases of primates deceiving social companions are very likely to reflect *past innovation*. We can, therefore, use these data to explore part of the range and nature of primate innovation.

A corpus of 253 records of tactical deception in primates has already been assembled, by survey and by literature search (Byrne and Whiten, 1990); this incorporates records collected in a previous, less systematic exercise (Whiten and Byrne, 1988b). Tactical deception was defined in purely functional terms, as 'Acts from the normal repertoire of the agent, deployed such that another individual is likely to misinterpret what the acts signify, to the advantage of the agent' (Byrne and Whiten, 1988, p. 271). Application of this definition was left to the primatologists who submitted the records, and in some cases its scope was probably stretched. (In particular, it is notoriously difficult to determine 'normal repertoire' in such flexible species, but some of the reported behaviours do not resemble actions described before for the species.) This does not diminish the usefulness of these records for examining innovation.

The distribution of deception across species was found to be non-random, even when a correction for observer effort was introduced (Byrne and Whiten, 1992). Thus, while it was certainly true that well-studied species like baboons, macaques and chimpanzees contributed disproportionately, the variation could not be wholly accounted for in terms of the number of long-term field studies that were capable of recording such subtle, close range behaviour in the species concerned. Rather, it seems that differences in brain structures affect the rate of use of deception, and consequently the rate of reporting. Byrne (1993, 1996b) showed that neocortical ratio is a good predictor of the frequency of deception, once the number of studies potentially recording it was taken into account. Although neo-cortical ratio scales with body size, and so is a less-than-ideal measure of a species' invest-ment in neocortex, it has been found to be the most sensitive of various brain size measures for identifying causes of variation in traits across primate species (Dunbar, 1992, 1995). Note that the widely used alternative of scaling against body size is inappropriate for exam-ining cognitive abilities: while it gives a good indication of the selection pressure that a species has experienced towards brain enlargement, it is unlikely to index the cognitive potential of the resulting brain structure (Byrne, 1996b). Indeed, some differences in rela-tive brain/body size may equally well be understood as selection acting primarily on body size (Shea, 1983; Deacon, 1990; though see Deaner and Nunn, 1999).

Most of the tactics were found to function by directing the focus of attention of another individual towards or away from a particular place or subject (Whiten and Byrne, 1988a), and the great majority could be understood as the result of associative learning (Byrne and Whiten, 1990; Byrne, 1997). This conclusion was based on consideration of the plausibility of the necessary circumstances, which could have allowed a particular individual primate to associate the behavioural tactic with effective acquisition of rewards, having occurred in its past life. In a few cases, however, the circumstances that would have to be envisaged were *so* implausible as to force serious consideration of alternative possibilities (Byrne and Whiten, 1991). There were 18 records of this type, and they were not distributed at random, nor did their distribution match that of the overall phylogenetic pattern of primate deception (Byrne and Whiten, 1992). Instead, the records hardest to explain as a result of associative learning were tightly clumped within the great apes, and spanned all species (orang-utan, gorilla, bonobo, chimpanzee). Byrne and Whiten (1992), therefore, suggested that it was most likely that the great apes had some ability to employ 'theory of mind' or 'mentalizing' abilities, so that they could thereby sometimes construct or anticipate deception. (More recently, Mitchell has attempted to account for even this class of great ape deception without invoking mentalizing, as 'strategic script violation': see Mitchell, 1999.) Great ape com-petence may not go as far as deliberately creating false beliefs; rather, it might involve an intention to create or prolong ignorance, or an understanding that other individuals' behaviour might be incorrectly aligned with reality. These early proposals seem to have been confirmed by more recent experimental studies (Boysen, 1998; Call and Tomasello, 1998; Hare *et al.*, 2001), although the history of claims and refutations in experimental testing of primate theory of mind suggests that caution is still advisable.

In this chapter, I focus on the *nature* of deceptive tactics, treating them as potential evid-ence of innovation in the social lives of individual primates. What aspects of primate behaviour are amenable to modification in the service of attaining goals by deception in a social context? Do patterns recur across species, or are there taxonomic variations in what aspects of behaviour are flexible in this way? And in what ways were the deceptive tactics 'innovative'—did they employ entirely novel actions, familiar actions in a novel context, or merely generalize actions slightly from their normal circumstances of use?

The approach taken was to sort all the 253 records of the published corpus into broad categories, on the basis of what feature of the usage was 'unusual'. Evidence from the rather few studies of deception in primates published since this corpus was collected (e.g. Mitchell, 1991) was also examined. Some features were relatively easy to describe and therefore to use in categorization: for instance, where a signal was used at an inappropriate intensity, or in an inappropriate context, or suppressed entirely in a context where signalling would have been appropriate. Other features were less clear-cut, and their use in categorization more difficult: for instance, sometimes a whole suite of signals was deployed inappropriately as if to create a single, misleading impression, and sometimes an apparently novel signal was invented for the purpose, an act not known to be in the species or individual repertoire. These cases mostly had to be examined on a case by case basis. The distribution across species of the records is shown in Table 11.1; all numbering, in both table and text, refers to the published corpus (Byrne and Whiten, 1990).

Exaggerate a signal

It has long been anticipated that animals might sometimes be exaggerating their signals (Dawkins and Krebs, 1978, 1979), especially in the domain of parent–infant conflict (Trivers, 1972, 1974), so it is no surprise that some records of deception were of this kind. As a typical case, David Chivers refers to 'the distress calls given by an infant siamang *Hylobates syndactylus* to the adult male, when it has been left by the male to travel on its own.... During the second year of life, once the male has taken over carriage of the infant when it is weaned from the female, the male progressively encourages the infant to travel after him. Often the infant protests at this, with squeals. It is deception to the extent that the infant is not really in distress, but is hoping for the easy way out—for the male to retrieve and carry him' (#150 in Byrne and Whiten, 1990). However, the difficulty of ever being sure about this last point—whether the infant was really in distress—most likely deterred many informants from even mentioning such cases.

The same caveat applies to the tactical use of tantrums, which may often be dismissed by researchers as functionless failures of normal control. However, Jane Goodall notes that a tantrum

> is observed most frequently in [chimpanzee *Pan troglodytes*] youngsters who are going through the peak of weaning, after they have begged, whimpered, and cajoled their mothers for an opportunity to suckle, but to no avail. In some ways it seems absurd to think that such a spontaneous outbreak could be a deliberate strategy for achieving a goal. Yet Yerkes (1943, p. 30) wrote, 'I have seen a youngster, in the midst of a tantrum, glance furtively at its mother ... as if to discover whether its action was attracting attention.' And de Waal (1982, p. 108) says, 'It is surprising (and suspicious) how abruptly [chimpanzee] children snap out of their tantrums if their mothers give in.' At Gombe a mother almost always does give in (#227, from Goodall, 1986, p. 576).

Temper tantrums are also reported in adult primates apparently functioning in the same way (e.g. #229).

Exaggeration was also reported in mating (e.g. a female exaggerating her copulation grunts to provoke interest from other males, #76), and in conflict (e.g. excessive screams in response

Table 11.1 A classification of means of deception used by primates. Numbers refer to the catalogue of records, arranged by species and by category, published as Byrne and Whiten (1990). Since some records afforded evidence of several forms of novel usage, a record may occur in more than one category

Classification	Strepsirhine	Callitrichid	Cebid	Cercopithecoid	Colobine	Hylobatid	Pongid	Hominid
Exaggerate				76. 99. 106.		150.		227. 229. 238.
Ignore signal				46. 64. 66.				207.
Modify own/other body				?28. 46.				205. 230.
Suppress signs of intentions			23. 25.	32. 40. 49. 53. 56. 63. 65. 71. 81. 83. 85. 86. 102.			154. 156.	160. 167. 169. 170. 177. 178. 179. 182. 183. 184. 185. 201. 206. 208. 209. 210. 211. 212. 213. 214. 215. 216. 217. 244. 252. 253.
Suppress signal			23. 24.	28. 54. 111. 123.	145.			161. 166. 168. 190. 191. 192. 193. 194. 195. 196. 197. 198. 199.
Hide (part of) body, or object in hand				29. 30. 31. 49. 55. 56. 57. 58. 59. 60. 61. 62. 112. 125.	135. 136. 137. 146.		155.	161. 162. 163. 164. 165. 166. 167. 177. 179. 180. 181. 182. 184. 200. 202. 203. 204. 205. 206. 250.

(Continued)

Table 11.1 (Continued)

Classification	Strepsirhine	Callitrichid	Cebid	Cercopithecoid	Colobine	Hylobatid	Pongid	Hominid
Show affiliation or neutral behaviour (for access)			20.	35. 41. 50. 51. 70. 82. 84. 87. 88. 89. 107. 110. 131. 132.	138. 139.			171. 172. 188. 233. 239. 240. 241. 242. 243. 245. 246.
Display or call in new context (e.g. at nothing, or at neutral stimulus)	5.	13. 14.	16. 17. 18. 19.	33. 34. 36. 37. 38. 39. 42. 43. 44. 45. 52. 67. 68. 72. 73. 74. 75. 77. 78. 90. 91. 92. 93. 94. 102. 103. 104. 105. 108. 113. 114. 115. 116. 126. 127. 128. 129. 130.		152	157. 158.	186. 218. 219. 220. 221. 223. 224. 235. 247. 249. 248. 251.
Lead others to/from	6.			69.	140.			189. 225. 226. 252.
Use 'displaced' actions								173. 228. 234. 236. 236.
Invent actions				80. 95. 124.		151.		160. 174. 224. 231. 252. 253.
Not categorized			21.	96. 97. 98. 100. 101. 109. 134.				187. 222. 235.

to minor attack, to solicit third-party aid, #106). Whether the behaviour in these cases was really exaggerated is a matter of judgement by skilled observers rather than verifiable fact, but sometimes exaggeration can be checked. De Waal, describes how a chimpanzee's pitiful hobbling was only evident when the original aggressor was present (#238).

Ignore a signal

As with exaggeration, the observational difficulties of distinguishing deliberate ignoring from accidental failure to notice mitigate against detection. Researchers are nevertheless convinced it occurs: '[Baboon *Papio cynocephalus*] mothers with weanlings will often "ignore" (avoid visual contact with) their milk-begging offspring. Ignoring as a social strategy...may be the most common form of deceptive behavior' (#64). The social tactic of ignoring may only become obvious to the scientist when it fails, as described by Fernando Colmenares, '...if the [baboon] female refuses to look in the male's eyes. The male may then move around the female trying to meet her eyes which she keeps away' (#66; see also #46).

Given the likelihood of underreporting, both this and the previous category of deception may not reflect idiosyncratic, learned tactics, but rather species-wide adaptations. Neither can usefully be discussed as evidence of innovation.

Modify body

When humans are frustrated by another's refusing to confront some sight, or deliberately ignoring their signals, they may physically force the issue. The same has been noted in primates. Jane Goodall describes a chimpanzees' tactic of deflecting unwanted attention to her infant, 'Twice she [the adult female Pom] reached out and touched his [her immature brother Prof's] face; once he responded and turned to groom mutually, but the second time he ignored the signal. Pom then put her hand under his chin—he was bent diligently over Pax—and forced his face up until their eyes met; he then groomed her' (#230, Goodall, 1986, pp. 571–2). Similarly, Suzanne Chevalier-Skolnikoff noted a stump-tailed macaque *Macaca arctoides* that 'looks directly at [her daughter, who had ignored her mother's signals], grasps her chin and turns her head towards[her]' (#46). These cases suggest that, in some sense, the primates may understand the connection between line-of-sight and efficient communication; the innovation lies in using a routine behaviour for an unusual purpose.

Other modifications appear to function by preventing anticipated behaviour. Kim Bauers describes a stump-tail macaque, engaged in mating a low rank male out of sight of dominant individuals, who 'suddenly turned and stared into his face, grasped his upper arm (still looking into his face), then covered his mouth with her hand briefly. The latter behaviour was most striking, as it was my strong impression that she had been afraid he would begin vocalising, and thus reveal their situation' (#28). The suggestion is that she understood the male's mental state, but nothing like that has been reported in any other monkey species. Since we cannot rule out the possibility of sheer accident—that is, this particular monkey might have just pushed at the male and hit his mouth—it would be unwise to hang theory upon this isolated case (however, see record #205, discussed under 'Hide from view', below).

Suppress signs of true intentions

Tactical *inhibition* of natural reactions is employed in more general circumstances than bodily modification. A common form of deception involves an unexpected lack of interest in some environmental object, with the frequent result that other individuals fail to discover the object. Marina Cords noted a long-tailed macaque *Macaca fascicularis*, who 'deliberately "hides" the fact that she is moving toward the peanut (1) by waiting a bit before moving once she is aware of the nut, (2) by climbing slowly and nonchalantly toward it, and (3) by stopping to eat a piece of carrot first' (#32). This tactic appears commonly among records of deception (e.g. #102, 210, 212, 213, 215, 216, 253). Suppression of interest may be quite prolonged, as noted by Birute Galdikas with an ex-captive orang-utan *Pongo pygmaeus*, restrained from drinking a medicine, who for half an hour 'just sat there and never once looked down at the medicine' (#156), yet immediately seized it when the researcher moved off. Deceptive inhibition has been noted in other contexts: sexual interest, in the context of dominant competitors (#65), sometimes prolonged over many hours (#208); aggressive intent, until a close approach is secured (e.g. #40, 81, 83, 85, 244); playful intent, by juveniles attempting to gain access to young infants (e.g. #214; Mitchell, 1991). In the groups from which these records were reported, not every individual is apparently able to suppress their interest, but whether this means the tactics are innovations, that is, novel, idiosyncratic tactics learned by experience, or that some individuals are just too impulsive, is not certain.

An 'arms-race' between two young chimpanzees, one of whom has been given privileged information about hidden food, was described by Emil Menzel (1974 and #252). The informed individual successively slowed her approach to the food, refrained from uncovering it, stopped going all the way to it, and eventually went in the opposite direction to the food—only racing for the food when the competitor was distracted elsewhere. It is not clear from the published descriptions over what timescale these changes took place. Menzel's experiment of giving one individual in a social group privileged information has been adapted for white-collared mangabeys *Cercocebus torquatus* (Coussi-Korbel, 1994) and domestic pigs, *Sus scrofa* (Held *et al.*, 2000, 2002), with very similar results. In the case of the pigs, for instance, informed individuals were found to take account of the several aspects of behaviour of the dominant competitor pig: its position and direction of movement relative to themselves, and relative to the food, its current direction of movement, and whether it was in potential line-of-sight contact with the site of the food. Each of these factors significantly affected the informed pig's likelihood of slowing approach, hanging back, or moving away from the hidden food. It is not difficult to imagine how these reactions might be conditioned by their effects on a dominant competitor into innovative tactics, which may function to deceive (as was found in chimpanzees and mangabeys). But that did not occur with pigs, where the competitor's behaviour was not significantly influenced, so conditioning by effect cannot be a general explanation of the origin of these (potentially useful) reactions. It is perhaps most likely that they reflect biological adaptations in socially foraging species.

Inhibition of interest can also involve innovative modification of actions, as in an adolescent female chimpanzee, Zwart, described by de Waal: 'she quietly moved in the general direction of the food and sat down a few meters away from it. She first looked around, with only a brief glance at the apple in the grass, before she moved closer. She sat down with one hand casually near the apple. Without ever looking down, blindly feeling around, she carefully closed her hand once she had reached the apple. Zwart then walked away, using the hand in which the apple was hidden just as if she were walking normally (which is unusual: chimpanzees usually walk on two or three extremities when carrying food)' (#209; see also #179, 184 and 211 for instances of objects held cryptically in foot or hand). The ability to seamlessly modify walking to suppress signs of holding an object has also been seen in a captive mangabey monkey (#49). Jane Goodall describes a male chimpanzee that had learnt a special skill of unscrewing a locking device, who 'quickly learned to modify his behaviour when adult males were in camp. Very nonchalantly, and with an apparent lack of purpose, he wandered to a handle. There he sat and performed the entire unscrewing operation with one hand, never so much as glancing at what he was doing' (#217, and Goodall, 1986, p. 578).

Locomotion may also be modified to conceal intentions. Howler monkeys *Alouatta palliata* aggressively contest group boundaries, and Jay Whitehead described two males that gained the advantage of surprise when they 'moved in an uncharacteristic manner that caused a minimum of noise from shaking branches' (#23; a rather similar use of silent approach in male-male agonism was noted in orang-utans, #154). Visual concealment can also be effected by slowness of movement, as Hans Kummer described in a female baboon *Papio hamadryas* that 'spent 20 min in gradually shifting in a seated position over a distance of about 2 m to a place behind a rock about 50 cm high where she began to groom the subadult male follower of the unit...The only aspect that made me doubt that the arrangement was accidental was the exceptionally slow, inch by inch shifting' (#56; see also #53 and #63). A Western lowland gorilla *Gorilla (g.) gorilla* in a zoo showed unusual as well as slow actions when, in order to attack a rival, he 'changed the nature of his approach, putting his feet down carefully, with his whole body adopting a tense demeanour. He gave every appearance of creeping up' (#160). It is tempting to attribute this innovative mode of movement to imaginative planning, but presumably young gorillas have extensive opportunities to learn the consequences of moving in different ways during social interactions, and the same may apply to a male spider monkey *Ateles geoffroyi*, noted by Katy Milton, that came to the ground when it followed the female it was consorting, evidently to avoid drawing the attention of competitor males (#25).

Although lumped as 'inhibition', the records discussed in this category therefore seem to mix cases where animals simply refrain from actions, an ability potentially explicable as biological adaptation, with innovative modifications of movement and actions employed in unusual and innovative ways.

Hide from view

Rather than hiding signs of interest while remaining in full view, a primate may also actively hide itself, or part of itself, or a detached object. In some cases, these actions suggest an understanding of line-of-sight visual perspective, as Nelly Menard's description of timid, unhabituated Barbary macaques *Macaca sylvanus* makes clear: 'When, without their knowing they suddenly found themselves too close to the observer, they initially had a barely perceptible reaction of surprise, then continued to walk in an apparently unconcerned manner in the original direction. They would then pass behind a tree but did not re-appear on the other side. After several minutes I realised that they had disappeared. I found them sometimes 25 to 30 metres away in a direction given by a line between three points: observer, tree, and macaque' (#29). Realizing that repeated cases could not be a coincidence, she made efforts to see what went on behind the tree, 'When it had got behind the trunk, it changed course abruptly, taking a direction which would purposefully keep it out of my field of view.' Such evidence of understanding *visual perspective* was categorized separately from evidence of mental perspective (i.e. theory of mind) in primate tactical deception, and found to be distributed widely across monkeys and apes (Byrne and Whiten, 1990, 1992).

More commonly, what is reported is that primate individuals seek places when they are not in sight of competitors, and there behave differently. Grey-Eaton's description of Japanese monkey *Macaca fuscata* behaviour is typical, 'A female will position herself behind a stump, and only solicit or allow a consort male to mount after she has scanned the surrounding area for other higher-ranking males' (#30; see also #31, 55, 57–59, 112, 135, 137, 146, 161–167, 180, 200, 250). The length to which some primates will go to hide is remarkable, as in Jim Moore's account of Hanuman langurs *Presbytis entellus* that insinuate themselves so deeply into spiny *Euphorbia* clumps that copulation takes 10 min instead of the usual fraction of a minute (#135). De Waal suggests that the effects of concealment are sometimes understood by the primates themselves, noting that occasionally a third party may draw attention to a secret consortship, 'The clearest instance was Dandy, witnessing a secret contact involving an oestrus female, Spin, with whom he had a very close friendship. Loudly barking, Dandy ran to the alpha male, who was far away, unaware of the contact, and led him to the scene where the two others were in the middle of mating. Such instances of "telling tales" are indicative of the chimpanzee's understanding of the concept of secrecy' (#250, de Waal, 1986).

Many species show understanding of the effect of visual concealment of objects. Colmenares description of a baboon is representative, 'When a subadult male happens to pick up a piece of food which, if discovered, would be taken by a dominant individual, the former may still be able to eat it by concealing it beneath a hand or foot' (#61, see also #60, 62, 125, 155). The concealment may be subtle, as in the captive bonobo *Pan paniscus* Kanzi's efforts to eat forbidden wild mushrooms, 'Often he will secretively conceal them between his thumb and forefinger as he walks, never breaking stride so that it is not obvious that he has picked a mushroom' (#181, see also #177, 182, 206). Although it seems that Kanzi had had practice, it may also be that bonobos are better able to suppress leakage of intent-betraying signals than the captive baboon that looked so conspicuously innocent that [the keeper] became suspicious (#86). Mitchell (1991, p. 525) reports several cases of captive western lowland gorillas 'minimizing body movement so as not to draw attention to [their] actions'. De Waal has reported cases of chimpanzees hiding parts of their bodies that might other-wise reveal their intentions; for instance, when a subordinate male with a prominent erection was suddenly confronted by an older male he 'immediately dropped his hands over his penis concealing it from view' (#203, and see also #202, 204; de Waal, 1982).

Great apes even seem to understand the effects of their own facial expressions, and sometimes pre-empt unwanted reactions. De Waal describes a male chimpanzee, showing a fear grimace when hearing a rival's call nearby, that 'quickly used his fingers to push his lips back over his teeth again. The manipulation occurred three times before the grin ceased to appear. After that, the male turned around to bluff back at his rival' (#205, de Waal, 1982). Similarly, Tanner and Byrne (1993) describe a young western lowland gorilla that was observed 'repeatedly to hide or inhibit her playface by placing one or both hands over the face,' thereby enabling her to approach close enough to launch playful attacks on her partner. Clearly these manipulations were intentional, and suggest that the apes had somehow comprehended the likely effects of their facial expressions, and dealt with that in an innovative way.

Most cases of primate hiding may reflect species-typical adaptations, although they are certainly valuable indications of visual perspective-taking abilities in primate cognition. However, some cases of great apes using routine behaviours to achieve concealment are certainly idiosyncratic, and seem to reflect innovative deployment of normal behaviours for unusual purposes.

Suppress signal

Vocal inhibition is frequently reported in primates as a means of deception (e.g. #23, already discussed above; see also #24, 123). As with any data that depend on the assertion that a behaviour 'normally occurs' or 'should be done', accepting that inhibition has occurred often requires us to rely on the assurance of experienced observers that the absence is unusual. However, some cases are very clear: for instance, when many species of monkey and ape mate, the emission of a distinctive copulation call is so reliable that its absence is striking—especially when the result is achievement of furtive mating, without discovery by dominant competitors. Robin Dunbar describes this in gelada *Theropithecus gelada*, where 'the animals do not give the shrill (in the case of the female) and loud (in the case of the male) post-copulation calls that are very characteristic of this species. Copulations in these circumstances are totally silent' (#111; see also #28, 54, 190). Monkeys regularly give loud, mobbing calls (often referred to as 'alarm calls') when they detect stealth predators like lions or leopards; the calls presumably impede effective hunting, in reaction to pursuit predators, silence is the normal reaction (e.g. when Guinea baboons *Papio papio* are confronted by African hunting dogs *Lycaon pictus*, Byrne, 1981). This might be a relatively hard-wired system, but Christophe Boesch describes colobus monkey *Colobus badius* response to chimpanzee pursuit, where 'only colobus far from the chimps will give alarm calls and those close to them remain totally silent' (#145), suggesting some flexibility. Flexibility in vocal inhibition has been studied experimentally in rhesus monkeys *Macaca mulatta* (Hauser, 1992; Hauser and Marler, 1993). The costs and benefits of food calling vs suppression were found to affect the monkeys' decisions in predictable ways.

In great apes, there is some evidence that voluntary control of vocalization must be learnt. In the mountain gorilla, suppression of copulation calls is often combined with hiding to effect secret matings. (Unlike western lowland gorillas, the mountain gorilla *G. (g.) beringei* lives commonly in multi-male groups, affording opportunities for female choice of mating partner.) A series of my own observations (#161–168) show that most mating between receptive females and subordinate adult males is carried out when the couple are out of sight of the dominant male; females will solicit males to accompany them out of sight; and in most cases the normal loud copulation calls are suppressed by both individuals. However, subadult males were generally unable to suppress these calls, and in one case a male who at one time could not keep silent, a few months later showed that he had learned the ability to suppress his copulation calls. Thus, on 26 August 1989, the young silver-back Shinda was several times with old female Flossie well behind the main group, 'On each occasion Shinda mates Flossie, not silently but with only quiet calls at the end of the copulation. We have the strong impression from his facial expression that Shinda is attempting (unsuccessfully) to inhibit his calling' (#163). However, by 14 November 'Shinda mates Kwiruka (for 2 min) in total silence: I was standing next to the two animals. This is the first time in which we have seen Shinda able to keep completely silent during a copulation. On a second occasion he again mated her (1.5 min) and was able to inhibit his copulation calls until the last 15 seconds, and when he did finally give copulation grunts' (#164). Similarly, Goodall describes a 9-year-old chimpanzee whose excited food barks caused others to take his food, but the following day in a similar circumstance, 'although we could hear faint, choking sounds in his throat, he remained virtually silent and ate his allotment undisturbed' (#195).

Fortunately, there is one circumstance that encourages vocal suppression sufficiently often to begin to make generalizations; this is when chimpanzees are involved in hunting and intergroup violence. Here, against the usual generalization, deception is not a rarity but the norm. Again, apes show *difficulty* in suppressing sounds. Goodall notes that 'sounds are suppressed during some hunts, although at other times a hunt is preceded by squeaks and even pant-hoots' (#191, Goodall, 1986, p. 137), and Boesch similarly reports that '[in some hunts] food calls and high excitement screams are emitted when seeing the prey; these immediately inform the prey of their intention and in some cases clearly negatively affect the hunt outcome' (#196). The problem of accidental discovery is far more acute when patrolling in areas of range shared with other communities, where error may be a matter of death rather than going hungry. Here, the efforts of chimpanzees to remain silent are sometimes remarkable. While hunting within such an area, Boesch records the reaction of a subordinate male whose prized meat was stolen from him, '[he] screamed with frustration for one minute with all the facial expression and movement that belong to such a situation; but, only the air expelled was audible, no sound was emitted' (#192). For animals commonly famed for noisiness, 'perhaps the most striking aspect of patrolling behavior is the silence of those taking part. They avoid treading on dry leaves and rustling the vegetation. On one occasion vocal silence was maintained for more than three hours. A male may perform a charging display during which he drums on a tree-trunk, but he does not utter pant-hoots. Copulation calls are suppressed by females, and if a youngster inadvertently makes a sound, he or she may be reprimanded.' (#197, Goodall, 1986, p. 490). This last feature, reprimanding careless individuals, may serve to teach appropriate behaviour. 'On two separate occasions the adolescent Goblin vocalized during patrols. Once, he was hit; the second time, embraced. During another period of noiseless travel an infant got loud hic-coughs; her mother became extremely agitated, repeatedly embracing the child until eventu-ally the sounds ceased. A human observer who is too noisy on such occasions may be threatened' (#198, Goodall, 1986, p. 579; males in secretive consort will also direct threats at females who vocalize loudly, see #194). Comfort is also used to silence a noisy youngster, 'on three occasions when he did scream, he was almost immediately embraced or touched by older males' (#199). It is clear that the *need* for silence can be appreciated, at least by adult chimpanzees, whereas the *means* of attaining silence does not come naturally to young chimpanzees but requires prolonged learning. Encouragement and punishment are used by adult chimpanzees in these circumstances to shape their youngsters' behaviour and acquire more appropriate actions: it is unclear why such behaviour has not been assessed as a candidate for active teaching (e.g. Caro and Hauser, 1992).

As with deceptive hiding, signal suppression may often reflect adaptation rather than innovative learning. Only in great apes is there clear evidence that learning is involved in acquiring innovations that require suppression of normal signalling and other activities. How easy this would be to learn varies between cases. Disadvantageous effects of vocalization may often be readily learnt by trial and error, for example, when mating close to a possessive and dominant individual; but during chimpanzee border patrols trial and error is not plausible (because mistakes would be lethal), and the use of threat and comfort shows remarkable ability to foresee future consequences (Byrne, 1998).

Distract by affiliation or other neutral activity

One of the commonest patterns of deception is where an individual directs affiliative or neutral behaviour at another, causing distraction just until its true goal can be attained. For the ethologist, the deceptive nature of the actions only becomes apparent afterwards, when the primate switches its attention to what is then presumed to have been its real goal. Shirley Strum describes such a tactic in olive baboons *Papio anubis*, where a female was not permitted to share meat by a male 'The female edged up to him and groomed him until he lolled back under her attentions. She then snatched the antelope carcass and ran' (#70, see also #20, 35, 41, 50, 51, 82, 84, 87, 110, 131, 132, 138, 139, 171, 172, 188, 233, 237, 239, 242, 243, 246; and Mitchell, 1991). In another case, again reported by Strum, a female in the same predicament 'left off grooming, and chased after [the male's] favourite female, attacking seemingly without provocation. He dithered, glancing between the aggression involving his friend and the carcass, but at length went to his friend's defence. The first female promptly doubled back to snatch the antelope' (#107). In some examples, the apparent purpose of affiliative or otherwise distracting actions is to launch an attack from close-range rather than to snatch food; de Waal (#237; 1986) records six cases of 'false reconciliation' by female chimpanzees, noting that this was rare compared with genuine reconciliations. Presumably the victims, whether of theft or violence, have no idea of the risk until it is too late. In contrast, the commonly observed behaviour of young primates towards the mothers of new babies is more ambiguous, as noted by Stuart Altmann of yellow baboons *P. cynocephalus*, 'Every day, one can see females approach mothers, pretend to be primarily interested in grooming the mother when what they are really after is an opportunity to sniff, touch, or even hold her infant. (As a result, mothers with small babies are often very well groomed.) But is the mother really deceived? Surely the multiparous ones know exactly what's going on!' (#88, see also #89, 240, 241, and see also #245). It may make more sense to view the operation of such tactics as more a matter of 'trade' than deception.

In all cases, circumstances are readily apparent that may serve to condition these behaviours as tactics. All that is necessary to suppose is that a primate agent, thwarted of its goal, engages in some unrelated activity, perhaps a lower priority goal. If this activity in fact distracts the problem individual then the agent will automatically be able to switch back to its original goal. The sequence of events, terminating in a reward, will act to condition the use of the unrelated activity as a tactic in future circumstances of the same kind. If so, these behaviours might then be genuine innovations for the agents involved. However, the patterns are sufficiently common for the alternative—that such distracting tactics are simply part of the species' adaptive social repertoire—to be hard to exclude.

Use vocal or visual display in novel context

One of the commoner tactics of primate deception, apparent misuse of a signal for other ends, is illustrated by Marc Hauser's account of a vervet *Cercopithecus aethiops*, which after making several attempts to copulate with a female, Borgia, was attacked by a coalition of the female and her relatives, 'After a few minutes of chasing, Tristan stopped and gave a "leopard alarm call" (confirmed by subsequent acoustic analyses). Borgia, all of her relatives, and the rest of the group quickly ran up into the tree, even though Tristan remained. My assistant and I searched the entire area for a leopard but did not find one. Moreover, no one else in the group gave an alarm call, including Tristan' (#128; similar misuse of alarm calls in #13, 16, 37, 33, 34, 126, 127, 129, 186, 218–220). A visual display may be used in the same way, as in a baboon *Papio ursinus*, pursued by attackers, which 'stands on hindlegs and stares into the distance across the valley. HL and the other newcomers stop and look in this direction; they do not threaten or attack ME. No predator or baboon troop can be seen through 10 × 40 binoculars' (#67; see also #130, 157, 221, 223, 224, 247).

In populations where predation is rare, the same ends are apparently achieved most commonly by threatening an uninvolved third party. This may often be referred to as redirected aggression, but it may serve a tactical purpose, as illustrated by a chacma baboon noted by Andrew Whiten, 'Adult male JG displaces female PK from her feeding patch. She responds by immediately enlisting JG, by characteristic rapid flicks of gaze from potential ally to target, to attack juvenile PA. PA (who is not PK's offspring) was merely feeding 2 m away, but is threatened and chased by JG. PK meanwhile continues feeding on her patch' (#93; see also #17, 43, 44, 72, 75, 90–94, 102–105, 113). How this tactic might have been learned associatively, as a result of readily imagined past circumstances, has been discussed by Byrne (1997). The victim of redirection is often the human observer (e.g. #18, 36, 42, 45, 68, 108, 152). Toshida Nishida describes a case where a wild chimpanzee embroiled him, 'Katabi came towards me, and began to screech/scream loudly, reaching his hand towards me (as if pointing). Then, he went round me repeatedly emitting screech/screaming loudly, while still reaching a hand to me. [His mother] Chausiku and her consort Kamemanfu at once glanced at me, with hair erect. I retreated a little bit away from Katabi, avoiding the possible attack from Chausiku or Kamemanfu or both. Undoubtedly both Kamemanfu and Chausiku misinterpreted that KB had been attacked or teased by me' (#251).

Whether these tactics always function by misleading others about a non-existent threat is doubtful; Dennis Rasmussen notes that in baboons 'stylized threatening away seemed to be a means of male-male alliance formation' (#90), and Fernando Colmenares concurs, 'Threatening a third individual...seems to be an excuse to re-initiate contact that often ends in greeting behaviour' (#91). This interpretation is strongly supported by the otherwise puzzling use of threats towards empty space, or an inanimate object, in much the same circumstances as threats are directed at other primates or the human observer. Dorothy Fragaszy describes this in capuchin monkeys *Cebus capucinus*, where an adult male intervenes in juvenile rough-and-tumble play, which had become violent, 'He looked toward the floor of the cage (about 3 ft below them) and threatened in that direction. The juveniles joined him in the threat. Nothing had moved on the floor in the threatened space; no animals had passed by there. It was an imaginary object; equally imaginary to all participants' (#19, see also #14, 38, 39, 52, 73, 74). In other cases, the threat is not only real, but incurred as a consequence of the apparent victim's behaviour. Robin Dunbar describes infant gelada *T. gelada*, refused suckling by its mother, deliberately approaching, hitting, and pulling the cape of an adult male, eventually incurring a mild attack; the result was that the mother retrieved the infant, and it was able to suckle (#114, see also #115, 116).

An unusual opportunity for misuse of a call, normally used exclusively in response to lost calls of group members, was noted in white sifakas *Propithecus verreauxi* by Alison Richard (#5). In the mating season, oestrus females give the same call in response to distant calls of solitary males, thereby 'summoning' potential mating partners.

Captivity presents much greater opportunities for primates to learn novel uses of signals, as in the Guinea baboon kept by Pfeiffer, that carried out the actions of defecating in order to avoid moving, and used calls at night to summon its owner (#77, 78), the orang-utan kept by Miles that used an ASL sign to gain access to an area for a forbidden purpose (#158), and the bonobo that solicited Savage-Rumbaugh's support against entirely innocent human carers (#249). In each case, associative learning from past, coincidental conjunctions of circumstances is plausible as an ontogenetic explanation, and the actions used tactically are normal ones in the species repertoire. Where captive apes have been taught artificial communication systems, the possibility of lying with them is introduced. Temerlin (1975, record #248) describes how the chimpanzee, Lucy, when asked who had made a mess, gave the name of a human carer when in fact it was her own. Whether or not she lied in the fully intentional sense, the means she employed was to use a familiar signal in an inappropriate context.

Although often it is possible to understand how innovative uses of visual or vocal displays may have been learnt, the facility with which primates acquire these tactics hints at rapid ability to learn in social circumstances. Monkeys and apes are known to learn more rapidly than other mammalian groups even under laboratory conditions (Passingham, 1981), and a difference in learning rate appears to be the most likely explanation for the lack of similar kinds of deception in many other highly social mammals. In captivity, where learning is made much easier by the predictable behaviour of humans, dogs and cats readily acquire innovative and deceptive ways of using their normal signals (Byrne, 1997).

Lead others

In deceptive leading, a primate combines a pattern of signals or bodily cues with its own movement in such a way as to cause others to follow—but in retrospect, it appears that the message conveyed was inappropriate. For example, Alison Richard has seen two oestrus female white sifakas precipitate conflict between males of their neighbouring groups: 'the two females repeatedly dashed at one another, still followed by their males, veering away only at the last moment. The follower males were thereby left face-to-face a few feet apart. For two-and-a-half hours the females mock-charged each other over and over again, and each time the males broke away and resumed their follower stations. Finally, however, the males did engage one another in what turned into a two-hour sequence of reciprocal chasing. During that time the females fed and rested' (#6; see #140 for a similar case in Hanuman langurs). Where the aim is to gain access to food by first leading dominant competitors away, it seems that a purposeful initiation of movement is sufficient to instigate following (e.g. chacma baboons, #69; chimpanzees, #225–6, 252, 189; in the last of these, ASL signs are used to support the interpretation). In all such cases, the possibility is very apparent that the useful effects of moving away were originally learnt by chance. In the informed forager experiments discussed already, chimpanzees, mangabeys, and pigs sometimes moved away from a food item they knew about, specifically when the dominant was close to it (Menzel, 1974; Coussi-Korbel, 1994; Held *et al.*, 2000). It is obvious how this response could be conditioned as a tactic for future use, and in chimpanzees and mangabeys that was observed to occur; the use of the behaviour is innovative, whereas the behaviour itself is part of the normal repertoire.

Use 'displaced' actions

Actions that might traditionally be termed displacement activity may sometimes be used tactically, as a distraction. This too may result from originally coincidental conjunctions of action and reward. Such tactical use of inappropriate actions has been frequently reported in chimpanzees: a mother may groom leaves and thus attract her infant to be groomed (#228), or she may tickle her infant to encourage it to follow (#234); during aggression, males may show great interest in trivial or non-existent objects, rather than pay attention to their opponents' displays (#236). In some cases, 'feigned attention' may also serve to break down tension and facilitate reconciliation, and de Waal notes that 'It is tempting to regard this face-saving tactic as a collective lie, that is one party deceiving, and the other acting as if deceived' (de Waal, 1986). Mountain gorillas, also, have been noted to use displacement activities in a tactical way, as in the juvenile that built a series of day-nests, each nearly next to the other, until the last nest allowed it to groom a new-born infant (#173). As with several categories of deception, it is only the tactical *usage* of displacement activities that is innovative: the behaviour itself is not new.

Use actions not expected for species

In this final category, we examine cases of tactical deception in which a primate seems genuinely to have 'invented' a new action. (This cautious phrasing reflects the fact that any claim of behavioural novelty is difficult to substantiate, because it relies on negative evidence.) I have divided this heterogeneous category into cases where the observer reporting the record apparently expected—a priori—that (a) the primate *would* not do what was seen, although the physical possibility was clear, or (b) the primate *could* not do what was seen, the action had not been thought part of the available repertoire. This division is perhaps an artificial one.

Actions considered highly improbable in context

Fernando Colmenares reports a baboon, PAY, that 'held her daughter Y by the tail and stood by her leader male NR. PAY was relatively far from NR but Y approaches and made physical contact with NR while he was feeding. Y was not very concerned with getting food but she grabbed and then dropped some food, or tasted it for a while. Sometimes after Y had grabbed some food and was playing with it PAY would pull her back by the tail and catch the food' (#95). The mother could presumably have learnt to move her daughter by tail-pulling in other circumstances, and might have already been in the habit of taking solid food from the daughter in order to eat it herself. Cyril Rosen describes a monkey *Cercopithecus mona* that 'pretended to drop and recover an item in order to secure a reward' (#124). However, this habit had developed in the context of Rosen's previous use of food to persuade the monkey to disgorge dangerous items from the mouth, so association of the reward with dropping actions as well as disgorging is to be expected. Pfeiffer kept a pet Guinea baboon that used several tactics to avoid going in certain locations, including 'in a lightning fast movement she bit her foot and was bleeding' (#80). Several times in the past, the discovery of a bleeding foot had terminated travel in this direction, and it was only by careful and surreptitious observation that the deliberate self-injury was discovered. While functionless self-injurious behaviour is known to occur in deprived captive conditions, its tactical use by a home-kept pet seems remarkable and hard to explain. De Waal notes a chimpanzee throwing a brick when pursuing a female, 'but almost without losing speed, so that he was able to catch his victim when she jumped back to the left in order to avoid the projectile' (#231). Since rock-throwing is often a part of chimpanzee aggressive displays, there are obvious prior contexts for discovering the effect on other individuals. Savage-Rumbaugh and McDonald (1988) describe a captive chimpanzee, Austin, that learned to use his dominant companion's fear of strange noises to his advantage, by covertly producing odd sounds. The researchers were unsure how these sounds were produced, although 'he was observed once making a very soft tapping noise on a pipe that was hollow. He then quickly scurried inside and looked back in that direction as though he had heard a very suspicious noise' (#224). Interpretation of this record is difficult, because the enriched regime that these chimpanzees experienced had already included experience with a 'scary monster' game.

In each of these cases, the reader's (and the original observer's) initial reaction is perhaps that the behaviour was novel and yet apparently crafted to purpose. However, careful examination in each case reveals opportunities for conventional, associative explanations to be feasible, and it is perhaps significant that all these cases come from captivity, often where the animals have extensive chance to interact with humans. Inevitably, such interaction makes inadvertent conditioning more likely (see e.g. Green, 1975; Byrne, 1997).

The quasi-experimental nature of Menzel's famous informed forager studies of chimpanzees (1974) gives more opportunity to follow the ontogeny of innovation. There, some tactics of an uninformed, dominant chimpanzee go beyond natural reactions to frustration, for example, 'On some occasions Rock started to wander off, only to wheel round suddenly precisely as Belle was about to uncover some food.' (#252). If Rock understood that he was being deceived, however, his tactical use of walking away and then turning round can be explained straightforwardly. The same applies to a highly unusual actions that Frans Plooij noted displayed by a chimpanzee, when confronted by a subordinate deceptively inhibiting its natural interest in hidden food. The apparent dupe 'left the feeding area again, but as soon as he was out of sight, he hid behind a tree and peered at the individual in the feeding area' (#253). Since chimpanzees are not stealth predators, and so not in the habit of peering from behind trees, an account in terms of instrumental conditioning of coincidentally performed activity is problematic. However, once it is allowed that a chimpanzee may *suspect deception* and understand the consequences of observation from a hidden position, then its effective counter-tactic becomes understandable (Byrne and Whiten, 1991; Hare *et al.*, 2000, 2001). Thus, if chimpanzees are considered to be able to compute with mental states, the innovation may be seen as an obvious, deduced solution. However, for both of these records, it may be possible to construct a simpler explanation in terms of detecting violation of a 'script' of highly predictable activities (Mitchell, 1999), an idea originally introduced to account for the effortless facility and speed with which humans deal with (logically complex) everyday situations (Schank and Abelson, 1977).

Actions considered to be invented by the individual using them
Turning now to actions which the observers had not believed were part of the species repertoire, each involves an ape. Warren Brockelman reports a female gibbon *Hylobates pileatus* that apparently lacked a mate, and 'gave a male-like solo in her territory,' quite unlike the normal notes on the end of the female call that 'sound like they are gasping for breath' (#151). It is perhaps conceivable that her call was simply at the extreme end of normal variation for the species, and occasioned by her unusual mood. The difficulty of being sure of the full extent of the 'normal repertoire' becomes even more acute with the two final cases, because they involve western lowland gorillas whose natural behaviour is very imperfectly known. In one, a gorilla approached a rival, but when about 5 m away, 'changed the nature of his approach, putting his feet down carefully, with his whole body adopting a tense demeanour. He gave every appearance of creeping up' (#160). The keeper who reported the incident did not recognize this as normal behaviour, but the full repertoire of the species in the wild is unknown. Moreover, the efforts of a young captive gorilla to conceal its playful intentions, described by Tanner and Byrne (1993), hint that stealth may be a normal part of play. Similarly, when a gorilla used hand-clapping in an interaction, Susanne Chevalier-Skolnikoff commented that 'hand clapping may not be considered part of the normal repertoire of gorillas' (#174); but in fact hand-clapping has subsequently been noted with some regularity in wild western lowland gorilla populations.

Discussion

Not all tactical deception by primates need reflect innovation of any sort. Where the precise form and context of deception is standardized, and the behaviour occurs widely between and within species, then treating the tactics as part of the species repertoire is more appropriate. This apparently applies to most reported cases of exaggeration or ignoring normal signals, suppression of attention, suppression of signalling, certain stylized forms of threat at inappropriate or non-existent objects, and various forms of hiding. This is not to say that these activities are independent of interesting cognitive capacities, and some of them have just such implications (e.g. visual perspective taking abilities in many species of monkey and ape); only that innovation is unnecessary.

Where innovation is clearly implied by the data, by far the most common case is that of *innovative use* of otherwise normal behaviours. Numerous behaviours are 'misused' in innovative ways: calls and displays are given in abnormal circumstances, displacement activities and behaviours that serve to lead others are used tactically in a fashion that clearly benefits the agent by manipulating competitors. These tactical uses are idiosyncratic, and normally it is clear how the tactics might have been acquired by rapid learning from natural but fortuitous conjunctions of circumstances (Byrne, 1997). The agents themselves may therefore have no conception of how and why the tactics work. The fact that such patterns occur widely in monkeys and apes but are far less common in other wild mammals (e.g. Hauser, 1997), may be explained as a consequence of the capacity for very rapid learning in those large-brained mammals. On this account, the enlarged neocortex found in many primates particularly facilitates more rapid learning of the details and circumstances of social interactions (Byrne, 1996a). Primates notice and remember the identities and association patterns of their companions, and to some extent their life histories and social relationships, more readily than do individuals of many other mammal species. With this enhanced social knowledge and an ability to learn rapidly (such that single-trial learning may be possible), primates often give the impression of understanding mental mechanisms and inventing novel behaviour to solve social problems. (And note that some other deceptive tactics, treated most parsimoniously on current evidence as adaptations, may prove in reality to reflect learning that is so rapid and widespread that it results in species-typical behaviours.) Nevertheless, changing species-typical behaviour may be hard to do. Even something as logically simple as suppressing a revealing vocalization is clearly sometimes quite difficult to perfect, as shown by the efforts of gorillas and chimpanzees to do so.

By some definitions, to learn a novel way of using a familiar display or signal might not even be called innovation, as no new behaviour pattern is added to the repertoire. This perhaps misses the point of innovation, since for the primate agents themselves the actions were (at some stage) novel solutions to problems. Nevertheless, in such an extensive survey of deception in primates, completed by many primatologists with extensive experience of one or many species, it is striking that there is so little evidence of entirely new motor patterns. Clearly, novel actions are invented with great rarity, and if this is the case for an order of mammals noted for large brain size and complexity of social living, it is likely to be true of most other terrestrial mammals. With this background, the cases where an animal modified its locomotion, manual actions, or facial appearance should repay careful study (discussed above under 'Modify body', 'Suppress signs of true intentions', 'Hide from view', and 'Use actions unexpected for species'). Some of these cases derive from captive circumstances, where there must always be concern at the possibility of inadvertent conditioning by humans. Where that is not the case, the most compelling data of innovative solutions comes from great apes. Notice in particular the evidence that behaviour is modified for purpose, with some understanding of the effect on other individuals, when facial expressions are modified or hidden (under 'Hide from view') and when the actions of others are shaped by threat or comfort during chimpanzee border patrols (under 'Suppress signal'). Confirmation of great ape innovative ability comes from the study of gestural communication in great apes, which shows remarkable plasticity and the development of many idiosyncratic patterns without human intervention, in contrast to the absence of any such findings in monkeys (chimpanzees: Tomasello *et al.*, 1985, 1989; gorillas: Tanner and Byrne, 1996;

Tanner, 1998; see also Miles, 1990, for evidence from a human-enculturated orang-utan). Great apes may be unusual both in the modifiability of their behavioural repertoire, associated with much greater extent of cortical control of motor activities even than in monkeys (Deacon, 1997), and the extent to which they can predict consequences of their own actions for the likely future behaviour of others.

The findings of this review must be regarded as preliminary, since they are based almost entirely on retrospective and often unsystematic survey. I hope they may form the basis for future, targeted study of the options available to non-human primates for innovative deception, and their means of operation. That this may be possible is suggested by the occasional finding of relatively high frequencies of deception in existing reports (e.g. in gorillas, 8 cases in 510 h focused on feeding in the wild, #161–8, and 21 cases in 81 h focused on social behaviour in captivity, Mitchell, 1991); what is now needed is to locate and direct research attention specifically at circumstances where deception is most frequent.

Summary

Primate tactical deception typically involves behaviour that is idiosyncratic to particular individuals, rather than species- or population-wide traits. As such, these tactics reflect innovation, although they may be used routinely by the time they are observed. Across a wide range of monkeys and apes, by far the most common deceptive innovation consists of using a familiar, normal part of the repertoire in an apparently inappropriate context. *Out-of-context usage* may constitute the bulk of innovation in animals: except in special circumstances like functional deception, this low key innovation may go unnoticed. In contrast, it is relatively rare for deceptive manipulation to involve wholly new actions: these few cases occur particularly in great apes rather than monkeys, and the novel action is usually a *modification* of an action in the normal repertoire. Innovation of manipulative tactics by means of modification of everyday actions implies considerable motor flexibility, also shown by great apes in their gestural communication, and is consistent with the ability to anticipate the likely effects upon others.

References

Adams, E. S. and Caldwell, R. L. (1990). Deceptive communication in asymmetric fights of the stomatopod crustacean *Gonodactylus bredini*. *Animal Behaviour*, **39**, 706.

Altmann, J. (1974). Observational study of behaviour: Sampling methods. *Behaviour*, **49**, 227–65.

Boysen, S. T. (1998). Attribution processes in chimpanzees. Heresy, hearsay or heuristic? Lecture given at *XVIIth Congress of the International Primatological Society*, University of Antananarivo, Madagascar, 10–14 August.

Byrne, R. W. (1981). Distance calls of guinea baboons (*Papio papio*) in Senegal: An analysis of function. *Behaviour*, **78**, 283–312.

Byrne, R. W. (1993). Do larger brains mean greater intelligence? *Behavioural and Brain Sciences*, **16**, 696–7.

Byrne, R. W. (1996a). Machiavellian intelligence. *Evolutionary Anthropology*, **5**, 172–80.

Byrne, R. W. (1996b). Relating brain size to intelligence in primates. In *Modelling the early human mind* (ed. P. A. Mellars and K. R. Gibson), pp. 49–56. Cambridge: Macdonald Institute for Archaeological Research.

Byrne, R. W. (1997). What's the use of anecdotes? Attempts to distinguish psychological mechanisms in primate tactical deception. In *Anthropomorphism, anecdotes, and animals: The emperor's new clothes?* (ed. R. W. Mitchell, N. S. Thompson, and L. Miles), pp. 134–50. New York: SUNY Press *Biology and Philosophy*.

Byrne, R. W. (1998). The early evolution of creative thinking: Evidence from monkeys and apes. In *Creativity in human evolution and prehistory* (ed. S. Mithen), pp. 110–24. London: Routledge.

Byrne, R. W. and Whiten, A. (1985). Tactical deception of familiar individuals in baboons (*Papio ursinus*). *Animal Behaviour*, **33**, 669–73.

Byrne, R. W. and Whiten, A. (1988). Towards the next generation in data quality: A new survey of primate tactical deception. *Behavioral and Brain Sciences*, **11**, 267–73.

Byrne, R. W. and Whiten, A. (1990). Tactical deception in primates: The 1990 database. *Primate Report*, **27**, 1–101.

Byrne, R. W. and Whiten, A. (1991). Computation and mindreading in primate tactical deception. In *Natural theories of mind* (ed. A. Whiten), pp. 127–41. Oxford: Basil Blackwell.

Byrne, R. W. and Whiten, A. (1992). Cognitive evolution in primates: Evidence from tactical deception. *Man*, **27**, 609–27.

Call, J. and Tomasello, M. (1998). Distinguishing intentional from accidental actions in orangutans (*Pongo pygmaeus*), chimpanzees (*Pan troglodytes*), and human children (*Homo sapiens*). *Journal of Comparative Psychology*, **112**, 192–206.

Caro, T. M. and Hauser, M. D. (1992). Is there teaching in non-human animals? *Quarterly Review of Biology*, **67**, 151–74.

Cheney, D. L. and Seyfarth, R. M. (1990). *How monkeys see the world: Inside the mind of another species.* Chicago, IL: University of Chicago Press.

Coussi-Korbel, S. (1994). Learning to outwit a competitor in mangabeys (*Cercocebus t. torquatus*). *Journal of Comparative Psychology*, **108**, 164–71.

Dawkins, R. and Krebs, J. R. (1978). Animal signals: Information or manipulation? In *Behavioural ecology: An evolutionary approach* (ed. J. R. Krebs and N. B. Davies), pp. 282–309. Oxford: Blackwell Scientific Publications.

Dawkins, R. and Krebs, J. R. (1979). Arms races between and within species. *Proceedings of the Royal Society of London, Series B*, **205**, 489–511.

de Waal, F. (1982). *Chimpanzee politics.* London: Jonathan Cape.

de Waal, F. B. M. (1992). Coalitions as part of reciprocal relations in the Arnhem chimpanzee colony. In *Coalitions and alliances in humans and other animals* (ed. A. H. Harcourt and F. B. M. de Waal), pp. 233–57. Oxford: Oxford University Press.

Deacon, T. W. (1990). Fallacies of progression in theories of brain-size evolution. *International Journal of Primatology*, **11**, 193–236.

Deacon, T. W. (1997). What makes the human brain different? *Annual Review of Anthropology*, **26**, 337–57.

Deaner, R. O. and Nunn, C. L. (1999). How quickly do brains catch up with bodies? A comparative method for detecting evolutionary lag. *Proceedings of the Royal Society of London, Series B*, **266**, 687–94.

Dunbar, R. I. M. (1992). Neocortex size as a constraint on group size in primates. *Journal of Human Evolution*, **20**, 469–93.

Dunbar, R. I. M. (1995). Neocortex size and group size in primates: A test of the hypothesis. *Journal of Human Evolution*, **28**, 287–96.

Goodall, J. (1971). *In the shadow of man.* London: Collins.

Goodall, J. (1986). *The chimpanzees of Gombe: Patterns of behavior.* Cambridge, MA: Harvard University Press.

Green, S. (1975). Dialects in Japanese monkeys: Vocal learning and cultural transmission of locale-specific vocal behaviour? *Zeitschrift fur Tierpschologie*, **38**, 304–14.

Gyger, M. and Marler, P. (1988). Food calling in the domestic fowl, Gallus gallus: The role of external referents and deception. *Animal Behaviour*, **36**, 358.

Harcourt, A. (1992). Coalitions and alliances: Are primates more complex than non-primates? In *Coalitions and alliances in humans and other animals* (ed. A. H. Harcourt and F. B. M. de Waal), pp. 445–71. Oxford: Oxford University Press.

Hare, B., Call, J., Agnetta, B., and Tomasello, M. (2000). Chimpanzees know what conspecifics do and do not see. *Animal Behaviour*, **59**, 771–85.

Hare, B., Call, J., and Tomasello, M. (2001). Do chimpanzees know what conspecifics know? *Animal Behaviour*, **61**, 139–51.

Hauser, M. D. (1992). Costs of deception: Cheaters are punished in rhesus monkeys (*Macaca mulatta*). *Proceedings of the National Academy of Science USA*, **89**, 12137–9.

Hauser, M. D. (1997). Minding the behaviour of deception. In *Machiavellian intelligence II: Extensions and evaluations* (ed. A. Whiten and R. W. Byrne), pp. 102–43. Cambridge: Cambridge University Press.

Hauser, M. D. and Marler, P. (1993). Food-associated calls in rhesus macaques (*Macaca mulatta*): II. Costs and benefits of call production and suppression. *Behavioral Ecology*, **4**, 206–12.

Held, S., Mendl, M., Devereux, C., and Byrne, R. W. (2000). Social tactics of pigs in a competitive foraging task: The 'informed forager' paradigm. *Animal Behaviour*, **59**, 569–76.

Held, S., Mendl, M., Devereux, C., and Byrne, R. W. (2002). Foraging pigs alter their behaviour in response to exploitation. *Animal Behaviour*, **64**, 157–66.

Hinde, R. A. (1973). On the design of check-sheets. *Primates*, **14**, 393–406.

Humphrey, N. K. (1976). The social function of intellect. In *Growing points in ethology* (ed. P. P. G. Bateson and R. A. Hinde), pp. 303–17. Cambridge: Cambridge University Press.

Jolly, A. (1966). Lemur social behaviour and primate intelligence. *Science*, **153**, 501–6.

Kummer, H. and Goodall, J. (1985). Conditions of innovative behaviour in primates. *Philosophical Transactions of the Royal Society of London, Series B*, **308**, 203–14.

Lefebvre, L., Whittle, P., Lascaris, P., and Finkelstein, A. (1997). Feeding innovations and forebrain size in birds. *Animal Behaviour*, **53**, 549–60.

Martin, P. and Bateson, P. (1986). *Measuring behaviour: An introductory guide*. Cambridge: Cambridge University Press.

Martin, R. D. (1990). *Primate origins and evolution*. London: Chapman and Hall.

Menzel, E. W. (1974). A group of chimpanzees in a 1-acre field: Leadership and communication. In *Behaviour of nonhuman primates* (ed. A. M. Schrier and F. Stollnitz), pp. 83–153. New York: Academic Press.

Miles, H. (1990). The cognitive foundations for reference in a signing orangutan. In *'Language' and intelligence in monkeys and apes* (ed. S. Parker and K. Gibson), pp. 511–39. Cambridge: Cambridge University Press.

Mitchell, R. (1999). Deception and concealment as strategic script violation in great apes and humans. In *The mentalities of gorillas and orangutans* (ed. S. Parker, R. Mitchell, and H. Mikes), pp. 295–315. Cambridge: Cambridge University Press.

Mitchell, R. W. (1991). Deception and hiding in captive lowland gorillas (*Gorilla gorilla gorilla*). *Primates*, **32**, 523–27.

Passingham, R. E. (1981). Primate specializations in brain and intelligence. *Symposia of the Zoological Society of London*, **46**, 361–88.

Reader, S. M. and **Laland, K. N.** (2002). Social intelligence, innovation and enhanced brain size in primates. *Proceedings of the National Academy of Sciences, USA,* **99**, 4436–41.

Savage-Rumbaugh, S. and **McDonald, K.** (1988). Deception and social manipulation in symbol-using apes. In *Machiavellian intelligence: social expertise and the evolution of intellect in monkeys, apes and humans* (ed. R. W. Byrne and A. Whiten), pp. 224–37. Oxford: Clarendon Press.

Schank, R. and **Abelson, R.** (1977). *Scripts, plans, goals, and understanding.* Hillsdale, NJ: Lawrence Erlbaum Associates.

Seyfarth, R., Smuts, B. B., and **Cheney, D.** (1986). Social relationships and social cognition in nonhuman primates. *Science,* **234**, 1361–6.

Shea, B. T. (1983). Phyletic size change and brain/body allometry: A consideration based on the African pongids and other primates. *International Journal of Primatology,* **4**, 33–61.

Tanner, J. (1998). Gestural communication in a group of zoo-living lowland gorillas. Ph.D. thesis, University of St Andrews.

Tanner, J. E. and **Byrne, R. W.** (1993). Concealing facial evidence of mood: Evidence for perspective-taking in a captive gorilla? *Primates,* **34**, 451–6.

Tanner, J. E. and **Byrne, R. W.** (1996). Representation of action through iconic gesture in a captive lowland gorilla. *Current Anthropology,* **37**, 162–73.

Temerlin, M. K. (1975). *Lucy: growing up human.* Palo Atto, California: Science and Behaviour Books.

Tomasello, M., George, B., Kruger, A., Farrar, J., and **Evans, E.** (1985). The development of gestural communication in young chimpanzees. *Journal of Human Evolution,* **14**, 175–86.

Tomasello, M., Gust, D., and **Frost, T. A.** (1989). A longitudinal investigation of gestural communication in young chimpanzees. *Primates,* **30**, 35–50.

Trivers, R. L. (1972). Parental investment and sexual selection. In *Sexual selection and the descent of man* (ed. B. Campbell), pp. 136–79. Chicago: Aldine.

Trivers, R. L. (1974). Parent-offspring conflict. *American Zoology,* **14**, 249–64.

Whiten, A. and **Byrne, R. W.** (1988a). The manipulation of attention in primate tactical deception. In *Machiavellian intelligence: Social expertise and the evolution of intellect in monkeys, apes and humans* (ed. R. W. Byrne and A. Whiten), pp. 211–23. Oxford: Clarendon Press.

Whiten, A. and **Byrne, R. W.** (1988b). Tactical deception in primates. *Behavioural and Brain Sciences,* **11**, 233–73.

Yerkes, R. M. (1943). *Chimpanzees.* New Haven, Connecticut: Yale University Press.

CHAPTER 12

INNOVATION AS A BEHAVIOURAL RESPONSE TO ENVIRONMENTAL CHALLENGES: A COST AND BENEFIT APPROACH

PHYLLIS C. LEE

Introduction

In this paper I attempt a general review of how non-human primates adjust to environmental variability via behavioural innovation, and place innovation into context as an adaptive strategy for coping with unpredictability. Innovation is defined here as a novel behaviour pattern or novel combinations of existing behaviours, which shift from being an unexpected one-off event to a more generally practised behaviour by individuals or groups. Examination of the environmental contexts or events that promote behavioural innovation suggests that unpredictable changes in the social or ecological context in the first instance are likely to elicit innovation and secondly such changes increase the adaptive value of innovations. In order for an innovative behaviour or technology to emerge and then become fixed in the repertoire, the risks associated with the new behaviour and the time, energy, and opportunity costs of the behaviour must be less than the benefits in terms of social or reproductive success, or an enhanced energy balance. Since males and females, and old and young, have different cost–benefit equations for established and novel behaviour, the effects of age and sex on the probability of innovation are also briefly explored. Life history plasticity and phylogenetic influences on cognitive capacities, social dynamics and physical skills may both enhance and constrain opportunities for innovation in different primate lineages.

Categories of innovation

The definition I use here is similar to that of Reader and Laland (see Chapter 1), and relates to the generation of a novel behaviour. A new behaviour may emerge in contexts for which a behavioural repertoire already exists, but where the established behaviour is no longer used in that specific context. The new behaviour appears and is extended as a novel solution to an existing problem. This is a classical perspective—the generation of novelty (see Kummer and Goodall, 1985, Chapter 10).

A second, and possibly more controversial, approach considers both the context of a behaviour and the behaviour itself. Thus an innovation can consist of an existing behaviour, which is applied to a novel context. Such a behaviour might be a manipulative skill that is used to solve a problem that the animal has not previously experienced. Thus the innovation is the translation of a skill that now enables individuals to exploit some new or previously unavailable resource or opportunity. The classical model of blue tits opening milk bottles to feed on cream is such an innovation. Here, the novelty is in the application of a behaviour—the tits use existing foraging skills and motor patterns. It requires the recognition of a new opportunity—the milk bottle containing high energy food—and the capacity to generalize and extend existing behaviour beyond prior experience. In this context, novelty may not be necessary in the behaviour, but is required in the capacity to make associations between behaviour and outcomes. Effectively, this is novelty in the cognitive as opposed to the performance domain, and as I have argued earlier, suggests that innovation is a process rather than simply a product of change (Lee, 1991).

Finally, there can be cases where both the behaviour and the context when it is used are novel—no prior learned response fulfils the new need or the new demand, and the animal has effectively invented a response to a completely unfamiliar situation. For example, when chimpanzees first began to crack open palm nuts using a wooden 'hammer and anvil', they were perhaps applying a new technique and new tools to a food that was previously unavailable. As such, this context would again fit the more classical definition of novelty (e.g. Kummer and Goodall 1985, Chapter 10).

While the generation of innovative behaviour itself is of considerable interest, the outcome of the innovation will primarily be considered here. By shifting behaviour between contexts, or by generating novel behaviour, there are consequences for individuals. Firstly, the innovation allows individuals to change between existing sets of outcomes, or to moderate the effects of an existing outcome, or to create a totally new outcome. In these contexts, the outcome is considered to be the consequence of a behaviour. Secondly, innovations lessen the predictability of cognitive or learned associations between a behaviour and its outcomes for individuals. This loss of the ability to predict outcomes potentially affects the participants in an interaction as well as other animals who are observers of interactions. Innovations which reduce predictability may be especially interesting or important in a social context or where social mediation or transactions between individuals are involved (see also Simonton, Chapter 14). Thirdly, an innovation can change the value of an interaction to one or both participants by altering its intensity or quality. Thus, innovations can be either technical and often associated with food extraction, foraging efficiency, or personal maintenance, or they can be social and strategic.

Why should innovation be of interest in the study of animal behaviour? As Reader and Laland (Chapter 1) note, innovation reveals capacities for learning, and that in the absence of learning, serendipity, exploration, and random 'discoveries' do not constitute innovation. Thus how individuals learn new skills or apply existing skills to new context, how a behaviour is developed, practised, applied, and altered over time are fundamental questions in understanding innovation. These are general issues in learning theory, and are not specific to innovation, but when an innovation does occur, it may provide new insights in a natural context for understanding general learning capacities. Innovations may also reveal the mechanisms for learning more directly, in that the relations between a novel behaviour, its generation, practice, and performance provide insights into how the process of learning occurs.

Problems in the study of innovation

While the definitions I give above are so general as to encompass many behavioural events, and thus potentially of limited usefulness (see Slater and Lachlan, Chapter 5), there is a further and more significant problem, which is that of distinguishing between the technical (tool use or manipulative foraging strategies—e.g. Byrne, 2001) and the social contexts of innovation. If strategies for manipulation of others take the form of more effective food extraction (which is competitive by nature), is this a technical innovation or a social innovation? Furthermore, technical innovation itself may require a social context for its effective learning, dissemination, and generalization, or spread to others. Again, the separation or categorization of forms of innovation appears to overlook the context— either social or ecological—in which the innovations arise and become established.

A further and significant issue is that if behavioural flexibility is the hallmark of the primate order, then how can we tell when an innovation in a behavioural sense has arisen? Are we, as observers, simply lacking the history of experiences for individuals and thus what appears 'new' may in fact be a behaviour that existed in previous generations or individuals no longer under observation. In long-lived and socially interactive species such as primates, it may never be possible in natural contexts to be sure that a behaviour is indeed completely novel. While much of the descriptive work on innovation comes from field observations of rare and unusual events, most detailed studies of innovative behaviour focus on experimentally induced behavioural alteration, such as the presentation of novel objects, foods, or extractive apparatus (e.g. Menzel and Menzel, 1979; Sumita *et al.*, 1985; Stammbach, 1988; Fragaszy and Visalberghi, 1990; Matsuzawa, 1994; Janson, 1996; Menzel, 1997; Visalberghi *et al.*, 1998), or rare cases where free-living individuals have been inadvertently 'introduced' by humans to genuinely unfamiliar objects (Kawai, 1965; Goodall, 1968).

Is it possible to distinguish how the use of an existing behaviour in a novel context differs from the generation of a completely novel behaviour applied to an existing context? Examples of non-technical novelty include communication and gestures, one being hand clasps during chimpanzee grooming (de Waal, 1994; Tomasello, 1994; Wrangham *et al.*, 1994; Seres and de Waal, 1997; Nakamura *et al.*, 2000). This was considered as a completely new behaviour, which emerged in a pre-existing context—grooming exchanges. Another is tandem buddy-walking, the waltzing hug seen in a number of captive chimpanzee colonies (Casanova, 2002). In these cases, the behaviour was apparently novel or at least spontaneously generated by specific individuals and its application to the social context was also unusual. Of interest in such contexts is whether the new behaviour had sufficient social value (a social or tactical innovation) for it to spread to other individuals, as noted for both hand clasp grooming and tandem buddy-walking, and for other novel forms of communication and gestures also seen in captive chimpanzees (e.g. Tomasello *et al.*, 1994).

Technical or tactical innovation?

Technical innovation can be thought of as having two components. The first is production of a novel behaviour in the physical domain (the use of physical objects in the environment, the making and use of tools) and the second is its dissemination in the social context. The classical example where the entire process was observed is that of the sweet-potato- and wheat-washing by Koshima Island Japanese macaques (see Kawai, 1965). Other examples have been described, including vervet monkeys obtaining water by dipping pods into tree boles during droughts (Hauser, 1988). Perhaps the most detailed examples are those from the use of tool technologies by chimpanzees, recently compiled by Whiten et al. (1999), and used as significant evidence for the existence of cultural variation between populations of chimpanzees.

Technical innovation appears to be widespread in vertebrates. Birds in particular excel at using novel technologies, and these are associated with complex memory or brain architecture (see Clayton and Dickinson, 1998; Hunt et al., 2001; Lefebvre et al., 2002). The recent work on New Caledonian crows demonstrating spontaneous but deliberate tool-making (Weir et al., 2002) suggests that problem solving in the physical domain is a major feature of this group of birds (see also Lefebvre and Bolhuis, Chapter 2).

Another modality consistently associated with the production of novelty is communication. The observation of dialects in killer whales (see Boran and Heimlich, 1999) and many passerines (Slater and Lachlan, Chapter 5) suggest that vocalizations are fertile grounds for innovation and thus innovation in communication should be generalized across a range of species. By contrast, innovations in social exchanges, which use novel behaviour to increase unpredictability or success in contests (social tactic innovation), may be more restricted in their distribution amongst animals. As such, compulsively social species such as carnivores, cetaceans, elephants, and primates might provide examples of tactical innovation. Tactical innovation is strategic; it relates to an intention to alter outcomes rather than being simply a by-product of general states of arousal. Here, I distinguish between social species where repeated interactions occur between known individuals and which have both the long-term recognition skills and memory to develop tactics or strategies for interaction, and colonies of obligate 'social' species. It should be noted again, that for socially living species, all innovation has a social dimension in its observation, perpetuation and dissemination. However, for non-social species, the generation of novelty to solve problems in the physical domain does not necessitate social transmission.

Tactical innovation which has a social context may be both especially difficult to perceive, and especially important in the study of the process of innovation. One possible approach is to consider the tactics used, for example, in contests (e.g. de Waal, 1989). The description of the 'new policies' adopted by a recently dominant male chimpanzee, of engaging in a controlling role by intervening in group disputes (de Waal, 1982) could be considered as a tactical innovation. What is key in determining whether an innovation has occurred is whether there is consistency within an individual using a new strategy during repeated contest with the same other animals, which can be contrasted with the novel behaviour being used consistently by different individuals in contests having similar content or qualities. Much of the work has focused on tactics used in deception and manipulation, where the ability to exploit novel contexts to achieve goals can be determined (see Byrne and Whiten, 1992; Whiten and Byrne, 1997; Byrne, Chapter 11). While deception is hardly a novel interaction, it is the means used for deception which can be open to innovation. Making yourself appear larger, more powerful or more threatening by using unfamiliar objects in the environment (e.g. the chimpanzee Mike's use of storage drums in his status displays: Goodall, 1968), is innovation with intent—an intention to deceive or manipulate another's perceptions.

Of equal interest are the tactics used in appeasement or reconciliation among primates, as this is the social 'glue' and needs to be responsive to highly variable contexts (Aureli and de Waal, 2000). Innovative behaviour in such contexts can be predicted to be highly variable, context-specific, and possibly specific to individual needs, which change both over time and from situation to situation depending on the value of the relationships (van Schaik and Aureli, 2000). Most analyses of reconciliation tend to focus on outcomes rather then to describe novel behaviour, but vocalizations, facial expressions, and new forms of body contacts in particular all have the potential to be used innovatively as part of a reconciliation or consolation strategy. de Waal (2000, p. 16) provides an example with his first description of reconciliation in the Arnhem chimpanzee colony: hooting, banging on metal drums, kissing, and embracing were all first seen associated together in this event. Exchanges of grooming or play in macaques, contacts with opponents' infants in baboons, the redirection of aggression towards other species such as birds in capuchins, vervet monkeys or even elephants (personal observation) could all be considered as innovations in the context of appeasement or relationship repair. Even a behaviour as dramatic as agonistic buffering, where a male uses a vulnerable infant as a flag to communicate his status and aggressive intent during an encounter, could be considered in its origin as an innovation for the reduction of tension.

In terms of outcomes, the introduction of a new behaviour or modes of behaviour in a strategic context can be proposed to have the following effects. Firstly it changes the outcome, which effectively lessens predictability between interactants. A male using novel gestures or combinations of gestures in a display represents an unfamiliar opponent, and thus the outcome of an interaction is more difficult to predict. In competitive contexts, this may be an especially critical feature of a capacity to innovate. Furthermore, the innovation may change the value of the interaction (its quality or intensity) to the interactants, and thus either confer an advantage on one participant, or alternatively ensure that both gain a mutalistic benefit. Males might be more likely to use novel tactics in contests, since for many species, male status or contest ability determines the outcome of male–male competition and is associated with reproductive success (e.g. baboons: Altmann et al., 1996). By contrast, novel tactics used in deception, or in the context of reconciliation and alliance formation, may be equally likely to occur in the interactions of both males and females. Here, an individual's needs to manipulate and restore the social fabric predisposes towards social innovation, and no differential advantage should accrue to either sex. Indeed, the likelihood of innovation in such contexts should be specific to the state of the individual at a moment in time, and a function of shifting needs and relationships over time. Species where needs and relationships are under constant adjustment (due to endogenous physical changes, reproductive states, or demographic variation) are those where there is a premium on the capacity to respond to social flux.

Cost–benefit approaches

As generally noted, innovation in a social context should follow a cost–benefit principle. Energy balance (as the currency for reproduction or survival in energy limited environments) should be enhanced through the use of a new tactic. Thus energy and time costs of the social engagement should be reduced, and the chances of a successful outcome increased. Failure in a social context carries with it physiological costs of hormonal changes (e.g. testosterone, oestrogen: Berkovitch, 2001), immunological depression (e.g. cortisol: Sapolsky and Ray, 1989), or other psychological stressors (behaviourally induced reproductive suppression: Abbott, 1993), not to mention constraints on growth or general physical fitness. Trade-offs against these costs relate to the benefits of maintenance of alliances, social status, or dominance rank, and strategic friendships with reproductive outcomes.

How can we determine a currency for cost–benefit analyses of innovations in the social domain? As noted above, the fitness payoff could be derived through enhanced hormonal, immunological, or psychological functions, which increase reproductive potential. Costs are those of stress, uncertainty, reduced growth, inhibition of participation in risky contexts, and these in turn lead to a loss of reproductive potential. Such costs in terms of fitness have yet to be measured, but the proximate costs (time costs, risk of injury) and benefits (hormonal status, nutritional state) can be explored.

In technical innovation, the change in the outcome may be that of increasing either the predictability or the magnitude of returns, and thus have specific benefits that can be weighed against the time and energy costs of nutrient extraction, or against the risks inherent in the use of novel actions, techniques or technologies. Again, the fitness returns to the individual resulting from a change in outcomes due to the introduction of novel behaviour are unknown, but proximate costs and benefits potentially can be more easily assessed. Obviously, the state and status of the individual will affect these payoffs. Young, growing individuals, reproductively active females, old, or ill animals all have different requirements for energy and physiological functioning. For example, the incorporation of novel medicinal plants in the diets of chimpanzees (e.g. Huffman and Seifu, 1989) has been suggested to function to reduce parasite infestations, which would improve condition. For the most part, however, such assessments have yet to be done, despite being noted as necessary some 15 years ago by Kummer and Goodall (1985, Chapter 10).

Contexts for innovation

If an innovation is considered in terms of the returns or outcomes, then we can ask whether there are specific social or ecological contexts that are more likely to promote novelty and innovation. As noted above, technical innovation may be easier to recognize and may be more closely linked to specific and measurable proximate costs and benefits of returns. However, there are few quantitative studies of ecological contexts for innovation, and thus a more general 'theoretical' approach will be used here. The aim is to develop some potential hypotheses for a cost–benefit model rather than to produce the model itself, given the limited data set at the moment.

The ecological context for most primates is one of variability rather than constancy. Seasonal cycles of food production, temperature, rainfall, and species distributions can be combined with unpredictable environmental hazards such as contact with predators, shifts of habitat occupation by competitors, extremes of rain or drought, cyclones, floods or other longer term, and thus unpredictable changes in dynamic ecosystems (see Table 12.1). Does a seasonal, predictable shift in the environment represent a context for innovation? At the level of an individual, the key mechanisms for coping with such predictable or cyclical changes are likely to be used repeatedly in each 'season' or cycle with success. These revolve around the reduction of energy costs and maximization of energy intake; the general optimization of energy balance which is closely associated with reproductive output in many species (Lee, 1996). There are times, however, when innovative responses to predictable cycles could be predicted and should be common. For example, a female facing late pregnancy or peak lactation during a hungry season might be expected to innovate in food choices or in balancing the costs of different activities. These are costs that can be modelled (e.g. Altmann, 1980; Barrett *et al.*, 1995). Indeed, since female reproduction is a function of energy balance, it can be predicted that females should be key innovators in the context of coping with seasonal fluctuations in energy balance, in general food competition and in food extraction. However, if the costs of failure as a result of an innovation also increase, then it may pay females to use tried and tested behaviour, rather than to innovate. This may explain why female chimpanzees are more likely to use tools in food extraction than are males, and why males, given their larger size and increased strength, exploit high risk—high return foods such as captured prey (McGrew, 1992; Stanford, 1998). Thus females may be less likely to be observed using novel tools or tools in novel contexts than are males (Reader, 2000). The balance of costs and benefits constrains opportunities for innovation

Table 12.1 Ecological variation and its potential effects on demography, sociality, and cognitive traits in primates

| | Scale of environmental change | | | |
| | Unpredictable | | Predictable | |
	Short term	Long term	Short term	Long term
Behavioural changes	Social flux ecological 'catastrophes' and opportunities	Dynamic relationships	Optimization of energy balance	Seasonal variation in relationships
Demographic changes	Relationship manipulation and repair	Changing group composition	Seasonal movements and breeding	Alteration of life history traits or kinship
Innovation potential	High: Social tactics; Experimentation with technology or novel foods	Low: Life history changes rather than strategies	High: Individual (age and sex specific) needs to innovate for energy balance and social contexts	High: Apprenticeship slippage and cultural variation

when the risks are high and gains low, but generates innovation when gains are also high. The distinction between shifting food types—experimentation with completely new foods—and novel strategies for exploiting known foods should reflect a similar cost–benefit relationship in the context of predictable cycles (Lee and Hauser, 1998).

Food type innovation may be an example where the cost–benefit ratios could be precisely modelled. In some contexts, early learned diets can be shown to enhance survival when infants closely follow or match their maternal diet (e.g. Hauser, 1993). If, however, the incorporation of novel food during periods of scarcity, either seasonal or unpredictable, results in an increase in dietary breadth and improves net energy balance, then survivorship should increase. Some species, which live in cyclical environments, or alternatively generally forage on unpredictable foods (sensu Ross, 1988) may be more likely to show the capacity to constantly sample novel foods. While many primates quickly learn about poisoned or unpalatable food (see Laska and Metzker, 1998), there are few studies that have assessed the loss of this response over repeated exposures. Aversive conditioning in baboons wanes quickly (e.g. Bergman and Glowa, 1986), suggesting that some species resist associating negative stimuli with specific food types over the long-term, as there may be an advantage to continually sampling foods for quality and quantity that outweighs a single (or even repeated) event of illness or horrible taste (see Hladik and Simmen, 1996). For other species, such as capuchins (Visalberghi *et al.*, 1998) and mandrills (Cambefort, 1981), novel foods are unlikely to be incorporated quickly into a diet, even when palatable, as there is more general dietary stability and constraint on choice. Capuchins appear to respond to short-term energy challenges less through the rapid incorporation of new foods but rather through manipulative innovation, increasing the efficiency of the exploitation of existing foods (Fragaszy and Visalberghi, 1989; Visalberghi, 1990; Panger *et al.*, 2002). However, for all these species repeated exposure and social acceptance are effective mechanisms for increasing the acceptance of novel foods (Visalberghi *et al.*, 1998; Visalberghi and Addessi, 2001).

Predictable cycles of changes in group composition may also promote innovation and the spread of innovations (e.g. influxes of males during a breeding season). A 'cultural trait' from one group is brought into a new group, and then disseminated. After diffusion, the novel behaviour pattern becomes established in both groups and without an observation of its initial occurrence, it would not be possible to determine how it had arisen. Such a behaviour was observed during a study of baboons in Ruaha National Park, Tanzania (Lee, personal observation). A recently transferred male began washing sedge roots that he had dug from the banks in the river. A number of juveniles, both males and females, began to use this cleaning technique (see Figure 12.1), and by the following dry season, most group members participated in the same behaviour (personal observation). Detailed observations were not made on the dissemination of the behaviour, although it was known that the behaviour originated with the male. Other examples come from Panger *et al.*'s work on capuchins (Panger *et al.*, 2002), where technical traditions in the use of stone tools for digging could be shown to move between social groups.

It can be suggested that technological innovation is a function of a market economy (e.g. Stanford, 2001), where the costs of producing a new behaviour must be outweighed by the benefits obtained through its use. If, however, the behaviour has minimal costs in

Figure 12.1 Baboon (*Papio cynocephalus*) root-washing on the edge of the Ruaha River, Tanzania. (Photo by P. C. Lee.)

terms of time, energy, or failure risks, then experimentation with a diverse array of repertoires might be expected, and this tendency to experiment for pleasure (such as stone handling among Japanese macaques, Huffman, 1996) would be difficult to categorize as an innovation. It is only when the trade-offs lead to stabilization and persistence of a behaviour in an individual or across a group, that innovation (as opposed to experimentation) can reliably be detected. When the surrounding behavioural traits are more effective, then an innovation is unlikely to persist beyond the initial trials, and the prevailing 'market' will ensure that marginally successful traits do not become established. But when market conditions change, then an experimental or innovative individual will gain advantages as a result of its tendencies for exploration, which may thus result in stabilization and dissemination of the behaviour.

Two conditions can be proposed as key to the retention of experiments leading to innovation by individuals. The first is that opportunity costs for experimentation with novelty must be low. The second is that the benefits can be expected to be high for the retention of a specific behaviour (it should be noted that the benefits of experimentation itself do not have to be high). Both of these may be more likely when animals are exposed to extremes of ecological variation, to unpredictable contexts, rather than to constant conditions or seasonally predictable variation, as discussed above (see also Reader and MacDonald, Chapter 4). Such conditions, it can be speculated, occur regularly over time—in the course of an individual's life, he or she is likely to encounter drought, super-abundances, hurricanes, disease outbreaks and so on, at some point—and over the course of the life of a social group, such events may be more common than has been thought, or indeed modelled. A generalized propensity to try new behaviour in new contexts may have significant selective advantages in contexts of unpredictable environmental variation, making individual tendencies to innovate more frequent in a population.

Does innovation require knowledge?

Both tactical and technical innovations have two components to their establishment. The individual must be capable of physically and cognitively generating the novel behaviour, or even if it is 'accidental', of holding some knowledge of its relative effectiveness in terms of changing outcomes. Thus animals with significant capacities for learning are expected, and indeed shown to, engage in more innovation (Reader and Laland, 2002; Reader and MacDonald, Chapter 4). The form that this 'knowledge' takes can be roughly partitioned into ecological knowledge, social knowledge, and the cognitive capacity to model outcomes (see Table 12.2).

If the capacity to learn, both as an immature and as an adult, underlies the generation of novelty, then one can ask why innovation in non-humans is relatively rare, at least to the human observer. With respect to technical innovation and opportunities for learning and dissemination, there may be significant differences between adults and immatures in their capacity to observe the innovations of others, as well as physical constraints such as size and strength, which limit opportunities for the expression of some behaviours. When innovations are produced by females or young animals, it can be predicted that these would

Table 12.2 Forms of knowledge in relation to learning and the capacity to innovate

Ecological knowledge
Food selection (assessment of relative quality between sources of the same foods, assessment of relative values between foods)
Food location (memory of past locations, tracking renewal rates, tracking abundance, and quality)
Food extraction (embedded foods, processing costs, technological vs cognitive extraction)

Social knowledge
Hypotheses about mental states (representation and intentionality)
Deception & manipulation (strategic relationship management in conflict)
Conflict resolution (relationship repair via reconciliation, consolation, and cooperation)

Goal-specific knowledge
Models of expected outcomes
Manipulation and management of outcomes by technical or social skill

be more likely to spread, since there is closer proximity and association between mothers and young, and between immatures themselves than is often the case with males, or males with other age–sex classes (Kawai, 1965; Russon, 1997; King, 1999). This lack of opportunity for observational learning may account for slowness in acquisition of novel behaviour or a lack of its dissemination. Inter-generational transmission of behaviour may both promote stability in technology and social tactics, as well as in food types—allowing for the generation of cultural consistency over time. However, inter-generational shifts in technology are also key components of modes of change in both knowledge and outputs (Parker and Russon, 1996). This is highlighted in the concept of 'apprenticeships' (Greenfield *et al.*, 2000). As knowledge is transmitted during training for tasks, the incorporation of error and novelty can lead to the production of new outputs—especially in technology. Whether these new outputs are the consequence of the random generation of changing techniques (slippage) or intentionally introduced changes that are more effective or perceived of as having a higher benefit, is open to debate. de Waal has recently (2000) argued that apprenticeships are a mechanism for consistency, in that the apprentice is allowed no flexibility in either the mode of production of an output or the output itself (see Table 12.3).

A final mechanism for the production of innovation may be associated specifically with the period of behavioural development. In species with a relatively extended developmental period (such as most higher primates: Joffe, 1997), demonstration, observation, and playful exploration all provide contexts for the production of novel behaviour. Again whether temporary modes of response or behaviour become fixed as a novel style, or incorporated into a range of contexts over the individual's lifespan is unknown. Species with a significant component of social learning may be more likely to generate novelty than those without 'teachers' or lacking a wider group context for social learning (Smith *et al.*, 2002). Thus species with short developmental periods and lifespan, but with social learning, as well as those with inter-generational exchange of information all should be capable of innovation, although this may be more marked for technical innovation.

Table 12.3 Apprenticeship concepts and their role in tradition and innovation (developed from Greenfield *et al.*, 2000; de Waal, 2001)

Innovation in the 'Apprentice' when
Guidelines are violated in response to self-generated perception of a new need
    Internally driven by shifts in individual state (developmental, hormonal, physiological)
    Externally driven as a response to changing external environments which change the value of existing traditions

Innovation in the 'Master' when
Guidelines are violated in response to changes in the cost/benefit ratio
    New opportunities in the market economy are presented—either immediate or with longer term payoffs
    Introduced by new individuals with alternative competing traditions which affect cost/benefit ratios of existing traditions

Conclusions and speculations

What are the conditions for innovation and how does attempting a cost–benefit approach increase our understanding of novel behaviour and novel outcomes? The following predictions can be made:

- Innovations will be likely to occur in immature individuals as a result of the processes of learning and acquisition of knowledge.
- They will be common in adults when there are significant benefits in terms of time, energy balance, or social outcomes under changing environmental contexts. Males are predicted to innovate in status seeking contexts, females in technology when it improves the energy available for reproduction and both sexes in contexts where the maintenance of a social group is vital to individual survival and reproductive success.
- Finally, both phylogeny and phenotypic variation in life history need to be taken into account in understanding patterns of innovation in response to environmental or social pressures. Some species may be constrained in their capacity to innovate, as they have relatively invariant life histories and thus may be unable to benefit from changes in the cost–benefit ratio, while other species with greater life history plasticity have more capacity for manipulating and balancing the risks or costs incurred due to innovations.

Thus innovation can be expected among the 'apprentice' rather than in the 'master' (e.g. de Waal, 2000), specifically when violations of established norms occur as a response to perceived new needs and new contexts. These can be internally generated by the innovator. Alternatively, when flexibility or the ability to respond to ecological, social, or demographic variation is at a premium, innovation is likely to occur among the 'master' as new opportunities are presented by environmental changes, or when social flux shifts the balance of costs and benefits against performing old behaviours and advantages new, more successful outcomes.

A developmental period associated with learned behaviour and specifically a social context for this learning may underlie the potential to innovate in species such as some birds, primates, carnivores, cetaceans, and elephants. However, whether innovation occurs is a function of the time and energy costs, balanced against the opportunities for benefits due to the production of the behaviour, and with the risks (costs) independently associated with failure of the new behaviour. Unpredictability in both an ecological and social context is key to setting the context for innovation, as has been widely noted. Under such conditions, there may be a low risk and low cost to the production of novelty, but a high cost and high risks associated with pursing a stable behavioural pattern. Why should unpredictability feature in innovation? With respect to ecological innovation, the ability to discriminate between similarities and differences in categories of food (Menzel, 1997), and to use a variety of food-handling techniques (Janson and Boinski, 1992) are associated with primates' knowledge of their habitats and how these change. The capacity to track and respond to changes in habitats over time (e.g. Robinson, 1986; Lee and Hauser, 1998) or to cognitively represent outcomes via the use of tools ('technical intelligence', Byrne, 1997), are all part of the capacity for technical or foraging innovation. Some of this capacity for technical intelligence may be present at birth, even in young inexperienced primates such as tamarins where tool use is rare (e.g. Hauser *et al.*, 2002).

Social innovation can also be seen as a response to social flux (Whiten and Byrne, 1988). Over the life of a primate, it will experience a constantly changing dynamic network of relationships, as a result of demographic changes within groups and as a result of manipulations of social contexts. Status or rank may shift over time, alliances are made and broken, new relationships are forged. Thus, stability in behaviour between generations may have costs, which are reduced by violations of the apprenticeship concept with respect to social behaviour. The concept of adaptive unpredictability or Protean behaviour (Driver and Humphries, 1988) has been explored in relation to deceptive signalling (Krebs and Dawkins, 1984; Hauser, 1997) as well as more generally applied to evasive actions in response to predation (Driver and Humphries, 1988). The cognitive implications of unpredictability with respect to courtship and mating have been developed by Miller (1997), who argues that preferring the unexpected (e.g. Small, 1993) plays a key role in selection for the capacity for social prediction. It is an old point, that adaptive unpredictability is vital to understanding the evolution of cognitive capacities, but it remains to be tested with respect to innovation. The extent of unpredictability may be difficult to assess, our knowledge of innovation is often anecdotal, and plasticity in life history traits also needs to be considered. In this respect, the search for general cost–benefit models of innovation may be in its infancy.

Summary

Primates as a group are renowned for their behavioural flexibility, their technical capacities, and for creating new contexts for social opportunities. Both phylogeny and life history underlie differences between species in their capacity to innovate, while within species, local ecological opportunities and constraints affect when, where, how often and among which age–sex classes innovations may arise and become fixed within behavioural repertoires. I have attempted to outline in a theoretical context how modelling costs and benefits could increase our understanding of innovation and dissemination of novel behaviour, but as I note above, we have only begun to have the data required to develop such models. Effectively such models are only heuristic tools at this stage, rather than predictive or testable. It is hoped, however, that they can generate further research, which will make the development of such models a reality in the future.

Acknowledgements

I thank Simon Reader and Kevin Laland for inviting and nurturing this contribution, Hilary Box for the initial stimulus to tackle these issues 10 years ago, and Elisabetta Visalberghi for her infinite patience in reading the first draft and for her insightful comments. Two referees made further very useful comments. Gillian Brown helped enormously throughout.

References

Abbott, D. (1993). Social conflict and reproductive suppression in marmoset and tamarin monkeys. In. *Primate social conflict* (ed. W. A. Mason and S. P. Mendoza), pp. 331–72. New York: SUNY Press.

Altmann, J. (1980). *Baboon mothers and infants.* Princeton, NJ: Princeton University Press.

Altmann, J., Alberts, S., Haines, S., Dubach, J., Muruthi, P., Coote, T. *et al.* (1996). Social structure predicts genetic structure in a wild primate group. *Proceedings of the National Academy of Sciences, USA,* **93**, 5797–801.

Aureli, F. and **de Waal, F. B. M.** (2000). Why natural conflict resolution? In *Natural conflict resolution* (ed. F. Aureli and F. B. M. de Waal), pp. 3–10. Berkeley, CA: University of California Press.

Barrett, L., Dunbar, R. I. M., and **Dunbar, P.** (1995). Mother-infant contact as contingent behaviour in gelada baboons. *Animal Behaviour,* **49**, 805–10.

Bercovitch, F. (2001). Reproductive ecology of Old World Monkeys. In *Reproductive ecology and human evolution* (ed. P. T. Ellison), pp. 369–96. New York: Aldine.

Bergman, J. and **Glowa, J. R.** (1986). Suppression of behaviour by food pellet-lithium chloride parings in squirrel monkeys. *Pharmacology, Biochemistry & Behaviour,* **25**, 973–8.

Boran, J. R. and **Heimlich, S. L.** (1999). Social learning in cetaceans: Hunting, hearing, and hierarchies. In *Mammalian social leaning* (ed. H. O. Box and K. L. Gibson), pp. 282–307. Cambridge: Cambridge University Press.

Byrne, R. W. (1997). The technical intelligence hypothesis: An additional evolutionary stimulus to intelligence? In *Machiavellian intelligence II* (ed. A. Whiten and R.W. Byrne), pp. 289–311. Cambridge: Cambridge University Press.

Byrne, R. W. (2001). Social and technical forms of primate intelligence. In *Tree of origin* (ed. by F. B. M. de Waal), pp. 146–72. Cambridge, MA: Harvard University Press.

Byrne, R. W. and **Whiten, A.** (1992). Cognitive evolution in primates: Evidence from tactical deception. *Man,* **27**, 609–27.

Cambefort, J. P. (1981). A comparative study of culturally transmitted patterns of feeding habits in the chacma baboon *Papio ursinus* and the vervet monkey *Cercopithecus aethiops. Folia Primatologica,* **36**, 243–63.

Casanova, C. C. N. (2002). Status and friendship in captive female chimpanzees (*Pan troglodytes*). Ph.D. Thesis, University of Cambridge.

Clayton, N. S. and **Dickinson, A.** (1998). Episodic-like memory during cache recovery by scrub jays. *Nature,* **395**, 272–4.

de Waal, F. M. B. (1982). *Chimpanzee politics.* London: Jonathan Cape.

de Waal, F. M. B. (1989). *Peacemaking among primates.* Cambridge, MA: Harvard University Press.

de Waal, F. M. B. (1994). Chimpanzees' adaptive potential: A comparison of social life under captive and wild conditions. In *Chimpanzee cultures* (ed. R. W. Wrangham, W. C. McGrew, F. M. B. de Waal, and P. G. Heltne), pp. 243–60. Cambridge: Cambridge University Press.

de Waal, F. M. B. (2000). The first kiss. In *Natural conflict resolution* (ed. F. Aureli and F. M. B. de Waal), pp. 15–33. Berkeley, CA: University of California Press.

de Waal, F. M. B. (2001). *The ape and the Sushi master: Cultural reflections of a primatologist.* New York: Basic Books.

Driver, P. M. and **Humphries, N.** (1988). *Protean behavior: The biology of unpredictability.* Oxford: Oxford University Press.

Fragaszy, D. M. and Visalberghi, E. (1989). Social influence on the acquisition of tool using behaviours in tufted capuchin monkeys (*Cebus apella*). *Journal of Comparative Psychology*, 103, 159–70.

Goodall, J. (1968). The behaviour of free-living chimpanzees of the Gombe Stream Reserve. *Animal Behaviour Monographs*, 1, 161–311.

Greenfield, P. M., Maynard, A. E., Boehm, C., and Schmidtling, E. Y. (2000). Cultural apprenticeship and cultural change. In *Biology, brains and behavior* (ed. S. Taylor Parker, J. Langer, and M. L. McKinney), pp. 237–78. Santa Fe: SAR Press.

Hauser, M. D. (1988). Invention and social transmission: New data from wild vervet monkeys. In *Machiavellian intelligence* (ed. R. W. Byrne and A. Whiten), pp. 327–43. Oxford: Oxford Science Publications.

Hauser, M. D. (1993). Ontogeny of foraging behaviour in wild vervet monkeys (*Cercopithecus aeithiops*); social interactions and survival. *Journal of Comparative Psychology*, 107, 276–82.

Hauser, M. D. (1997). Minding the behaviour of deception. In *Machiavellian intelligence II* (ed. A. Whiten and R. W. Byrne), pp. 112–43. Oxford: Oxford University Press.

Hauser, M., Pearson, H., and Seelig, D. (2002). Ontogeny of tool use in cottontop tamarins, *Saguinus oedipus*: Innate recognition of functionally relevant features. *Animal Behaviour*, 64, 299–311.

Hladik, C. M. and Simmen, B. (1996). Taste perception and feeding behaviour in nonhuman primates and human populations. *Evolutionary Anthropology*, 5, 58–71.

Huffman, M. A. (1996). Acquisition of innovative cultural behaviour in nonhuman primates: A case study of stone handling as socially transmitted behaviour in Japanese macaques. In *Social learning in animals: The role of culture* (ed. C. M. Heyes and B. G. Galef), pp. 267–89. London: Academic Press.

Huffman, M. A. and Seifu, M. (1989). Observations on the illness and consumption of a possibly medicinal plant, *Veronia amygdalina* (Del.), by a wild chimpanzee in the Mahale Mountains National Park, Tanzania. *Primates*, 30, 51–60.

Hunt, G. R., Corballis, M. C., and Gray, R. D. (2001). Laterality in tool manufacture by crows. *Nature*, 414, 707.

Janson, C. H. and Boinski, S. (1992). Morphological and behavioural adaptations for foraging in generalist primates: The case of the cebines. *American Journal of Physical Anthropology*, 88, 483–98.

Janson, C. R. (1996). Towards an experimental socioecology of primates: Examples from Argentine brown capuchin monkeys (*Cebus apella nigritus*). In *Adaptive radiations of neotropical primates* (ed. M. Norconk, P. Garber, and A. Rosenberger), pp. 309–25. New York: Plenum Press.

Joffe, T. (1997). Social pressures have selected for an extended juvenile period in primates. *Journal of Human Evolution*, 32, 593–605.

Kawai, M. (1965). Newly-acquired pre-cultural behavior of the natural troop of Japanese monkeys on Koshima Islet. *Primates*, 6, 1–30.

King, B. J. (1999). New directions in the study of primate learning. In *Mammalian social learning* (ed. H. O. Box and K. L. Gibson), pp. 17–32. Cambridge: Cambridge University Press.

Krebs, J. and Dawkins, R. (1984). Animal signals: Mindreading and manipulation. In *Behavioural ecology* (ed. J. Krebs and N. B. Davis), 2nd edn, pp. 380–402. Oxford: Blackwell.

Kummer, H. and Goodall, J. (1985). Conditions of innovative behaviour in primates. *Philosophical Transactions of the Royal Society of London, Series B*, 308, 203–14.

Laska, M. and Metzker, K. (1998). Food avoidance learning in squirrel monkeys and common marmosets. *Learning & Memory*, 5, 193–203.

Lee, P. C. (1991). Adaptations to environmental change: An evolutionary perspective. In *Primate responses to environmental change* (ed. H. O. Box), pp. 39–56. London: Chapman & Hall.

Lee, P. C. (1996). Lactation, condition and sociality: Constraints on fertility of non-human mammals. In *Variability in human fertility: A biological anthropological approach* (ed. L. Rosetta and C. G. N. Mascie-Taylor), pp. 25–45. Cambridge: Cambridge University Press.

Lee, P. C. and **Hauser, M. D.** (1998). Long-term consequences of changes in territory quality on feeding and reproductive strategies of vervet monkeys. *Journal of Animal Ecology*, **67**, 347–58.

Lefebvre, L., Nicolakakis, N., and **Boire, D.** (2002). Tools and brains in birds. *Behaviour*, **139**, 939–73.

Matsuzawa, T. (1994). Field experiments on the use of stone tools by chimpanzees in the wild. In. *Chimpanzee cultures* (ed. R. W. Wrangham, W. C. McGrew, F. B. M. de Waal, and P. G. Heltne), pp. 351–70. Cambridge, MA: Harvard University Press.

McGrew, W. C. (1992). *Chimpanzee material culture*. Cambridge: Cambridge University Press.

Menzel, E. W. and **Menzel, C. R.** (1979). Cognitive, developmental and social aspects of responsiveness to novel objects in a family group of marmosets (*Saguinus fuscicollis*). *Behaviour*, **70**, 251–79.

Menzel, C. R. (1997). Primates' knowledge of their natural habitat: As indicated in foraging. In. *Machiavellian intelligence II* (ed. A. Whiten and R. W. Byrne), pp. 207–39. Cambridge: Cambridge University Press.

Miller, G. (1997). Protean primates; the evolution of adaptive unpredictability in competition and courtship. In *Machiavellian intelligence II* (ed. A. Whiten and R. W. Byrne), pp. 312–40. Cambridge: Cambridge University Press.

Nakamura, M., McGrew, W. C., Marchant, L. F., and **Nishida, T.** (2000). Social scratch: Another custom in wild chimpanzees. *Primates*, **41**, 237–48.

Panger, M. A., Perry, S., Rose, L., Gros-Louis, J., Vogel, E., Mackinnon, K. C. *et al.* (2002). Cross-site differences in foraging behavior of white-faced capuchins (*Cebus capucinus*). *American Journal of Physical Anthropology*, **119**, 52–66.

Parker, S. T. and **Russon, A. E.** (1996). On the wild side of culture and cognition in the great apes. In *Reaching into thought: The minds of the great apes* (ed. A. E. Russon, K. Bard, and S. T. Parker), pp. 430–50. Cambridge: Cambridge University Press.

Reader, S. M. (2000). Social learning and innovation: Individual differences, diffusion dynamics and evolutionary issues. Ph.D. Thesis, University of Cambridge.

Reader, S. M. and **Laland, K. N.** (2002). Social intelligence, innovation and enhanced brain size in primates. *Proceedings of the National Academy of Sciences, USA*, **99**, 4436–41.

Robinson, J. G. (1986). Seasonal variation in use of time and space by wedge-capped capuchin monkeys, *Cebus olivaceus*: Implications for foraging theory. *Smithsonian Contributions to Zoology*, **431**, 1–60.

Ross, C. (1988). The intrinsic rate of natural increase and reproductive effort in primates. *Journal of Zoology, London*, **214**, 199–219.

Russon, A. E. (1997). Exploiting the expertise of others. In *Machiavellian intelligence II* (ed. A. Whiten and R. W. Byrne), pp. 174–206. Cambridge: Cambridge University Press.

Sapolsky, R. and **Ray, J.** (1989). Styles of dominance and their endocrine correlates among wild olive baboons. *American Journal of Primatology*, **18**, 1–13.

Seres, M. and **de Waal, F. M. B.** (1997). Propagation of hand clasp grooming among captive chimpanzees. *American Journal of Primatology*, **16**, 935–69.

Small, M. F. (1993). *Female choices*. Ithaca: Cornell University Press.

Smith, V. A., King, A. P., and **West, M. J.** (2002). The context of social learning: Association patterns in a captive flock of brown-headed cowbirds. *Animal Behaviour*, **63**, 23–35.

Stammbach, E. (1988). An experimental study of social knowledge: Adaptation to the special manipulative skills of single individuals in a *Macaca fascicularis* group. In *Machiavellian intelligence* (ed. R. W. Byrne and A. Whiten), pp. 309–26. Oxford: Oxford University Press.

Stanford, C. (1998). *Chimpanzees and red colobus.* Cambridge, MA: Harvard University Press.

Stanford, C. (2001). *Significant others.* New York: Basic Books.

Sumita, K., Kitahara-Frisch, J., and Norikoshi, K. (1985). The acquisition of stone tool use in captive chimpanzees. *Primates*, **20**, 513–24.

Tomasello, M. (1994). The question of chimpanzee culture. In *Chimpanzee cultures* (ed. R. W. Wrangham, W. C. McGrew, F. B. M de Waal, and P. G. Heltne), pp. 301–17. Cambridge, MA: Harvard University Press.

Tomasello, M., Call, J., Nagell, K., Oliguin, R., and Carpenter, M. (1994). The learning and use of gestural signals by young chimpanzees: A trans-generational study. *Primates*, **35**, 137–54.

van Schaik, C. P. and Aureli, F. (2000). The natural history of valuable relationships in primates. In *Natural conflict resolution* (ed. F. Aureli and F. B. M. de Waal), pp. 307–33. Berkeley, CA: University of California Press.

Visalberghi, E. (1990). The acquisition of nut-cracking in two capuchin monkey groups (*Cebus apella*). *Folia Primatologica*, **49**, 168–81.

Visalberghi, E. and Addessi, E. (2001). Acceptance of novel foods in capuchin monkeys: Do specific social facilitation and visual stimulus enhancement play a role? *Animal Behaviour*, **62**, 567–76.

Visalberghi, E., Valente, M., and Fragaszy, D. (1998). Social context and consumption of unfamiliar foods by capuchin monkeys (*Cebus apella*) over repeated encounters. *American Journal of Primatology*, **45**, 367–80.

Weir, A. S., Chappell, J., and Kacelnick, A. (2002). Shaping of hooks in New Caledonian crows. *Science*, **297**, 981.

Whiten, A. and Byrne, R. W. (1988). Taking (Machiavellian) intelligence apart. In *Machiavellian intelligence* (ed. R. W. Byrne and A. Whiten), pp. 50–65. Oxford: Oxford University Press.

Whiten, A. and Byrne, R. W. (1997). *Machiavellian intelligence II.* Cambridge: Cambridge University Press.

Whiten, A., Goodall, J., McGrew, W. C., Nishida, T., Reynolds, V., Sugiyama, Y. *et al.* (1999). Chimpanzee cultures. *Nature*, **399**, 682–5.

Wrangham, R. W., de Waal, F. M. B., and McGrew W. C. (1994). The challenge of behavioural diversity. In *Chimpanzee cultures* (ed. R. W. Wrangham, W. C. McGrew, F. B. M. de Waal, and P. G. Heltne), pp. 1–18. Cambridge, MA: Harvard University Press.

CHAPTER 13

INNOVATION AND CREATIVITY IN FOREST-LIVING REHABILITANT ORANG-UTANS

ANNE E. RUSSON

Introduction

One recent source of interest in innovation is evidence that in many non-human species, social learning and cultural processes probably underpin expertise, such as knowledge and skills (Box and Gibson, 1999). Even when expertise is so widely acquired socially that conformity is the norm, it remains to explain how it was innovated (i.e. invented or modified to meet changing demands or novel circumstances), and how innovative and cultural processes mesh (i.e. how innovations enter into a community's repertoire).

Great apes are interesting in this regard because they may have a high potential for both innovation and culture. As primates, they rely on life-long learning to acquire and update most of their expertise (Fleagle, 1999; Parker and McKinney, 1999). They show powerful social acquisition processes in the form of sophisticated social learning mechanisms, apprenticeship, and cooperative problem solving (e.g. Tomasello and Call, 1997; Russon *et al.*, 1998; Whiten *et al.*, 1999; Boesch and Boesch-Achermann, 2000; van Schaik and Knott, 2001). They also rank high on characteristics linked with innovation (e.g. van Schaik *et al.*, 1999; Huffman and Hirata, in press). They face cycles of food scarcity, which can compel invention, and superabundance, which encourages exploration (Knott, 1998; van Schaik *et al.*, in press; Yamagiwa, in press). They are opportunistic foragers with broad, generalized diets (Rodman, 2000; Yamagiwa, in press). Socially, they live within flexible societies that allow working alone as well as working socially, the latter under conditions of very low as well as very high social tolerance (van Schaik *et al.*, 1999, in press; Reader and Laland, 2001). Development and intelligence are also probable contributors to a propensity to innovate. Intelligence is prominent among factors likely to favour innovation; the role of development may be less obvious.

Great apes also stand out among non-human primates for extended life histories with disproportionately prolonged immaturity, extremely high adult body mass, and large brains (Fleagle, 1999). In terms of acquiring expertise, prolonged immaturity extends their period of intensive learning past infancy through the juvenile period. Massive adult bodies and brains, prolonged immaturity, and long lives amplify the changes actors experience over their lifetime compared to other non-human primates—in their needs, physical capabilities, social opportunities and limitations, and abilities to absorb and apply new information (Parker and Gibson, 1990; King, 1994; Parker and McKinney, 1999; Russon, in press a)—correspondingly amplifying the changes needed to their expertise. Their prolonged development may favour innovation because emerging characteristics can force actors to change their expertise and/or open new opportunities for doing so.

Great apes also stand out for their intelligence, which most experts now accept as more sophisticated than that of other non-human primates (Byrne, 1995; Russon *et al.*, 1996; Parker and McKinney, 1999; Matsuzawa, 2001; Suddendorf and Whiten, 2001). The cognitive mechanisms that support their problem-solving and expertise acquisition extend beyond the associationist ones typically attributed to non-human species. Abilities credited to them include cause–effect and spatial cognition up to and including understanding relations and relations-between-relations, conceptual classification, mental representation, mental problem solving, multiple representations, imitation, self-concept, intentional tactical deception, and analogical reasoning (e.g. Rumbaugh *et al.*, 1996; Russon *et al.*, 1996; Tomasello and Call, 1997; Byrne and Russon, 1998; Langer, 1998, 2000; Parker and McKinney, 1999; Thompson and Oden, 2000; Suddendorf and Whiten, 2001). Their cognitive systems appear to be constructed during development, on the basis of experience, by combining and recombining existing cognitive structures to create new ensembles then hierarchically integrating ensembles into unified, higher-level structures (Gibson, 1993; Inoue-Nakamura and Matsuzawa, 1997; Parker and McKinney, 1999; Langer, 2000). Hierarchical cognitive systems may be intrinsically generative (Gibson, 1990; Rumbaugh *et al.*, 1996).

Probable reasons for great apes' cognitive prowess concern their primate phylogenetic heritage, tropical forest habitat, frugivorous preferences, extremely large bodies, and large brains. These culminated in extremely broad diets, intermittent reliance on difficult foods, and unusually flexible social systems—all of which require complex and flexible problem-solving capacities (Russon and Begun, in press). Their challenging diet has repeatedly been singled out as a key evolutionary force behind great apes' enhanced mentality (Parker and Gibson, 1977; Byrne, 1997; Russon, 1998). Living great apes' techniques for obtaining difficult foods confirm great cognitive complexity, in the form of lengthy tool-assisted and manipulative sequences that combine and recombine multiple skills into hierarchically organized structures (Byrne and Byrne, 1991; McGrew, 1992; Russon, 1998, 2002a; Byrne *et al.*, 2001; Matsuzawa, 2001). Acquisition can occupy the first 7–10 years of life (Parker, 1996; Boesch and Boesch-Achermann, 2000).

Cognitive perspectives on great ape innovation are then particularly appropriate, especially as regards foraging. If great apes stumble on useful innovations accidentally, they are especially well equipped to represent the problem, 'notice' the innovative solution, and voluntarily recruit it into their functional repertoire (Byrne, 1995, 1999a). Combining, recombining, and hierarchization, may offer generative properties that render great apes especially apt at invention and improvisation. Coupled with their manual dexterity, their high intellect may allow them to invent certain kinds of tools easily, like sponges or probes, under diverse living conditions (Huffman and Hirata, in press).

For great apes, how cognition affects innovation, how innovative and social processes balance one another in generating complex expertise, and how this balance shifts over time have yet to be worked out. My observational studies of cognition on ex-captive orang-utans under rehabilitation to forest life in Indonesian Borneo offer some insights. My focus has been foraging, a type of expertise considered likely to foster innovation (Reader and Laland, 2001), but observations also open windows on other types of behaviour. Here I discuss a selection of innovations noted in the course of these studies. My interests lie primarily with innovation as the process by which individuals invent new or modified behavioural variants, so my aim is to dissect the processes generating innovative behaviours with emphasis on the cognitive mechanisms involved.

Innovation in rehabilitant orang-utans

Rehabilitant orang-utans' circumstances suggest they will have an unusually high innovative potential. Virtually all orang-utan rehabilitants were orphaned and captured from the wild as young infants under 2–3 years old (Swan and Warren, 2000). Captivity varies in quality from abysmal cages to luxury hotels and Bornean villages to Taiwanese apartments, and in duration from a few days to over 15 years. On resuming forest life, typically as juveniles or adolescents, ex-captives' forest expertise ranges from virtually naïve (if captured less than 1 year old) to infant levels (if captured late in infancy). Since 1995, regulations require that ex-captives be reintroduced into forests that support no wild orang-utans (Smits et al., 1995); some programs release them in pre-formed groups for social support, typically peer groups, into areas without resident rehabilitants. All programs offer daily supplementary provisions while needed. In short, newly released ex-captives typically have little if any relevant forest expertise and desperately need to acquire it. Newly released ex-captives experience unusually rich social contact and relatively benign social tolerance, given their own immaturity and their release group companions, but have little access to knowledgeable conspecifics. Few if any experienced residents are available and release groups provide few if any experts. These circumstances should call forth as much inventiveness as ex-captives can muster, as one of their only routes to success.

Cases I offer derive from my observations of ex-captives under rehabilitation at Camp Leakey (CL), Tanjung Puting National Park, Central Kalimantan and the Wanariset Orang-utan Reintroduction Project (WORP), East Kalimantan. CL and WORP have cared for ex-captives since 1971 and 1991, respectively. During the period that my observations represent CL ex-captives ranged freely through the forest and base camp. WORP ex-captives are group-housed in large socialization cages pre-release and released into isolated blocks of protected forest, Sungai Wain (SW) and Meratus (M). WORP follows the release regulations and protocols described above; CL released ex-captives to a forest with resident wild orang-utans, although the two communities do not integrate readily (Galdikas, personal communication; Rijksen, 1978). My observations total 2500 h over 14 years, focused on adult females (CL, 1989–1993) and juveniles and adolescents (WORP, 1994–2001) (for details, see Russon and Galdikas, 1993, 1995; Russon, 1998, 2002b).

I used Reader and Laland's (Chapter 1) definition of innovation as either new or modified individually learned behaviour patterns or the process that generates them, interpreting learning broadly to mean individual acquisition that is contingent on appropriate experience. I did not exclude idiosyncratic (particular to one actor), insightful, or improvisational (one-off, opportunistic) behaviour. Idiosyncrasy is a valuable indicator of a behaviour's individual vs social origins. In cases of apparent insight, experience has typically paved the way. Improvisations epitomize the creativity that is the essence of the innovative process; they show the inventiveness needed to handle highly variable problems and, especially when they achieve immediate pay-offs deftly, they cannot be dismissed as random. Repetition or re-use may not be useful for improvisations in complex situations insofar as the same situational configuration is highly unlikely to recur. Which behavioural modifications produce departures from established patterns that are valuable enough to spread to conspecifics and contribute to between-group diversity is of great importance, but this cannot be determined without looking widely at the options.

Operationally, innovations were behaviours I saw only once in my 2500-h observation, or behaviours that appeared newly in one actor or in several actors with no opportunity for social exchange. On novelty, I included variants that differed visibly from the common pattern; for functional behaviour, I focused on variants that altered effectiveness. I identified innovations against the backdrop of systematic data on imitation (CL) and foraging (SW, M) so cases involving foraging, at least, rest on relatively firm ground. Typically, I questioned knowledgeable rehabilitation project staff about unusual behaviours. Some of the expertise I consider here has been discussed elsewhere, primarily relative to socio-cultural processes (Russon and Susilo, 1999; Russon, 2002a, b, 2003; Russon, in press a). My focus here is features likely to involve innovation; together, the two analyses may offer some insights into how social and individual processes mesh in advancing expertise at individual and community levels.

Note that these cases were not generated by formal study of innovation. Like many cases currently used as evidence of innovation or other rare behaviours, they owe to opportunistic observations made in the course of other studies and my evaluation, as a researcher experienced with these orang-utans in these forests, that they were rare, unusual, or never seen before. Others who rely on cases like these (e.g. Whiten and Byrne, 1988; Lefebvre *et al.*, 1997; Reader and Laland, 2001) consider that such cases do not represent uninformed casual observations or anecdotes because they come from experienced scientists familiar with their subjects, and they are valuable where the state of scientific study is embryonic, as it is currently for innovation. The 2500-h observation is low compared to long-term field studies of wild orang-utans but rehabilitants are visible almost 100 per cent of the time from as close as 5–10 m; wild orang-utans are less visible from farther away. Except as noted, observations are all my own, rather than my assistants', so my basis for identifying innovations is direct knowledge of the whole database.

An orang-utan innovation sampler

I offer cases beyond foraging to suggest orang-utans' innovative scope, in light of suggestions that innovation may be more common in some kinds of problems than others. From a cognitive perspective, great apes face three broad problem domains: physical (contingent, probabilistic relations in the physical world, such as space, cause–effect), logico-mathematical (necessary relations, such as features, categories, number of items in a set), and social (relations with social or animate partners) (Parker and McKinney, 1999; Langer, 2000). I chose illustrative cases in all domains to suggest the range of situations in which orang-utans can innovate and the qualities of their innovations. I identified these cases based on my general knowledge of these rehabilitants' behaviour and living conditions, not systematic observations of these behaviour categories. They are therefore less secure than foraging cases and quantification is not meaningful. I dissected the more complex cases in efforts to infer the process producing them.

Physical

Foraging is probably rich terrain for innovation because it concerns physical problems, which vary probabilistically (Langer, 2000). To ensure their effectiveness and efficiency, actors need to adjust their foraging skills to the particulars of each instance (Russon, 1999a, in press b; Gosselain, 2000; Byrne *et al.*, 2001). Mastering complex foraging skills requires major investments in time and energy, so being able to make repairs and adjustments may help ensure that energies are not wasted. For example, great apes make tools to suit the task and can adjust them to suit task changes as work progresses (Parker and Gibson, 1977; Fox *et al.*, 1999). Skills for other physical problems, like nest-building or arboreal locomotion, should operate in similar fashion. Orang-utans' tree swaying, for instance, must allow improvisation because it operates in conditions that vary widely in spatial configuration (tree layout), vegetation compliance (by species and size), surrounding vegetation (obstructions, connections), etc.

Rehabilitants' innovations appeared in cause–effect and spatial object manipulations. Two orang-utans used other orang-utans' heads to crack open hard items: Supinah, an adult female (CL), used a juvenile's head as an anvil; Tono, a juvenile male (SW), hammered a termite nest against an adolescent's skull. Kinoi, a juvenile male (WORP), made rings from 1 m lengths of garden hose, joining the ends by carefully forcing one into the other (3 cases). Enggong (SW), a juvenile male, invented a new way to make a 'workseat' (i.e. a palm leaf petiole deliberately bent horizontal) to sit on while extracting a palm's shoot from its crown. He pulled an immature leaf towards and over the crown's centre then sat on it above the shoot (Russon, 1998) (2 cases); normally, he and others pushed a mature leaf away from the crown's centre then sat on it beside the shoot.

Concerning space, locomotion offers considerable innovative scope, especially in arboreal contexts where discontinuity and changeability put premiums on improvisation. Orang-utan arboreal locomotion is known for its great flexibility and its non-stereotyped, work-it-out-as-you go quality; some suggest it may qualify among orang-utans' greatest cognitive challenges and may have been the evolutionary pressure behind one of their representational abilities, a self-concept (Chevalier-Skolnikoff *et al.*, 1982; Povinelli and Cant, 1995). Witness to this locomotor flexibility are these innovations: Siti, a juvenile female, standing or walking on her hands (SW, 2 cases); an infant hanging and spinning from a liana by his teeth (WORP, 1 case); and Davida, a juvenile female, positioning herself supine on a path, planting both feet near her buttocks and both hands behind her shoulders, lifting her body into a yoga-like crab position, and walking backwards several steps up the path (CL, 1 case). I have observed most proto-tool improvisations in arboreal contexts:

(1) *Vehicle trees.*   Orang-utans characteristically transfer between trees by deliberately swaying a vehicle tree across the gap (Bard, 1993; Povinelli and Cant, 1995). I noted one innovative tree-swaying tactic (2 cases): keep hold of the vehicle tree after transferring into the target tree instead of letting go immediately as usual. Two immatures (CL) were in a

chase through the trees; the fleeing partner tree-swayed across a gap and on reaching the target tree, kept hold of the vehicle tree. This stopped the chase because it left the pursuer no way to cross the gap. Enggong, a juvenile male (SW), kept hold of his vehicle tree after transferring into a palm. He walked to the palm's crown holding his vehicle tree, sat down, removed some of the vehicle tree's leafy branches, put them on his head as a hat (it was raining heavily), then released the vehicle tree (Russon, 1998).

(2) *Arboreal supports.* Two rehabilitants improvised arboreal supports. Circumstances forced Aming, an adolescent male (SW), to transfer out of one tree fast, up through the crown. To reach the next tree, he had to climb onto tiny terminal branches likely to break under his weight. Aming rapidly bunched several terminal branches together, shifted his weight onto his 'rope', and then swung into the next tree (1 case). Bento, as a heavy-bodied subadult (M), was climbing along two branches from the trunk of an *Anthropcephalus chinensis* tree to its terminal branches, for fruit. The two branches were sturdy and close together near the trunk but progressively, they splayed wider apart towards the edge of the tree's crown; they also narrowed, becoming less able to support his weight. Bento kept hold of both branches, positioned his body between them, and edged out towards the fruit. The effect was zipper-like: hands and feet progressively pulled the branches together and his torso followed in the middle.

(3) *Ladders.* Bento and Luter, adolescent males (M), each climbed a pair of tiny saplings, 0.75 m apart, like a rung-less ladder. Standing upright, each grabbed an overhead support with their hands (to take weight off the saplings, which were much too small to support either climber), seized one sapling in each foot, and inched up into a food tree. Luter left the food tree the same way, in reverse: he descended feet first onto the same two saplings, inched half way down, and released his overhead hold; as the saplings bent under his weight, he held on and rode them to the ground. Inept orang-utans, by mistake, ride vegetation that bends under their weight and crash to the ground; the more skilled deliberately ride one slender tree to the ground as it bends. Only in these two cases have I seen orang-utans use two trees this way.

(4) *Chairs.* Uce, an adolescent female (SW), made chairs by pulling two slender trees across each other into an 'X' (2 cases). She climbed a pair of slender trees 1 m apart to a height of 2 m, moved onto one and bent the second across the first, then moved onto the second and pulled the first with her, so the two trees crossed each other. Uce sat on the outer side of one lower leg of the 'X' and for balance, held the upper ipsilateral arm of the 'X' above her head. From her chair, she extracted rattan hearts.

(5) *Handles.* Enggong, in example 1, could not sway his vehicle tree far enough to reach the palm despite repeatedly pumping and re-aiming. He finally stopped swaying, cracked a branch on the vehicle tree's trunk so it dangled loosely, grabbed the loose end like a handle, and swayed toward the palm. This handle still did not bring him close enough to reach the palm, so he made another the same way (Russon, 1998). The second handle extended his reach enough to grab the palm and climb in. Juveniles in WORP cages also made handles. Several threaded short lengths of hose through cage mesh or hollow pipes, bent the hose double, and dangled from it. A juvenile female, Koko, threaded pilfered wire through a link of a horizontal chain then bent the wire double, like a hairpin. She then took the wire's doubled ends, threaded them together through a second chain link

about 10 cm from the first, and wove them to fasten them securely. This made a handle, fastened at each end to the chain, from which she dangled and twirled. WORP handles used operations described as weaving or knotting elsewhere (Maple, 1980). By implication, orang-utans readily invent handles (Enggong had no contact with the cage-living ex-captives and weaving/knotting reports are from foreign zoos) using a variety of materials.

These cases suggest several avenues to innovation. Vehicle tree innovations suggest new combinations, notably by interconnecting objects or activities that are normally unrelated (vehicle tree with rain hat, handle, and partner vs self). All entail multiple representations of one item (e.g. tree as vehicle and rain hat or handle material, vehicle for self and for partner). Most innovative supports (chair, ladder, rope, 'zipper', WORP handles) reflect known orang-utan patterns of distributing weight over multiple arboreal supports and deliberately manipulating habitat compliance (Cant, 1987; Povinelli and Cant, 1995). All also manipulate one object—object relation to create a support structure, *join* or *bring together* several items, which orang-utans also use in other contexts (join fibres to make rope; Maple, 1980). What is, innovative, then, are the behavioural tactics for creating that relation, constrained by materials at hand. Enggong's handle comes close to a meta-tool because it modifies a (proto-) tool, the vehicle tree (Russon, 1998). Meta-tools, first reported in wild chimpanzees, represent the most cognitively sophisticated tool use known in great apes (Matsuzawa, 1991).

Social

Social problems share the probabilistic contingencies of physical problems, probably to a greater degree given that they involve animate entities, so they should be equally rich ground for innovation.

(1) *Slaves.*   Cabrik was one of two females in a cage-housed group of 10–12 adolescents (WORP). She was exceptionally thin, probably because she had little success competing for food at feeding times; she frequently withdrew and scrounged leftovers. Whatever food Cabrik obtained, however, the males quickly stole. At one feeding, a male approached her, grabbed her arm, and led her down to the grill floor of the cage. The cage had a double floor, grill work 1 m above tiles, to reduce orang-utans' contact with debris. The male held Cabrik while she reached through the grill and retrieved food fallen on the tiles, then took it from her. I saw this manoeuvre once, but it had the air of a well-practised routine. Its likely basis is: (1) Cabrik adopted scrounging as her feeding strategy; (2) she scrounged food from the tile floor, and was perhaps the only orang-utan who could because she had extremely long arms; (3) opportunistically, after Cabrik retrieved food from the tile floor, males stole it from her; (4) some male(s) noticed the pattern and recreated it deliberately, including leading Cabrik into place.

(2) *Iconic sign.*   Princess, an adult female (CL) trained in sign language (Shapiro, 1982), invented a new vocal sign. Like other nursing mothers I observed, she made a 'huffing' vocalization *in reaction to* her infant's suckling, as if it tickled or hurt. When her infant was about a year old, I observed Princess huff voluntarily *before* nursing: once she huffed then sucked her own nipple, once she huffed then her infant immediately moved from her mother's back and began suckling. Princess created a new iconic sign from a spontaneous

functional action, by recreating it voluntarily; similar cases have been reported in other great apes (Parker and McKinney, 1999).

(3) *Deception.*   One innovative deception grew out of an elaborate interaction between a forest assistant and Unyuk, an adult female (CL). It began with Unyuk's studying and imitating the assistant's Swiss Army knife use, shifted to a complex grooming interaction using the knife, and culminated in her deceptively grooming to cover preparing to steal a backpack (for details, see Russon, 2002a). Unyuk behaved as if this were her first close encounter with a Swiss Army knife; given local conditions, it probably was. Deception began 14 min into the interaction, during grooming, when Unyuk first spotted backpacks behind the assistant, unguarded. Thereafter her grooming behaviour changed visibly. While continuing to accept and even solicit grooming, she casually repositioned twice; this looked like part of the grooming exchange and raised no suspicion, but it put her in position to reach past the assistant for the packs. Once in position, she grabbed a pack and ran. Her deception suggests calculated altering and blending of theft and grooming scripts (scripts are representations of the sequence of elements that constitute events; Mitchell, 1999). If she built her deception from scripts she knew, she had to improvise to mesh with the interactive flow, her dual purposes, and a situation that was novel or nearly so.

(4) *Request.*   Siti, as an adolescent female (SW), was gnawing and tearing off fibrous husk to expose a wild coconut's three 'eyes'. Eyes are visible on the shell and easily punctured, as in domestic coconuts, but each accesses a separate jelly filled cavity. Siti exposed one eye and poked through it with a finger, then turned aside, made a probe tool, and used it to poke out jelly (this is the only tool use on coconuts I have seen). After emptying one cavity, Siti handed her coconut to a forest assistant. He held it briefly then gave it back, gesturing to her to work on it herself. She took the coconut back but a moment later handed it to him again. Again, he held it without acting. Siti took a stick and chopped repeatedly at the coconut as he held it, as one would with a machete. Everyone who has watched the event, on video, interprets it as Siti's twice asking the assistant to chop her coconut open with his parang (Indonesian machete-like knives), his refusing twice, then her reiterating and clarifying her request by acting out what she wanted done. Given the complex conjunction of circumstances and the specificity of her request, Siti's arrangement of these skills must have been unique even if she was familiar with all constituent skills.

Several patterns appear in these social cases. Slave use and the iconic sign owe to auto-cuing, recreating behaviour originally provoked by external cues using internal (voluntary) cues (Donald, 1991). Both require first noticing that a valuable contingency has occurred; to this extent, they resemble accident-spawned innovations. The deception, beyond the novel blend of two scripts, involves simultaneously entertaining two representations of an event; it qualifies as pretending, with signs of symbolic pretending (Russon, 2002a).

Siti's coconut request, beyond its unique arrangement of skills, involved parang skills beyond Siti's own capabilities. Forest staff regularly use parangs; Siti sometimes watched closely and once reportedly pilfered a parang then chased staff 10–15 m with it before they could retrieve it. Staff had evidently chopped open coconuts in the past (old coconuts, split open, lay nearby), probably in Siti's presence (she foraged in this area often), but Siti was unlikely to have functional parang skills herself. Chopping coconuts open is difficult for humans and letting parangs fall into Siti's hands is clearly risky. Therefore, some of Siti's parang knowledge must have been observational. Further, Siti appeared to interpret the assistant's not acting on the coconut as his not understanding her request, because she altered her message in the direction of clarifying by showing him what she wanted done. Her demonstration then involved imitation (enacting behaviour she was unlikely to have learned experientially) using a substitute tool (stick representing parang), used for instructional vs functional purposes. It drew on skills from three cognitive domains-physical (food skills, tool use), logical (tool substitution), and social (imitate, mime-demonstrate). Events this complex are rare and probably unique, so creative interweaving of diverse forms of expertise to suit situation-specific purposes must have occurred.

Logic

Logical tasks concern necessary relations (e.g. the number of items in a set never changes with spatial arrangement) (Langer, 2000) so innovation may be less prevalent in logical expertise. I have a few candidates for innovations involving logic. Several rehabilitants created ordered sets. Princess, an adult female (CL), broke 6–8 sticks of like diameter to similar length and arranged all on the ground, carefully parallel to each other. Siswi, an adolescent female (CL), repeatedly pulled lengths of bark from a pile created by splitting a log for firewood, then replaced each on the pile–after turning each so that its inner surface (orange) faced up and bark surface (brown) down.

(1) *Body part ordering.*   Paul, as an adolescent (SW), used an idiosyncratic system for ordering fragments of termite nests he was opening for termites inside (Russon, 2002b). He held a main chunk in one hand and fragments he cracked from it in the other. He held the first fragment between the tips of his thumb and one finger, pinned fragments of the first to the base of his palm with his third fingertip, balanced fragments of the second on his wrist, and once balanced fragments of the third higher on his wrist. He sucked termites from fragments in reverse order, wrist to fingers, discarding the smallest fragment as he emptied it then returning to its direct parent. His fragment ordering resembles rudiments of human body part counting systems (Saxe, 1981). The utility of such a system is obvious. Cracking breaks termite nests to smithereens; as debris accumulates, which fragments are full vs empty gets confused. Several rehabilitants used simpler place-based systems for organizing chunks (e.g. hold chunks to be cracked later in their feet). Paul's system, perhaps, extended such simpler ones by placing each next good fragment on the next available spot near the last one.

I offered this sampler primarily to show the kinds of innovations and avenues to innovation that stand out even without systematic searching, not as an exhaustive list. It also shows that innovation can occur well beyond foraging in orang-utans; their innovation may in fact be relatively generalized across problem domains.

Foraging

I have focused on palm foods, a class that poses difficult problems requiring highly complex expertise. Orang-utans eat many palm items, including fruit, flowers, flower stalks, young and mature leaves, heart (meristem), and leaf petiole pith (parenchyma). From some species, they eat all these items. Palm foods other than flowers and fruits are available year-round (i.e. permanent). Permanent palm foods are especially interesting because many are armed with anti-predator defences (e.g. heart and pith are embedded and sometimes protected by spines or sharp-edged petioles), non-consumption cannot be attributed to non-availability or seasonal changes in the foods themselves, and they comprise a set of interrelated, complex foraging problems. These features affect how palm expertise is acquired, innovation included.

Identifying foods

Wild orang-utans probably learn to identify forest foods and much of that learning probably occurs in infancy, socially, through mothers (Box and Gibson, 1999; Russon, 2002b). Ex-captives, almost all orphaned in infancy, are therefore likely to be ignorant of many forest foods. Many behave as if they are when newly released: They ignore and even refuse foods that knowledgeable orang-utans prefer—even foods they later come to prefer (Russon, 2002b).

Change over time in rehabilitants' food repertoire in Sungai Wain echoes this pattern, especially for palm foods: many edible items were ignored for years. The established repertoire included 114 species by 05/1995 and 241 species by 07/2000; of about 40 palm species available, 4 were eaten by 05/1995 (5 items) and 26 by 07/2000 (56 items) (Peters, 1995; Russon, 2002b). Some of this change may reflect new opportunities created by residents' range expansion, knowledge introduced by new ex-captives released into the forest, sampling artefacts, etc. Some of it, however, probably reflects residents' new discoveries: Permanent palm foods added after 1995 are widely available in Sungai Wain and became popular after researchers first recorded their consumption, so it is implausible that rehabilitants lacked opportunities to try them or were eating them and researchers failed to notice. Evidence for discovery is strong for palm foods added after 1997. Sungai Wain received no new ex-captives after 1997, it has no wild orang-utans, and it is physically isolated from other forests with orang-utans, so new food knowledge could not have been introduced socially.

Three of five orang-utans I followed in 1998 (Tuti, Paul, Panjul) ate three new palm foods: two permanent foods, pith of two tree palms (*Pholidocarpus* sp. and *Licuala spinosa*, locally *serdang* and *daun biru*), and *Pholidocarpus* sp. fruit stalks. All three foods were first recorded on the first day of my yearly observation session and frequently thereafter, having never been recorded in several hundred hours' observation over several previous years that included these same orang-utans.

These foods were first recorded at the end of the prolonged fires and drought that afflicted Borneo in 1997–1998, so severe conditions probably forced rehabilitants to find new foods (Russon and Susilo, 1999). Serdang and daun biru probably became valuable as food sources because they remained available; both palm species favour areas that retain moisture (Whitmore, 1985). Similar food innovations have been reported in wild orang-utans in the wake of ecological disasters (Suzuki, 1992). Why rehabilitants did not eat daun biru pith beforehand is puzzling. It is easy to obtain, and they had eaten other items from daun biru and pith from other common tree palms for years. They may have ignored daun biru pith simply because preferred items were available concurrently.

Ecological conditions do not explain discovery of other new palm foods, however, or the acquisition process involved. Discovering palm foods other than fruits or flowers by trial and error exploration is improbable because of the prevalence of anti-predator defences that encourage would-be consumers to avoid palms, not explore them. Rehabilitants probably bypassed serdang, for instance, because of its formidable size and spines.

Rehabilitants could have discovered these new palm foods by generalizing based on similarities with known palm foods (see Russon, 2002b). All three new foods resembled items already in their repertoire and the first rehabilitants seen eating these new foods were among the most knowledgeable in Sungai Wain, which suggests that discovery owed to and built upon existing knowledge. Palms form a complex family of about 215 genera and 2600 species that vary widely in appearance (e.g. single vs multi-stemmed, crowned vs sheathed, tree vs climber, spiny vs smooth, edible vs toxic; Jones, 1995). All share a few basic structural features (e.g. perennial, woody, single meristem embedded in stem tip) and some share features at the genera or subfamily level (e.g. *Korthalsia* vs. *Calamus* sp.; climbing palms (rattan)). This makes it unlikely that orang-utans identify palms as strictly individual species or as an undifferentiated group (i.e. they probably recognize both species similarities and differences).

Consistent similarities and differences in rehabilitants' techniques for obtaining palm heart suggest they group palm plants into four categories, based on structural features associated with palm biology (robustness, defences, meristem access point; Russon, 2002b). Categories are: (1) slender rosettes (extract shoot from rosette centre with one pull, bite meristem off shoot's base); (2) robust, non-climbing rattans (tear open rosette base from side, bite/tear out meristem); (3) crowned tree palms (extract shoot from crown's centre, pulling it section by section, bite meristem off each section's base); (4) climbing rattans and sheathed tree palms (tear open rattan stem or sheath below apex from side, bite/tear out meristem).

Some distinguishing features are structural and not directly perceptible (e.g. meristem location) so these categories may be cognitively vs perceptually defined. Cognitive categorization supports rapid identification of unfamiliar items by generalization, based on the new item's similarity to a known category. Cognitive categorization is within orang-utans' reach; the rate at which rehabilitants expanded their range of palm foods and the sophistication of their skill with new palm foods are consistent with this possibility, so it could explain how rehabilitants identified new palm foods (for details, see Russon, 2002b).

Food processing techniques

Food processing techniques and potential innovations depend on the food problems presented. Palm foods present multiple defences. Palm heart (meristem) lies within the stem's apex, but this can be located on the forest floor or inaccessible 30 m above and protected by a leathery sheath, massive crown, spines, and/or spiny or sharp-edged petioles, depending on the species. Palms also grow, so some foods become problem sets naturally graded in difficulty: for instance, heart can be found in tiny seedlings, robust rosettes or massive trees.

Accordingly, palm techniques are typically arrangements of many tactics, or operations, each of which handles a specific defence. Many defences protect other foods as well (e.g. spines, shells, chemical repellents), so techniques typically draw operations from an established kit and operations may be reused for various foods. Importantly, then, orang-utans build palm techniques by combining and recombining sets of operations. Functional techniques require applying operations in organized fashion and handling their interplay (i.e. a strategy to orchestrate the set). Evidence shows that the strategies behind great apes' food processing techniques can be hierarchically organized products of hierarchical cognition (Byrne and Byrne, 1991; Matsuzawa, 1991; Russon, 1998). These qualities should affect how and where innovation occurs.

I focus on two tree palms, *Borassodendron borneensis* and *Oncosperma horridum* (locally, *bandang* and *nibung*), on which rehabilitants (SW, M) forage regularly for several foods. Their processing techniques are the most complex I know and they vary with age, experience, and sociality, so they offer good terrain for exploring innovation.

(1) *Bandang heart.* One principle underpins rehabilitants' bandang heart techniques: Extract the new shoot as it emerges from the palm's apex then bite meristem from its base. Bandang growth turns obtaining heart into a set of problems graded in difficulty by palm age. In immature bandang, the shoot is small, grass-like, in the centre of a slender rosette on the ground; in mature bandang, it is stout, spear-like, embedded in a massive crown, encircled by 50+ leaves with razor-edge petioles atop a 15-m trunk. Rehabilitants handle the variation with two strategies, acquired in order: a basic one for shoots in small rosettes (grab then pull the whole shoot at once) and a mature one for spears in large rosettes or trees (subsection the spear then pull and eat sections one by one) (Figure 13.1). The mature strategy works because leaves are fans of many finger-like laminae that are separated at the tips, even as shoots. Table 13.1 summarizes evidence on these techniques and possible innovations.

One juvenile male, Paul, advanced from *naiveté* to a mature strategy in four observable steps, in enough detail to suggest how progress occurred (Russon in press a; Table 13.1).

Figure 13.1 Paul pulls the new leaf from an immature bandang (robust rosette), for its heart (apical meristem) tissue. Note: the new leaf is in the centre of what will be the palm's crown; Paul is pulling a section of the new leaf's fan that he carefully selected and has bent the section's tip (sections of few laminae often snap when pulled); Paul is bracing against a nearby small tree to enhance his pulling strength. Drawing by Priyono, from a photograph by A. Russon (1996).

Table 13.1. Innovations in bandang heart techniques. Column headings: Obs. (h) = total hours observation; Loc'n = forest location (SW = Sungai Wain, M = Meratus); No. of cases = number of bandang heart cases observed (>1 sect = no. cases in which the orang-utan subdivided the bandang shoot into multiple sections then pulled each separately); Consec. pulls = no. of cases in which the orang-utan pulled several shoot sections consecutively, without pausing to eat; Tree brace = no. of cases in which the orang-utan braced against a tree to facilitate pulling out the shoot; 2-step pull = no. of cases in which the orang-utan pulled out the shoot in two steps, first pulling it down through the side of the crown and then pulling it out; First section = no. of cases in which the orang-utan selected the first section(s) of the shoot to pull in an unusual way (i.e. from the middle or both ends of the fan of laminae instead of from one end)

Name	Age (Year)	Obs. (h)	Loc'n	Total[1]	>1 sect	Consec. pulls	Tree brace	2-step pull	First section
Males									
Aming	9 (96)	42	SW	10	0				Mid
Bento	6 (95)	52	SW	7	0				
	7 (96)	38	SW	21	8				
	8 (97)	28	SW	7	3				
	10 (99)	54	M	3	3				
	11 (00)	22	M	3	2				
	12 (01)	40	M	18	16				
	13 (02)	37	M	4	2	1			
Charlie	10 (96)	57	SW	5	1				
	11 (97)	14	SW	16	11				
Enggong	5 (95)	48	SW	1	0				
	6 (96)	63	SW	64	35		1		
	7 (97)	28	SW	5	5				
Luter	8 (00)	14	M	7	4				Mid
Mono	5 (99)	23	M	5	3				
	7 (01)	36	M	16	10	4			
Paul	5 (95)	50	SW	3	0				
	6 (96)	87	SW	40	8				
	7 (97)	70	SW	42	24	1	15	5	
	8 (98)	9	SW	6	5				
Females									
Ida	4 (96)	25	SW	16	0		2		
Imelda	8 (96)	21	SW	3	0				
Maya	6 (99)	12	M	7	2		2		
	8 (01)	7	M	4	1		1		
	9[2] (02)	31	M	6	3				Each end
Pasaran	7 (01)	7	M	1	0				
Sariyem	5 (95)	31	SW	11	1				
Seni	8 (01)	8	M	1	0				
Siti	5 (96)	15	SW	4	0				
	9 (00)	73	SW	15	14				

(Continued)

Table 13.1. (*Continued*)

Name	Age (Year)	Obs. (h)	Loc'n	No. of cases Total[1]	>1 sect	Consec. pulls	Tree brace	2-step pull	First section
Tuti	8 (97)	7	SW	3	1				
	9 (98)	26	SW	1	1				
	11[2] (00)	18	SW	5	5				
Uce	11[2] (99)	35	SW	18	18	2	1		
Total		452		180					

[1] Included are periods when orang-utans were observed eating bandang heart.

[2] Female with first infant (<1 year old), so formally adult, but age-wise adolescent.

Paul was first seen eating bandang heart at about 5 years old, after 3–5 months forest experience, with Sariyem; he *scrounged* her discarded leftovers then tried a *flawed basic strategy* (pull shoot from small rosette, incorrectly eat tip). Within 6 months, he had mastered a *basic strategy plus two tactical enhancements* (two-step vs simple pull; brace against tree while pulling). Paul probably invented both tactics; both were idiosyncratic and he had few sources of social input because others resident in his area (Enggong, Bento) did not tolerate his proximity. A year later, Paul had acquired the *mature* strategy. Social learning likely contributed because Paul often foraged with the more knowledgeable Enggong in the interim, but he could have discovered the key to the mature strategy independently. Laminae sometimes slip free when Paul or others pull a shoot (Figure 13.1); noticing this accidental subdividing then recreating it deliberately leads to the mature strategy. It may also have been the additional year of cognitive development that enabled Paul to master his strategy; other data suggest the youngest age at which orang-utans master this strategy is 5 years (Russon, 1999b).

Four other tactical innovations stand out (Table 13.1). (1) Four rehabilitants with mature strategies (Paul, Uce, Mono, Bento) acquired similar economizations: pull several shoot sections consecutively, without eating any, then carry them aside and eat the lot. The first three must each have invented this because they had no contact. (2) Five rehabilitants (Maya, Paul, Ida, Uce, Enggong) braced themselves against a nearby tree to increase their ability to pull robust shoots. The first four must each have invented this because they had no contact with one another. (3) Most rehabilitants pulled their first section from one side of the shoot's fan then worked their way across to finish on the other side. The first section appears to be the hardest to extract (often snaps during pulling, will not budge, etc.) and it typically comprises only a few laminae, probably because the shoot fits very tightly in its socket. It affects extracting other sections because its removal reduces pressure, so the rest of the shoot can be extracted in larger sections and more easily. Three orang-utans invented new approaches to making the first section: Aming and Luter each took it from the middle of the shoot's fan; Maya took her first two sections from opposite sides of the fan, thereby loosening the rest of the shoot on both sides. (4) Bento, as a skilful subadult, enhanced his technique for other palm hearts by applying the subsectioning strategy to a new task, pulling petioles. After twice trying and failing to pull out a mature petiole that interfered with his obtaining *daun biru* and *nibung* heart, he subdivided the petiole in half lengthways and pulled each half separately.

Mature bandang heart techniques are hierarchically organized (Russon, 1998); these observations suggest more innovation at detail (tactical) than strategy levels. Many of these innovations could have been accidentally spawned. While Paul may have acquired the mature strategy socially, that it could have been discovered fortuitously is interesting with respect to how this strategy may first have emerged. Tactics (1), (3), and (4) suggest innovations generated by understanding: (1) and (3) concern organizational features that do not affect the technique's functionality, only its efficiency; (4) suggests transferring an operation from its original task to a related one.

(2) *Bandang pith.*   Rehabilitants also eat bandang pith (parenchyma). Their strategy is to bite open the petiole sheath lengthways, tear out strips of pith, and chew the pith for juice. Palm growth changes petiole size, toughness, and accessibility. Evidence on pith techniques comprises 156 cases on 15 orang-utans (SW) in 1996 and 1997.

I was able to trace Siti's progress from naïve to a mature strategy over 18 months, beginning at her release into Sungai Wain as a 5-year-old (7 cases 1996, 35 cases 1997). Immediately post-release Siti associated with Kiki and Ida, two female peer friends. Staff, who monitored daily, saw no one in Siti's area eat bandang pith until Kiki discovered it independently after a month. Kiki chose tiny rosettes at ground level; initially she chewed the whole petiole but within 3–4 days removed the sheath and chewed only pith. Siti and Ida were first seen eating bandang pith as leftovers they scrounged from Kiki; within a month both could obtain it themselves. Within another week Siti improved her technique to handle mature petioles (i.e. robust, arboreal).

Over the next 17 months Siti added tactical refinements. One was idiosyncratic, so probably innovative. Siti chose bandang with a liana or branch running diagonally through the crown and mature petioles arching over the liana/branch. Hanging from or sitting on the liana/branch, she bit into one such petiole; the petiole typically cracked and flopped over the liana/branch, which then acted as a 'hanger' anchoring the petiole while she tore it apart (5 cases; Figure 13.2). The hanger function could have resulted accidentally from her spatial choice. The palm-liana/branch layouts offer structures for hanging or sitting while working, so she needed layouts that would allow her to reach a petiole's underside. In such layouts, once positioned, she would face a petiole arching upwards, over her head—and also over the liana/branch. Siti's biting and tearing pulled the petiole towards her so when it cracked, it would fall towards her and, coincidentally, over the liana/branch. This hanger function is unlikely to have remained a fortuitous side effect because it was reused, and rehabilitants use and shift vegetation elsewhere to brace, anchor, and support plants when they apply force.

Two more refinements suggest coinnovation with Judi, a near-adult female who became Siti's regular companion. The two frequently ate bandang pith together, even working the same petiole. They repeatedly chose the second newest petiole for pith (5 cases). Others always took mature petioles for pith and shoots for heart, systematically pushing the second newest leaf aside. Each palm has some 50 mature petioles to one shoot and one second-newest leaf, so competition cannot explain Siti and Judi's choice. Also, these two regularly made their first bite into the petiole near the leaf then tore the petiole open towards its base. All others regularly made their first bite about a third of the way down from the leaf then tore the petiole open in both directions. Judi and Siti were both broadly

Figure 13.2 Siti uses a 'hanger' to support a bandang petiole she is tearing open for its pith. Her tearing open the petiole has cracked it and it has flopped over the tree from which she dangles. Farther up the hanger tree are dried remains of three other petioles cracked and supported the same way. This implies that she or someone else used the same hanger technique on this palm in the past. Photograph by A. Russon (1997).

naïve at release, Judi learned painfully slowly, and Siti had only the aptitude of a young juvenile. One learning from the other is therefore highly unlikely.

(3) *Nibung heart.*   Nibung heart is the hardest to obtain but, local people say, the tastiest. Nibung is a multi-stemmed tree palm that grows over 15 m tall (Whitmore, 1985). Immature stems are completely enshrouded with 10 cm long fierce spines. Stems lose their spines and become relatively smooth as they mature but then develop a tubular crown shaft, a tough, leathery sheath below the leafy crown, which surrounds the meristem. Techniques for nibung heart require, variously, tactics for foiling spines, sheath, and location. Nibung heart, like bandang heart, poses a set of problems graded in difficulty relative to the stem's age. Differently, nibung grow into multi-stem clusters by basal suckers, so stems of many different sizes can occur in a single cluster. Table 13.2 suggests innovations in orang-utans' nibung heart techniques.

In Sungai Wain, we first recorded nibung heart as a food in 1996, four years after rehabilitants were first reintroduced there. In total, four adolescents (two female, two male) and one subadult male have been seen eating it, all with extensive forest experience. All five chose stems less than 8 m tall (most less than 2 m) adjacent to an access structure, for instance,

Table 13.2. Changes over time in techniques for obtaining Nibung heart. Column headings: M/F = sex (M = male, F = female); Age = age in years (J = juvenile, A = adolescent); Exper. = amount of experience with forest life, post-release (<4 = less than 4 years, ≥4 = 4 years or more); ## = number of nibung palm stems the orang-utan opened for heart; Stem ht. = height (m) of nibung stems the orang-utan attacked (≤2 = 2 m tall or less, <8 = 2–8 m tall, ≥8 = 8 m tall or greater); Access = location from which the orang-utan tore apart the nibung stem (G = ground, A = on an 'access tree' adjacent to the nibung stem, N = on a nibung stem); Bite tip of stem, Pull stem over, Husk sheath, Other = unusual tactics used in extracting heart from within the stem's tip (bite and tear open the tip, pull the whole stem to the ground before tearing the tip open, tear open the sheath protecting the tip in adult stems, other = other unusual tactic, see text for description). Cell values indicate the number of cases in each category (values bracketed when based on WORP staff reports vs my direct observation)

Year	Name	M/F	Age J	Age A	Exper. <4	Exper. ≥4	##	Stem ht. ≤2	Stem ht. <8	Stem ht. ≥8	Access G	Access A	Access N	Bite tip of stem	Pull stem over	Husk sheath	Other
Sungai Wain																	
96	Panjul	M		9		>4	2	2			1	1		2			
	Aming	M		9	4		1					1		?			
98	Tuti	F		10		5	1		1		1	1		1			
00	Siti	F		9		4	1	1			1			1			
	Charlie	M		14		8	1		(1)			(1)		(1)			
Meratus																	
99	Bento	M		10		6	17	10	7		10	8		17	5		
	Luter	M	7		1		1*	(1)			(1)			1			
	Mono	M	6		2		2*	1			1			1			
00	Luter	M		8	2		15	14		(1)	14	1	(1)	1	6	(1)	
01	Bento	M		12		8	11	7	3	1	9	1	1	11		1	
	Mono	M		8	4		3		3	3		3	3	3		3	Split petiole
02	Seni	F		8		4	1*		1*		1*			1*			
	Bento	M		13		9	10	10			10	3		7	3		

* = also ate stem leftovers. Boxes highlight potentially innovative behaviour.

a log or tree from which they could reach the stem's tip. Positioned on the access structure, they leaned toward the stem's tip, bit and tore open its side, then bit and tore out heart.

In Meratus, nibung heart was an established food by 1999, only 2 years after the first reintroductions there. In 1999, Bento, an adolescent male about 10 years old, was the major known consumer. He was transferred to Meratus in 1997 after 4 years in Sungai Wain. In Sungai Wain he was not known to eat nibung but did forage heavily on other palms and was highly skilled in palm techniques. Mono and Luter, juvenile males 6–7 years old, also consumed nibung heart and often scrounged Bento's nibung leftovers. We observed a first innovation in nibung techniques in Meratus, in 1999: Bento pulled the whole nibung stem over via a mature leaf, then tore its tip open while seated on the ground (this works for immature stems yet to develop woody trunks) (5 cases). The juveniles only attacked stems less than 2 m tall, which they tore apart in place, from the ground.

In 2000, we found indirect evidence of rehabilitants newly attacking mature nibung 10–15 m tall. Observers found mature nibung stems (i.e. trees) standing but dead, crowns simply gone, in areas where rehabilitants regularly ranged and ate immature nibung heart. The culprits were never spotted but everyone seeing the damage agreed, only orang-utans could have done it. Destroying a nibung crown requires foragers with the strength and dexterity to climb to the palm's apex and tear it apart while arboreally positioned. No other resident species qualify. The only scenario imaginable was an orang-utan climbing an access tree to a point near a nibung crown, and then tearing its sheath open from positions in the access tree or the nibung itself. WORP forest staff twice saw orang-utans atop a mature nibung, one smallish, unidentified individual and Luter, who tried to tear the tip but failed, unable to bite through the sheath. In 2001, Bento and Mono were seen obtaining heart of mature nibung in this fashion (Figure 13.3).

The late appearance of nibung heart techniques and subsequent variants again suggest a process of building expertise in repeated, small steps. If new expertise built on old and if new palm foods may be identified by generalizing from existing palm categories, themselves defined by techniques, then known palm heart techniques may be relevant. Before 1996, known palm heart techniques differed for climbing (i.e. rattans) vs tree palms. For tree palms known at that time, daun biru and bandang, rehabilitants pull the shoot out of the *top* of the stem, from the *centre* of the crown; both are single-stemmed and unsheathed (Russon, 1998; Russon and Susilo, 1999). For climbing rattans, they tear into the *side* of the stem below the apex; most are spine-covered the length of the stem, many are multi-stemmed, and all lack sheaths. From this perspective, nibung is ambiguous. It is a tree palm but in most stages of growth it does not resemble rehabilitants' tree palm category (multi-stem clusters, immatures uncrowned, adults sheathed). Rehabilitants have never been seen using the tree palm technique for adult crowned nibung; for immatures, it is inappropriate. Nibung, especially immatures, better resemble climbing rattans (spiny, clustering, uncrowned, unsheathed). The formidable defences, orang-utans' food conservatism, and the difficulty of identifying mature nibung all suggest that rehabilitants would initially bypass nibung for foods easier to identify and obtain, first attack immature nibung using the rattan technique, and have greatest difficulty with mature nibung trees. This is what we observed.

Figure 13.3 Mono, adolescent male, obtaining heart tissue from a mature Nibung stem (*Oncosperma horridum*). Mono is in the process of tearing open the tough sheath that protects the heart, just below the crown; the job resembles husking a cob of corn. His repeated biting into the sheath has weakened the stem to the extent that the whole tip plus crown has broken off and fallen to the ground. Photograph by A. Russon (2001).

The order in which variants appeared suggests how rehabilitants generated their techniques: (1) for immature stems less than 8 m, bite and tear open the tip from the side, positioned on an access structure or (for stems less than 1–2 m) the ground; (2) for immature stems less than 8 m, pull the stem over then bite and tear open its tip from the side, on the ground; (3) for mature trees over 8 m, bite and tear open the tip from the side, and strip off the sheath (as in husking corn), on a nearby access structure or the nibung trunk itself.

Variant (1) is simply the climbing rattan technique. Variant (2) suggests an innovation (Bento, Table 13.2), perhaps spawned by fortuitous accident. While tearing apart a nibung stem from an access structure, rehabilitants often pull against a mature leaf, to bring the stem closer, to help support their weight as they lean to bite, or for balance (personal observation). The force can crack the stem, accidentally, mostly near the tip where biting and tearing have weakened it. If tips crack, they fall off or orang-utans pull them free.

Orang-utans then descend and open the tip on the ground, which simplifies the job and saves energy by eliminating tasks of supporting and balancing oneself arboreally and of counterbalancing the forces applied. Bento's pulling the stem over amounts to the same operation, enacted voluntarily and refined. This operation cannot be a product of associationist learning alone because Bento enhanced it to handle a harder task (whole stems, not just loose tips) and altered its sequential position within the technique (pull stems down before, not during, biting, and tearing).

Variant (3) probably arose by adjusting variant (1) for new defences presented by mature nibung (sheath, woody trunk) and old defences that disappear (spiny stem). It first appeared in 2000 so it qualifies as innovative, although who invented it is unclear. Variant (3) maintains the rattan strategy (bite/tear into the stem's tip from the side) and adds new tactics (strip sheath, climb nibung trunk). Other possibilities seem unlikely. It is unlikely that variant (3) was acquired as a completely distinct technique, for instance, because the three rehabilitants seen taking heart from mature nibung (or trying to) are the same three who take it from immature nibung (Table 13.2). It is also unlikely that variant (3) owes to trial-and-error experimentation because of the fierce defences, the inaccessible location, and the amount of time and energy required to gain rewards (e.g. Mono took >30 min to obtain one mature heart). It is possible that this clique of rehabilitants acquired variant (3) socially, from some outsider. By 2000, WORP had reintroduced over 300 ex-captives plus a small number of rescued wild orang-utans into Meratus and Bento, at least, ranged widely enough to meet orang-utans outside his clique. My sense is that importation is unlikely. Ex-captives beyond this clique were mostly younger, too small, and less experienced (most are released as juveniles less than 5–6 years old; WORP personal communication). Second, social learning of manipulative skills probably requires close proximity; that requires social tolerance (Coussi-Korbel and Fragaszy, 1995), which commonly requires established, positive relationships (Russon, 2002b). Bento, Mono, and Luter were maturing males, so more subject to competition than affiliation and ex-captives do not integrate well with wild orang-utans (Rijksen, 1978). The area in which mature nibung were heavily destroyed is near the project post and within the habitual range of this clique of 7–8 rehabilitants; it is so regularly monitored that outsiders would have been reported.

Discussion

Many of the factors proposed to favour innovation likely operated in these rehabilitant orang-utans. Extreme living conditions, ecological and social, probably forced or afforded some innovations. Behavioural propensities common to great apes (e.g. nest, probe, extract) and specific to orang-utans (e.g. join items) likely spawned innovations that resemble skills known elsewhere. Developmental changes in strength, weight, needs, etc. probably provoked age-related innovations. Rehabilitants appeared very adept at noticing and taking advantage of fortuitous accidents and on-the-spot improvisation.

Innovation rates

To my knowledge, no reliable quantitative data exist on innovation rates in orang-utans, wild or rehabilitant. The view that innovation is rare in wild orang-utans is an inference based on a proxy indicator: tool-assisted foraging is habitual in some wild communities but has never been observed in others, despite comparable opportunities for invention (van Schaik *et al.*, 1999; van Schaik and Knott, 2001).

Similar inferences can be drawn for rehabilitants. The innovation sampler indicates that rehabilitants innovate in a very broad range of problems, improvise effectively, and improvise for highly variable problems. By inference, their innovation rates should be high relative to species or populations that innovate less broadly, face less variable problems, or cannot improvise effectively. Systematic evidence on foraging innovations has further implications. I considered three of several hundred items these rehabilitants consume, so innovation rates are well beyond what these items indicate. If food availability affects innovation, orang-utan innovation rates may be paced by El Niño, which governs the climatic cycles causing the extreme food scarcities and abundances that orang-utans experience. Innovations in techniques for accessing permanent foods imply that innovation rates may also be paced by developmental changes, especially major ones (e.g. male adolescent growth spurt, critical phases of brain development). Cognitive considerations predict considerable innovation in physical problem solving because these problems can vary from one instance to the next; similar considerations likely apply to social problems, perhaps to an even greater degree. Innovation rates may be relatively uninteresting in these domains because innovation should be an intrinsic feature.

Innovation rates implied for rehabilitants, juvenile to subadult, seem at odds with Reader and Laland's (2001) finding that immature primates may be less innovative than adults. These authors suggest innovation frequently builds on other skills, so it requires levels of understanding more typical of adults than immatures. Abnormal lives and living conditions almost certainly played a major role in enhancing innovation in rehabilitants. Development, however, may be an important consideration. Adolescence brings cognition to near adult levels in great apes, alters morphology in ways that seriously affect arboreality and foraging, and changes sociality in ways that affect ranging and relationships. These changes could promote extensive innovation.

Evidence on rehabilitant orang-utan innovation also runs counter to the view that animal motor skill is usually not generative, that is, does not involve categorical analysis and recombination of elementary parts as human praxis does (e.g. Corballis, 1989). If great apes have generative cognitive systems, even in rudimentary form, we may have to turn the question around and ask if they ever *don't* innovate. For complex physical problems especially, innovation should be an everyday event in great apes.

Development

Because my foraging studies spanned several years, they offer insights into development and innovation. Innovation in rehabilitants' techniques for obtaining permanent palm foods were not associated with fluctuating food availability. Palm heart and pith are always available to these rehabilitants, in palms of all sizes. Ecological or social pressures may affect the extent of reliance on permanent foods and which species or items are consumed, but less clearly affect which individuals within a species are used (e.g. mature or immature palms). The nibung and bandang innovations identified were more clearly related to developmental changes, cognitive and physical, in orang-utans themselves. Developmental changes create new constraints (e.g. small branches may no longer support one's weight, pregnant/nursing females experience changed nutritional and travel constraints) and new opportunities (e.g. greater strength, more experience, and advanced cognition enable new skills). Developmentally created constraints and opportunities can drive innovation because they induce actors to change their expertise in response to individual factors.

Cognition

Virtually all these innovations amount to small variations on a theme or small advances to expertise (e.g. in foraging, most were tactical modifications to established strategies). Some derived from applying known expertise to new problems. Some were achieved by combined or rearranging known elements in new ways. A chimpanzee innovation at Gombe, Mike's use of fuel drums in his display, was similarly attributed to combining a number of existing perceptions and understandings in novel ways (Donald, 1991). Others resulted from recruiting accidental patterns into the voluntary repertoire (Kummer and Goodall, 1985, Chapter 10). These patterns are consistent with hierarchical models of great ape cognition and a hypothesis derived from it, that great apes may typically acquire behavioural strategies socially but their details, individually (Byrne and Russon, 1998).

Hierarchical cognitive systems may be intrinsically generative, as some have pointed out (Gibson, 1990; Rumbaugh *et al.*, 1996). They can create emergents, new competencies produced without specific training by generative operations of the actor's brain (Rumbaugh *et al.*, 1996). Some emergents may owe to bringing together expertise across cognitive domains, perhaps by jumping to higher levels of abstraction, interconnecting different cognitive abilities, or analogical reasoning (e.g. Premack, 1983; Parker, 1996; Thompson and Oden, 2000; Russon, in press c). Whatever the process, bringing together expertise from different domains makes it possible to address tasks that involve multiple, heterogeneous problems. Some problems facing free-living great apes have this heterogeneous quality. A single foraging task may present an interrelated set of physical, social, and logical problems (Russon, 1998; Stokes, 1999). For such problems, the particular mix of expertise needed and its organization can be unique to each case. Accordingly, each behavioural solution may have novel features. That great apes are good innovators is then consistent with critical survival problems in their natural habitat.

Grain

Whether innovations occur, and at what rate, may depend on the level at which behaviour is analysed because the steps by which advances are made define the shape and pace of innovations. These steps can be thought of as the 'grain' of an individual's behaviour and cognition. Grain determines the proper level at which behaviour should be analysed, which Byrne (1999b) describes as 'carving at the joints'. Great apes' low cognitive ceilings and slow cognitive development compared to humans, for instance, suggests relatively finer grain, and small, hard-gained innovative steps.

Grain is especially difficult to establish in cognitive systems that build structures through combination and hierarchization, which can organize behaviour at multiple levels. In great apes, grain may range from simple voluntary motor actions on the world to operations on object–object relations and arrangements or organized combinations of these components (Byrne *et al.*, 2002; Russon, 2002b). Grain is also a function of development in such systems (Case and Okamoto, 1996), especially during immaturity. In great apes, then, no common grain is appropriate for parsing behaviour: grain should vary between and within individuals as a function of age and experience. In particular, great ape immatures and learners operate at lower-level grains than adults and experts (Inoue-Nakamura and Matsuzawa, 1997). Infants are probably limited to simple voluntary actions on the world, while juveniles and older individuals are limited to operations on object–object relations and arrangements of these (Parker and McKinney, 1999). The same behaviour may even represent different grains depending on the actor's development and/or experience (Case, 1985; Byrne *et al.*, 2001; Russon, in press c).

Some current differences of opinion about great ape behaviour, innovation included, may owe to differing grains of analysis. Behavioural repertoires that treat simple actions (e.g. grip, hold) as units equivalent to operations on object–object relations (e.g. scrape off, twist apart) do not distinguish levels of organization relevant to great apes, so they may fail to show up important differences. It is only at the roughest grain that some claims make sense: for instance, that chimpanzee cultural transmission is represented by two behaviours, tool use and gestures (Tomasello, 1999). Some assessments relevant to great ape innovation and innovative rates discuss great ape expertise at rougher grain, for instance using tools or not (van Schaik *et al.*, 1999), and others use finer grains, for instance kiss-squeaking with or without leaves (Peters, 2001), ant dipping two-handed with long sticks or one-handed with short sticks (McGrew, 1998), or opening nibung stems in place or after pulling them down. These two grains of analysis suggest, respectively, low vs high rates of innovation. For complex expertise organized into multilevel ensembles of sets of skills, innovation could affect either grain or both. Without establishing the grain at which actors parse and compose behaviour, it is impossible to determine which of the two assessments is correct, if either. For great ape innovation, rate-related issues are likely to remain unresolved until grain issues have been factored into conceptual and methodological equations. The solution must involve carving expertise as the joints, where great apes themselves define the joints. If their expertise normally advances at lower levels, as evidence suggests, lower level analyses may often be the better choice.

Current evidence leaves a clouded picture of innovation in orang-utans and other great apes. Beyond the problem of how to model innovation in species with these cognitive capacities lies the problem of identifying it empirically. Innovation is defined against the backdrop of what an individual already does and its community already does. Both require extensive knowledge from the field that can only be gained by long-term study. In this context, rehabilitants may offer a valuable window on innovation. Ecological skills could be compared between orphaned ex-captives newly reintroduced to forest life and wild peers, for instance, to approximate independent vs socially supported problem solving. This would require assuring that the two groups remain separate yet tackle equivalent ecological problems. For orang-utans, no situations meeting these criteria currently exist; should they arise, they may offer the best opportunity possible for studying innovation in the field.

Summary

This chapter considers innovation from a cognitive perspective in ex-captive orang-utans readapting to forest life. Orang-utans are likely to be of high innovative potential partly because of their sophisticated cognition and prolonged development. Ex-captives offer an exceptional window on orang-utan innovation: most, orphaned and captive-reared, must acquire feral expertise without normal social support. I described innovations identified in the course of studying ex-captives' foraging behaviour to suggest how innovation occurs in foraging. I also described opportunistically identified innovations in non-foraging domains to suggest the scope of orang-utan innovation. Several recognized conditions appeared to foster innovation in ex-captives (e.g. extreme living conditions, behavioural propensities, coinnovation) and many predictable forms appeared (e.g. new combinations, voluntarily recreating fortuitous patterns, minor 'tactical' enhancements to behaviour programs). In addition, developmental change appeared to create an important impetus for innovation by altering behavioural constraints and opportunities. Moreover, complex cognition, including generativity, appeared to underpin cases of deft improvisation and creatively, interweaving multiple types of expertise in complex situations. These cases suggest orang-utan innovation is relatively generalized rather than focused on specific types of problems, and relatively common. They also show the importance of the grain of analysis for species like great apes whose behaviour can be organized into multi-level programs. In these species, innovation may occur at some levels but not others, so detecting its frequency and even its occurrence depends on probing behaviour at the appropriate level.

Acknowledgements

Studies of the ex-captive orang-utans were sponsored by the Research Institute of the Indonesian Ministry of Forestry and Estate Crops, Samarinda, E. Kalimantan, Indonesia, the TROPENBOS Foundation of the Netherlands, and the Wanariset Orang-utan Reintroduction Project. Glendon College and York University, Toronto, and The Natural Sciences and Engineering Research Council of Canada provided funding to support the research.

References

Bard, K. A. (1993). Cognitive competence underlying tool use in free-ranging orang-utans. In. *The use of tools by human and non-human primates* (ed. A. Berthelet and J. Chavaillon), pp. 103–113. Oxford: Clarendon Press.

Boesch, C. and Boesch-Achermann, H. (2000). *The chimpanzees of the Taï forest.* Oxford: Oxford University Press.

Box, H. O. and Gibson, K. R. (ed.) (1999). *Mammalian social learning: Comparative and biological perspectives.* Cambridge: Cambridge University Press.

Byrne, R. W. (1995). *The thinking ape.* Oxford: Oxford University Press.

Byrne, R. W. (1997). The technical intelligence hypothesis: An alternative evolutionary stimulus to intelligence? In *Machiavellian intelligence II* (ed. A. Whiten and R. W. Byrne), pp. 289–311, Cambridge: Cambridge University Press.

Byrne, R. W. (1999a). Cognition in great ape ecology: Skill-learning ability opens up foraging opportunities. In *Mammalian social learning: Comparative and ecological perspectives* (ed. H. O. Box and K. R. Gibson), pp. 333–50. Cambridge: Cambridge University Press.

Byrne, R. W. (1999b). Imitation without intentionality: Using string parsing to copy the organization of behaviour. *Animal Cognition*, **2**, 63–72.

Byrne, R. W. and Byrne, J. M. E. (1991). Hand preferences in the skilled gathering tasks of mountain gorillas (*Gorilla g. beringei*). *Cortex*, **27**, 521–46.

Byrne, R. W. and Russon, A. E. (1998). Learning by imitation: A hierarchical approach. *Behavioural and Brain Sciences*, **21**, 667–721.

Byrne, R. W., Corps, N., and Byrne, J. M. E. (2001). Estimating the complexity of animal behaviour: How mountain gorillas eat thistles. *Behaviour*, **138**, 525–57.

Cant, J. G. H. (1987). Positional behavior of female Bornean orang-utans (*Pongo pygmaeus*). *American Journal of Primatology*, **12**, 71–90.

Case, R. (1985). *Intellectual Development from Birth to Adulthood*. Orlando, FL: Academic Press.

Case, R. and Okamoto, Y. (ed.) (1996). *The role of central conceptual structures in the development of children's thought* (series no. 246, Vol. 61). Chicago, IL: University of Chicago Press.

Chevalier-Skolnikoff, S., Galdikas, B. M. F., and Skolnikoff, A. (1982). The adaptive significance of higher intelligence in wild orang-utans: A preliminary report. *Journal of Human Evolution*, **11**, 639–52.

Corballis, M. C. (1989). Laterality and human evolution. *Psychological Review*, **96**, 492–505.

Coussi-Korbel, S. and Fragaszy, D. M. (1995). On the relation between social dynamics and social learning. *Animal Behaviour*, **50**, 1441–53.

Donald, M. (1991). *Origins of the modern mind*. Cambridge, MA: Harvard University Press.

Fleagle, J. G. (1999). *Primate adaptation and evolution*, 2nd edn. San Diego: Academic.

Fox, E. A., Sitompul, A. F., and van Schaik, C. P. (1999). Intelligent tool use in wild Sumatran orang-utans. In *The mentalities of gorillas and orangutans* (ed. S. T. Parker, R. W. Mitchell, and H. L. Miles), pp. 99–116. Cambridge: Cambridge University Press.

Gibson, K. R. (1990). New perspectives on instincts and intelligence: Brain size and the emergence of hierarchical mental construction skills. In *'Language' and intelligence in monkeys and apes: Comparative developmental perspectives* (ed. S. T. Parker and K. R. Gibson), pp. 97–128. New York: Cambridge University Press.

Gibson, K. R. (1993). Tool use, language and social behavior in relationship to information processing capacities. In *Tools, language and cognition in human evolution* (ed. K. R. Gibson and T. Ingold), pp. 251–69. Cambridge: Cambridge University Press.

Gosselain, O. P. (2000). Materializing identities: An African perspective. *Journal of Archaeological Method and Theory*, **7**, 187–217.

Huffman, M. A. and Hirata, S. (in press). Biological and ecological foundations of primate Behavioral traditions. In *Towards a biology of traditions: Models and evidence* (ed. D. Fragaszy and S. Perry). Cambridge: Cambridge University Press.

Inoue-Nakamura, N. and Matsuzawa, T. (1997). Development of stone tool use by wild chimpanzees (*Pan troglodytes*). *Journal of Comparative Psychology*, **111**, 159–73.

Jones, D. L. (1995). *Palms throughout the world*. Washington: Smithsonian Institute.

King, B. J. (1994). *The information continuum*. Santa Fe, NM: School of American Research.

Knott, C. D. (1998). Changes in orang-utan caloric intake, energy balance, and ketones in response to fluctuating fruit availability. *International Journal of Primatology*, **19**, 1061–79.

Kummer, H. and Goodall, J. (1985). Conditions of innovative behaviour in primates. *Philosophical Transactions of the Royal Society of London, Series B*, **308**, 203–14.

Langer, J. (1998). Phylogenetic and ontogenetic origins of cognition: Classification. In *Piaget, evolution, and development* (ed. J. Langer and M. Killen), pp. 33–54. Mahwah, NJ: Lawrence Erlbaum Assoc.

Langer, J. (2000). The descent of cognitive development. *Developmental Science*, **3**, 361–88.

Lefebvre, L., Whittle, P., Lascaris, E., and Finkelstein, A. (1997). Feeding innovations and forebrain size in birds. *Animal Behaviour*, **53**, 549–60.

Maple, T. L. (1980). *Orang-utan Behaviour*. New York: Van Nostrand Reinhold Co.

Matsuzawa, T. (1991). Nesting cups and metatools in chimpanzees. *Behavioural and Brain Sciences*, **14**, 570–71.

Matsuzawa, T. (2001). Primate foundations of human intelligence: A view of tool use in nonhuman primates and fossil hominids. In *Primate origins of human cognition and behavior* (ed. T. Matsuzawa), pp. 3–25. Tokyo: Springer-Verlag.

McGrew, W. C. (1992). *Chimpanzee material culture: Implications for human evolution.* Cambridge: Cambridge University Press.

McGrew, W. C. (1998). Culture in nonhuman primates? *Annual Review of Anthropology*, **27**, 301–28.

Mitchell, R. W. (1999). Deception and concealment as strategic script violation in great apes and humans. In *The mentalities of gorillas and orangutans* (ed. S. T. Parker, R. W. Mitchell, and H. L. Miles), pp. 295–315. Cambridge: Cambridge University Press.

Parker, S. T. (1996). Apprenticeship in tool-mediated extractive foraging: The origins of imitation, teaching and self-awareness in great apes. In *Reaching into thought: The minds of the great apes* (ed. A. E. Russon, K. A. Bard, and S. T. Parker), pp. 348–70. Cambridge: Cambridge University Press.

Parker, S. T. and Gibson, K. R. (1977). Object manipulation, tool use and sensorimotor intelligence as feeding adaptations in cebus monkeys and great apes. *Journal of Human Evolution*, **6**, 623–41.

Parker, S. T. and Gibson, K. R. (ed.) (1990). *'Language' and intelligence in monkeys and apes: Comparative developmental perspectives.* New York: Cambridge University Press.

Parker, S. T. and McKinney, M. L. (1999). *Origins of intelligence: The evolution of cognitive development in monkeys, apes, and humans.* Baltimore: The Johns Hopkins University Press.

Peters, H. H. (1995). *Orangutan reintroduction? Development, use and evaluation of a new method: Reintroduction.* M.Sc. Thesis, Groningen University, Netherlands.

Peters, H. (2001). Tool use to modify vocalisations by wild orang-utans. *Folia Primatologica*, **72**, 242–44.

Povinelli, D. and Cant, J. G. H. (1995). Arboreal clambering and the evolution of self-conception. *Quarterly Review of Biology*, **70**, 393–421.

Premack, D. (1983). The codes of man and beast. *Behavioural and Brain Sciences*, **6**, 125–37.

Reader, S. M. and Laland, K. N. (2001). Primate innovation: Sex, age and social rank differences. *International Journal of Primatology*, **22**, 787–805.

Rijksen, H. D. (1978). *A field study on sumatran orang utans (Pongo pygmaeus abelii Lesson (1827).* Wageningen: H. Veenman & Zonen B. V.

Rodman, P. S. (2000). *Great ape models for the evolution of human diet.* Poster presented at the International Congress of Anthropological and Ethnological Sciences, Williamsburg, Virginia: http://www.cast.uark.edu/local/icaes/conferences/wburg/posters/psrodman/GAMHD.htm

Rumbaugh, D. M., Washburn, D. A., and Hillix, W. A. (1996). Respondents, operants, and *emergents*: Toward an integrated perspective on behavior. In *Learning as a self-organizing process* (ed. K. Pribram and J. King), pp. 57–73. Hillsdale, NJ: Lawrence Erlbaum Associates.

Russon, A. E. (1998). The nature and evolution of orang-utan intelligence. *Primates*, **39**, 485–503.

Russon, A. E. (1999a). Orang-utans' imitation of tool use: A cognitive interpretation. In *The mentalities of gorillas and orangutans* (ed. S. T. Parker, R. W. Mitchell, and H. L. Miles), pp. 119–45. Cambridge: Cambridge University Press.

Russon, A. E. (1999b). *Acquisition of food processing expertise in free-ranging orangutans.* Presented at Gesellschaft für Primatologie, Utrecht, The Netherlands, August 17–22.

Russon, A. E. (2002a). Pretending in free-ranging rehabilitant orang-utans. In *Pretending in animals, children, and adult humans* (ed. R. W. Mitchell), pp. 229–40. Cambridge: Cambridge University Press.

Russon, A. E. (2002b). Return of the native: Cognition and site-specific knowledge in orang-utan rehabilitation. *International Journal of Primatology*, **23**, 461–78.

Russon, A. E. (2003). Comparative developmental perspectives on culture: The great apes. *Between biology and culture: Perspectives on ontogenetic development* (ed. H. Keller, Y. H. Poortinga, and A. Schölmerich), pp. 30–56. Cambridge: Cambridge University Press.

Russon, A. E. (in press a). Developmental perspectives on great ape traditions. In *Towards a biology of traditions: Models and evidence* (ed. D. Fragaszy and S. Perry). Cambridge: Cambridge University Press.

Russon, A. E. (in press b). The role of enculturation in great ape foraging expertise. In *Culture and meanings among apes, ancient humans and modern humans* (ed. F. Joulian). Kluwer/Plenum Press.

Russon, A. E. (in press c). The nature of great ape intelligence. In *The evolution of great ape intelligence* (ed. A. E. Russon and D. Begun). Cambridge: Cambridge University Press.

Russon, A. E. and Begun, D. R. (ed.) (in press). *The evolutionary origins of great ape intelligence.* Cambridge: Cambridge University Press.

Russon, A. E. and Galdikas, B. M. F. (1993). Imitation in free-ranging rehabilitant orang-utans. *Journal of Comparative Psychology*, **107**, 147–61.

Russon, A. E. and Galdikas, B. M. F. (1995). Constraints on great apes' imitation: Model and action selectivity in rehabilitant orang-utan imitation (*Pongo pygmaeus*). *Journal of Comparative Psychology*, **109**, 5–17.

Russon, A. E. and Susilo, A. (1999). The effects of the 1997–98 droughts and fires on orang-utans in Sungai Wain Protection Forest, E. Kalimantan, Indonesia. In *Impacts of fire and human activities on forest ecosystems in the tropics: Proceedings of the Third International Symposium on Asian Tropical Forest Management* (ed. H. Suhartoyo and T. Toma), pp. 348–72. Samarinda, Indonesia: Tropical Forest Research Center, Mulawarman University and Japan International Cooperation Agency.

Russon, A. E., Bard, K. A., and Parker, S. T. (ed.) (1996). *Reaching into thought: The minds of the great apes.* Cambridge: Cambridge University Press.

Russon, A. E., Mitchell, R. W., Lefebvre, L., and Abravanel, E. (1998). The comparative evolution of imitation. In *Piaget, evolution, and development* (ed. J. Langer and M. Killen) pp. 103–43. Mahwah, NJ: Lawrence Erlbaum Associates.

Saxe, G. B. (1981). Body parts as numerals: A developmental analysis of numeration among the Oksapmin of Papua New Guinea. *Child Development*, **52**, 306–16.

Shapiro, G. L. (1982). Sign acquisition in a home-reared/free-ranging orang-utan: Comparisons with other signing apes. *American Journal of Primatology*, **3**, 121–9.

Smits, W. T. M., Heriyanto, and Ramono, W. S. (1995). A new method for rehabilitation of orangutans in Indonesia: A first overview. In *The neglected ape* (ed. R. D. Nadler, B. F. M. Galdikas, L. K. Sheeran, and N. Rosen), pp. 69–77. New York: Plenum.

Stokes, E. J. (1999). *Feeding skills and the effect of injury on wild chimpanzees.* Ph.D. Thesis, University of St Andrews, Scotland.

Suddendorf, T. and **Whiten, A.** (2001). Mental evolution and development: Evidence for secondary representation in children, great apes, and other animals. *Psychological Bulletin,* **127**, 629–50.

Suzuki, A. (1992). The population of orang-utans and other non-human primates and the forest conditions after the 1982–83's fires and droughts in Kutai National Park, East Kalimantan, Indonesia. In *Forest biology and conservation in Borneo* (ed. G. Ismail, M. Mohamed, and S. Omar), pp. 190–205. Center for Borneo Studies Publ. No. 2, Yayasan Sabah, Kota Kinabalu, Sabah, Malaysia.

Swan, R. A. and **Warren, K. S.** (2000). Health, management, and disease factors affecting orang-utans in a reintroduction centre in Indonesia. Presented at *The Apes: Challenges for the 21st Century.* Chicago, May 10–13.

Thompson, R. K. R. and **Oden, D. L.** (2000). Categorical perception and conceptual judgements by nonhuman primates: The paleological monkey and the analogical ape. *Cognitive Science,* **24**(3), 363–96.

Tomasello, M. (1999). The human adaptation for culture. *Annual Review of Anthropology,* **28**, 509–29.

Tomasello, M. and **Call, J.** (1997). *Primate cognition.* New York: Oxford University Press.

van Schaik, C. P. and **Knott, C. D.** (2001). Geographic variation in tool use on *Neesia* fruits in orang-utans. *American Journal of Physical Anthropology,* **114**, 331–42.

van Schaik, C. P., Fox, E. A., and **Sitompul, A. F.** (1996). Manufacture and use of tools in wild Sumatran orang-utans. *Naturwissenschaften,* **83**, 186–8.

van Schaik, C. P., Deaner, R. O., and **Merrill, M. Y.** (1999). The conditions for tool use in primates: implications for the evolution of material culture. *Journal of Human Evolution,* **36**, 719–41.

van Schaik, C. P., Preuschoft, S., and **Watts, D. P.** (in press). Great ape social systems. In *The evolutionary origins of great ape intelligence* (ed. A. E. Russon and D. Begun). Cambridge: Cambridge University Press.

Yamagiwa, J. (in press). Diet and foraging of the great apes: Ecological constraints on their social organizations and implications for their divergence. In *The evolutionary origins of great ape intelligence* (ed. A. E. Russon and D. Begun). Cambridge: Cambridge University Press.

Whiten, A. and **Byrne, R. W.** (1988). Taking (Machiavellian) intelligence apart: Editorial. In *Machiavellian intelligence: Social expertise and the evolution of intellect in monkeys, apes, and humans* (ed. R. W. Byrne and A. Whiten), pp. 50–65. Oxford: Oxford University Press.

Whiten, A., Goodall, J., McGrew, W. C., Nishida, T., Reynolds, V., and **Sugiyama, Y.** (1999). Culture in chimpanzees. *Nature,* **399**, 682–5.

Whitmore, T. (1985). *Palms of Malaya.* Singapore: Oxford University Press.

HUMAN INNOVATION

CHAPTER 14

HUMAN CREATIVITY: TWO DARWINIAN ANALYSES

DEAN KEITH SIMONTON

Introduction

Of all species that evolved on this planet, *Homo sapiens* is without doubt the one that acquired the greatest capacity for creativity. This human ability not only exceeds that of any known animal, primate or otherwise, but even surpasses the capacity of other hominids, no matter how closely placed on the evolutionary tree. Indeed, judging from both technology and art, the creativity demonstrated by the Cro-Magnons may even represent a quantum jump over that displayed by the Neanderthals—notwithstanding their assignment to the same genus and even species. It has even been argued that the human ability to generate creative ideas vastly exceeds evolution's capacity to generate novel life forms (Basalla, 1988). After all, human creativity has generated all of the diverse cultures and civilizations on this globe, each replete with inventions and techniques, mythologies and philosophies, works of art and musical compositions, customs and laws, fashions, games and sports, and a host of other artefacts having no clear counterparts in the animal world.

Not surprisingly, given the wealth of products generated by the creative mind, the subject of creativity has attracted a huge amount of scientific research. The earliest empirical investigations on the subject date back to the days of Quételet (1835/1968) and Galton (1869), and the pace of research accelerated appreciably in the last half of the twentieth century. The research literature has become so vast and rich that creativity has become the focus of several hefty handbooks (e.g. Glover *et al.*, 1989; Sternberg, 1999) and, most recently, the two-volume *Encyclopedia of Creativity* (Runco and Pritzker, 1999). Accordingly, it is impossible to provide anything more than a superficial overview of the research findings within the confines of a single book chapter or journal article (see, e.g. Simonton, 2000b). Therefore, rather than do so here, I wish to take a more restricted approach to the phenomenon of human creativity. In particular, I will examine the literature from a single theoretical perspective, namely that of Darwinism. Because Darwinian theory plays such a substantial role in the scientific study of animal behaviour, this theoretical discussion will help define the connections, if any, between human creativity and animal innovation. In addition, Darwinian theory currently represents the only comprehensive explanatory framework that accommodates most major empirical findings in the field (Simonton, 1999b, in press). But before this theoretical treatment can begin, it is first necessary to make two distinctions, the first between creativity and innovation and the second between primary and secondary Darwinism.

Human creativity and animal innovation

Creativity and innovation are closely related but not absolutely identical concepts. On the one hand, creativity is defined as producing an idea or behaviour that is simultaneously novel and useful (Simonton, 1999b). It is most often used in the context of problem solving, where a solution to a problem is deemed creative when it is original and at the same time satisfies some criterion or set of criteria for what counts as a successful solution. Solutions that work but that are not novel are considered algorithmic or routine, whereas solutions that are highly original but are totally useless are deemed non-creative, even crazy. Thus, accountants are not considered creative when they complete financial reports or tax returns because they are simply applying well-known techniques. At the other extreme, paranoid psychotics may devise extremely creative 'solutions' to life problems—such as the fantasy world depicted in the film *Beautiful Mind* based on the experiences of John Nash—and yet these solutions are maladaptive rather than adaptive. Rather than enabling these persons to hold jobs and to maintain relationships, their delusions and hallucinations serve to alienate them from meaningful activity.

On the other hand, innovation is usually defined as the act of introducing something new, whether a novel idea or an original behaviour. Although the criterion of usefulness or adaptiveness is not an explicit part of the definition, this standard is often implied. This implication holds because a new idea or behaviour would seldom be introduced unless it served some useful function (e.g. new weapons, agricultural practices, metallurgical techniques, fashions that indicate status in social hierarchies, religious beliefs serving as coping mechanisms or vehicles of social control). However, unlike in the concept of creativity, innovation does not necessarily stipulate that the innovator also be the *creator* of the innovation. Instead, the innovator may simply be disseminating the original and adaptive idea or behaviour acquired from someone who gets credit for the actual creativity. Hence, a farmer who adopts a new hybrid seed or a biomedical researcher who adopts a new staining technique would be styled an innovator, but not a creator. Even so, this distinction is not absolute, because the definition of innovation also does not exclude the possibility that the innovator and creator is one and the same person. Indeed, in most domains of human creativity, this distinction often breaks down. For instance, Albert Einstein's paper on the photoelectric effect applied Max Planck's quantum solution to the black body problem to a different phenomenon. Because Einstein had adopted Planck's new theory, he can count as an innovator. Yet because he was applying quantum theory to a phenomenon distinct from Planck's original application, Einstein can also be considered a creator—so much so that it was this work that explicitly won him the Nobel Prize for physics.

These ambiguities notwithstanding, one point is clear: Creativity is logically prior to innovation. Whether the creator and the innovator are the same person or different persons, one cannot innovate without an act of creation first taking place. It is largely for this reason that research on human creativity seldom uses the word 'innovation,' finding the latter process more secondary. Indeed, the term innovation is largely confined to extremely narrow applications, most often to work and organizational settings (West and Rickards, 1999). This situation contrasts greatly with practice in the research on animal behaviour where the word 'innovation' is used more or less the same way as 'creativity' is utilized in human research (as is apparent in most of the chapters in this volume). For example, when Imo, the young female Japanese macaque, innovated the practice of washing of sweet potatoes (Kawai, 1965), she was displaying creative behaviour, because what she did was both novel and adaptive. Therefore, even though the review that follows the usage favoured in the human literature, it must be recognized that the discussion can also be said to treat innovative behaviour in the human species.

Primary and secondary Darwinism

By the definition of creativity just given, Charles Darwin's *Origin of Species* certainly must be considered an exemplary creative product. First, it was novel, so much so that it sparked considerable controversy, a not uncommon response to original ideas. Even Alfred Wallace, the reputed cofounder of the theory of evolution by natural selection, assigned Darwin primary credit for this creative achievement. Second, it was adaptive—in the scientific sense that it explained a tremendous diversity of biological facts using a highly parsimonious scientific theory. Moreover, by making a convincing case that evolution was a scientific fact, and by hypothesizing a process by which evolution could take place, Darwin provided a comprehensive framework for understanding all of life.

In effect, Darwin's *Origin of Species* offered an implicit theory of creativity. That is, Darwin proposed a process by which new life forms could emerge and evolve. This process consists of only two fundamental steps: (a) the production of spontaneous variations and (b) the selective retention of those variations that are the most fit. A significant feature of the variation process is what may be called its 'blindness' (Campbell, 1960; Simonton, 1999b). According to Darwin, the variations exhibited no foresight, nor were they guided by some presumably divine purpose. Although he was not able to specify exactly how these blind variations came about, later advances enabled subsequent evolutionary theorists to ground the mechanism in genetic recombination and non-directed mutation. The advent of genetics also provided the basis for understanding how adaptive variations were retained. The final product of these developments was what we will call a 'blind-variation and selective-retention' or BVSR model (Cziko, 1998).

So impressive was the explanatory power of Darwinian theory that the same BVSR model began to be applied to other phenomena. These applications can be considered as examples of *secondary* Darwinism, to distinguish them from *primary* Darwinism, which is confined to biological evolution (Simonton, 1999b). For instance, both the acquisition of immunity and neurological development in complex nervous systems have been described as BVSR mechanisms (Cziko, 1995). B. F. Skinner's conception of operant conditioning provides another explicit example of secondary Darwinism (Dennett, 1995; Epstein, 1991). The organism emits behavioural operants that are either reinforced or punished by environmental conditions. Even more strikingly, Skinner and his students soon extended the operant conditioning paradigm to creative behaviour (Epstein, 1991). This was by no means the first attempt to explicit creativity in BVSR terms. Only a score years after the *Origin* appeared William James (1880) proposed that the creative process might operate in an analogous fashion to biological evolution. Of special pertinence here is Campbell's (1960) article advocating an expressly BVSR model of the creative process (see also

Campbell, 1965, 1974). Moreover, for the past dozen years I have been actively engaged in developing my own elaborations and extensions of the same framework (e.g. Simonton 1988c, 1997a, 1999a, b).

Given that the theory has received such extensive treatment, it is impossible to provide a complete presentation in the limited space allotted in this chapter. Consequently, I will only provide an abstract of the general framework and documentation. Furthermore, in presenting the BVSR model, I will confine discussion to the blind variation stage of the theory, ignoring the stage of selective retention. The reason for doing so is that the latter part of the model is the least distinct from alternative theories. In problem solving, for instance, the problem is already given along with the criteria for what constitutes an acceptable solution. Hence, it comes as no surprise that the BVSR model does not substantially differ from rival accounts of the creative process at this stage of problem solving. In a sense, the circumstance is comparable for what holds for primary Darwinism. Alternative theories of the origin of species all agreed that life forms were well adapted to their respective environments (Cziko, 1995). Where they differed is how that adaptation originated. Creationists believed that God pre-selected the adaptations, even arguing that the adaptations proved God's existence. Lamarckians, in contrast, held that adult organisms actively acquired adaptive characteristics that were then passed down to their offspring. Only Darwinism attributed the adaptations to the selection of those spontaneous variations that yielded higher fitness. In the case of secondary Darwinian theory, this contrast dovetails with the definition of creativity—the two-fold components of novelty and utility. The variation portion of the model is concerned with the generation of novelty, while the selection portion of the model is concerned with the determination of utility. The former process has primacy over the latter.

Readers who seek more information about the full model can consult the more detailed articles and monographs, especially my book *Origins of Genius: Darwinian Perspectives on Creativity* (Simonton, 1999b).

Human creativity and secondary Darwinism

According to Campbell's (1960) theory, creativity begins with the generation of behavioural or ideational variations. Moreover, this process almost invariably requires, at some crucial juncture, the generation of ideational variations that are to a certain degree 'blind.' Campbell's use of this particular term was perhaps unfortunate because it has caused numerous misunderstandings about the nature of his theory (Cziko, 1998; Simonton, 1999b). Some of these misconceptions will be discussed later in this chapter. At this point, let it suffice to say that variations are blind to the extent that the creator cannot completely anticipate what idea will work and what will not—or what idea will succeed and what will fail. As a consequence, the creative mind must engage in a certain amount of free association, primary process, defocused attention, exploratory play, tinkering, trial-and-error, or some other relatively unrestricted activity. Significantly, the BVSR model does not require that the variation procedure always operate according to the same mechanism. Indeed, one of the most fascinating aspects of creativity is that it results not from one process but many, as illustrated in the following three examples.

First, creativity can certainly work in a fashion hardly distinguishable from operant conditioning. The individual may generate various permutations of established behaviours—often through playful manipulation of objects in the environment—and thereby encounter a combination of acquired behaviours that serves to solve some problem. That is, behavioural creativity is the consequence of subjecting the components of a repertoire of acts to some exploratory combinatorial process. Köhler (1925) provided a classic illustration of such behavioural insights in his observations of how Sultan the common chimpanzee was able to join two sticks to retrieve a banana placed just out of reach. Epstein (1991) has proposed a model that explicates Sultan's insight behaviour in totally operant terms. Furthermore, Epstein has shown that this model does an excellent job predicting 'insightful' behaviour in pigeons that are given problems comparable to those that Köhler provided his apes. Although one might be inclined to dismiss this behavioural BVSR as too primitive to support major acts of human creativity, this behavioural process actually has considerable importance. Kantorovich (1993), for instance, has argued that playful 'tinkering' often provides the basis for breakthrough discoveries in science.

Second, for an organism that is endowed with sufficient cognitive complexity, the BVSR process can be rendered more efficient. As Campbell (1960) pointed out, both the generation of variations and the testing of those variations can occur covertly rather than overtly. Dennett (1995) styled creatures that engage in this kind of problem solving 'Popperian,' in contrast to the 'Skinnerian' creatures of the previous example. Such organisms take advantage of the fact that they contain internal representations of the external world, along with representations of various ways of acting on that world. The internal representation, for instance, might be a 'cognitive map' of the physical environment, which the organism can then use to conceive alternative routes should the normal pathway be obstructed. By engaging in such internalized 'trial-and-error,' the organism increases the odds that when it finally emits an overt behaviour, that action will be successful.

Third and last, although there is no doubt that much creativity operates in the above manner—especially in everyday problem-solving situations—many of the more impressive acts of human creativity rely on a more sophisticated BVSR process. Human beings have minds that contain not just images of themselves and the outer world, but also 'cultural artefacts' that can be used in lieu of those mental images. Those artefacts include language, logic, mathematics, graphics, symbols, and various tools and devices, whether mechanical or electronic. Dennett (1995) called creatures that can exploit these means to problem solving 'Gregorian' (a term inspired by some ideas of Richard Gregory, the British psychologist). The only Gregorian creature we currently know of is *Homo sapiens*. Probably the supreme vehicle for this highly abstract form of BVSR is mathematics. Once a correspondence has been established between mathematical symbols and the external world, the symbols can undergo efficient manipulations to yield discoveries that then can be tested against the world, and new discoveries thus made. Sometimes these predictions will be derived in a systematic fashion from the mathematical representations, but other times the predictions will be the serendipitous result of playful tinkering with the abstractions.

How plausible is the notion that human creativity represents another type of secondary Darwinian process? To address this question, I first provide a brief overview of the empirical evidence, and then turn to objections that are often raised against this view.

Evidence

A vast amount of data can be marshalled in support of a BVSR theory of creativity (Simonton, 1999b). Below I can do no more than provide a tiny sample of the empirical evidence. This evidence can be grouped into the following five categories: cognitive processes, individual differences, developmental influences, creative careers, and socio-cultural phenomena.

Cognitive processes

The BVSR model provides a strong theoretical basis for understanding the significance of serendipity. In fact, serendipitous discoveries have a role in cultural evolution similar to that of the mutation in biological evolution (Kantorovich and Ne'eman, 1989). Both are unexpected events that can set the course of historical change in new directions. More significantly, the theory explicates the results emerging from laboratory experiments on problem solving and insight. For instance, creative problem solving appears to be stimulated when individuals are presented with unpredictable or incongruous juxtapositions of stimuli (Finke *et al.*, 1992; Rothenberg, 1986). Such exposure evidently evokes more diverse and unconstrained ideational variations.

Individual differences

Secondary Darwinism offers a theoretical foundation for the psychometric instruments that assess the cognitive processes underlying creativity (Simonton, 1999a). For instance, the tests of divergent thinking that originated with Guilford (1967) and the Remote Associates Test of Mednick (1962) both assess the capacity to generate numerous and diverse variations, the bulk of which turn out to be useless and hence 'blind'. Indeed, adaptive variations are a positive function of the number of non-adaptive variations that are generated (e.g. Derks and Hervas, 1988). The theory also dovetails quite well with the personality traits associated with creativity (Simonton, 1999b), especially with the tendency for highly creative individuals to exhibit a certain amount of psychopathology (Eysenck, 1995; Ludwig, 1995). So long as incapacitating mental breakdowns are avoided, certain seemingly psychopathological symptoms can facilitate Darwinian creativity by increasing the number and scope of variations generated. The association between creativity and psychopathology may largely reflect their common foundation in a deficiency regarding the cognitive capacity for excluding 'irrelevant' information (Eysenck, 1995). Although this deficiency is less pronounced in the creative than in the psychotic, it enables the former to generate variations that would otherwise be excluded from consideration. Even so, Darwinian theory predicts that the rates of these psychopathological symptoms should vary across disciplines according to the magnitude of 'blindness' each domain requires for successful creativity (Simonton, 1999b). For instance, artistic creators should exhibit higher levels of psychopathological traits than do scientific creators, and that is in fact the case (Ludwig, 1995; Feist, 1998).

Developmental antecedents

Any developmental factor that enhances the capacity of an individual to generate numerous and diverse variations should have a positive impact on the development of creative potential. Consequently, the development of creative talent should include events and circumstances that encourage nonconformity, independence, appreciation of diverse perspectives, a variety of interests, and other favourable qualities. That indeed appears to be the case. For instance, eminent creators are more likely to have come from unconventional family backgrounds (Simonton, 1987), to have been subjected to multiple and diverse role models and mentors (Simonton, 1977b, 1984a, 1992), and to have had diversifying and atypical educational experiences and professional training (Simonton, 1976a, 1983, 1984b, c). The distinction between artistic and scientific creativity is also relevant here. Developmental events that tend to nurture originality are prone to be much more frequent or intense in the lives of artistic creators relative to scientific creators (Simonton, 1984b, 1986, 1999b).

Creative careers

The Darwinian view of creativity makes a striking prediction about the relation between quantity and quality of output. In biological evolution, those individuals who produce the most total offspring will usually have more offspring survive to reproduce themselves. But the more prolific organisms will also tend to produce the most progeny who die before reaching maturity. Thus, reproductive success is often associated with reproductive failure. A similar pattern is observed in the careers of eminent creators (Simonton, 1997a). Those who are the most prolific will have the most successful works, but they will also have the most unsuccessful works. So, quality is strongly associated with pure quantity. Produce more variations, and the odds will be increased that some variations will survive. Even more remarkably, this same relationship holds *within* careers, not just across careers. The mathematical function that describes the changes in creative output across the life span is the same for successful and unsuccessful products (Simonton, 1988a, 1997a). Those periods in which the creator produces the most total works will be those in which the most outstanding works appear, including the single best contribution (Simonton, 1991a, b). In fact, the ratio of successful products to total output stays more or less stable throughout the career (Simonton, 1977a, 1984b). In other words, the expected probability of success stays constant regardless of the creator's age, yielding what has been called the 'equal-odds rule' (Simonton, 1997a). Because of this principle, creative individuals are not able to increase their hit rates, nor do the hit rates decline with age, nor will they even exhibit some curvilinear form (Over, 1989; Simonton, 1997a; Huber, 2000). The fascinating aspect of this principle is that it be would predicted from the Darwinian viewpoint. To the extent that the variation process is blind, good and bad ideas will appear more or less randomly across careers, just as happens for genetic mutations and recombinations.

Socio-cultural phenomena

It has long been known that creative personalities are not randomly distributed across either cultures or historical periods, but rather such individuals will cluster into 'Golden Ages' separated by 'Dark Ages' (Simonton, 1988b). This fact suggests that there are special political, cultural, economic, and societal circumstances that may serve to either encourage or repress the development and manifestation of the individual capacity to generate variations (Simonton, 1999b). In the category of negative factors is international war, a condition that even has an adverse impact on creativity in science and technology (Fernberger, 1946; Price, 1978; Simonton, 1980). More generally, threatening conditions of various kinds, whether political or economic, tend to restrict the range of ideational variations that are emitted or permitted in a given time and place, and thus undermine creative behaviour (Sales, 1973; Doty *et al.*, 1991; Simonton, 1999b). With respect to positive factors, creative individuals are most likely to appear when a multi-ethnic civilization is fragmented into a large number of separate nations, which would presumably enhance the cultural heterogeneity while at the same time permitting cross-fertilization of ideas (Simonton, 1975, 1976d). In addition, when a civilization is characterized by conspicuous ideological diversity—the presence of numerous rival philosophical schools—then creativity tends to increase, even in those domains that have relatively little to do with intellectual trends (Simonton, 1976c). Finally, after a civilization opens itself up to foreign influences, it tends to become the site for a revival of creative activity (Simonton, 1997b). The process operating here has certain parallels with the role of hybridization in biological evolution (Harrison, 1993).

The foregoing comments should be qualified by the recognition that artistic and scientific creativity require somewhat different circumstances (Simonton, 1976b). Although both require a Zeitgeist that supports the free exploration of ideas, scientific creators generally require more stable socio-cultural settings than do artistic creators (Simonton, 1975), a finding that parallels the contrast mentioned earlier between the biographical backgrounds of creative scientists vs creative artists.

Objections

Although the secondary Darwinian theory of creativity has attracted many criticisms, these objections are almost entirely based on incorrect conceptions of what is and is not claimed by a BVSR model of creativity (Cziko, 1998; Simonton, 1999b). Thus, the theory is often taken to task for arguing that creativity is totally random and that the creator does not take advantage of acquired expertise in the field—when the theory claims nothing of the sort. In the first place, the theory assumes that accumulated knowledge and skill provide a necessary but not sufficient basis for creativity (Simonton, 1991b, 1996, 2000a). Something more must be added to take the creative mind beyond the limitations and constraints of that expertise, to generate truly original ideas that go beyond what has worked before. Moreover, although occasionally this additional step requires the introduction of truly random processes—James Watson's (1968) identification of the specific genetic codings providing a case in point—randomness in any strict sense is the exception rather than the rule. Much more often the creative act requires that the creator rely on the 'quasi-random' processes that are a natural by-product of free association, primary-process thinking, defocused attention, deficient perceptual filtering, playful tinkering, and the like (Simonton, in press).

Just as significant, the critics of the Darwinian theory repeatedly overlook the fact that the ideational variations that are proposed to feed the creative process are also expected to vary greatly in the degree of 'blindness' displayed (Cziko, 1998; Simonton, 1999b, in press). Some variations are highly constrained, almost algorithmically so, and others are almost completely unguided, with an unbroken continuum connecting these two extremes. This continuous dimension is analogous to what holds in primary Darwinism, where the variations can vary from the 'clones' generated by asexual reproduction (the absolute zero point of the blindness scale), through genetic recombinations governed by varying degrees of chromosomal linkage and epistasis, to mutations, which can also exhibit varying levels of a priori constraints (including differential mutation rates at distinct loci). In addition, the relative level of variational blindness will vary across types of creative products and even within various portions of a single product (Simonton, 2002, in press).

Secondary Darwinism has received unanticipated support from computer programs that exhibit genuine creativity. So far, the only successful simulations of human creative behaviour do so only by introducing some form of stochastic mechanism, sometimes even in the form of a random-number generator (Boden, 1991). Of special interest to the Darwinian theory of creativity is the emergence of the set of programs known as genetic algorithms and genetic programming (Goldberg, 1989; Koza, 1992). Not only have they produced original and adaptive solutions to real problems, but in addition these programs operate in an outright Darwinian manner, via the technique of blind-variation and selective-retention.

Human creativity and primary Darwinism

A Darwinian treatment of human creativity cannot be complete without embedding the behaviour in primary Darwinism. To do this requires that we establish the selection pressures that might support the evolution of creativity as a characteristic human activity. These selection pressures may be of two kinds: natural and sexual.

Natural selection

Superficially, it might seem to be an easy task to attribute human creativity to the fitness it confers upon each individual member of our species. After all, it is obvious that the supreme adaptive success of *Homo sapiens* is immediately apparent in the impressive technology that we have acquired over the past few thousand years—a technology that must be cumulative consequence of human ingenuity. Even at the individual level of day-to-day living, creativity often comes in handy, enabling us to solve the problems that frequently throw themselves in our path. Such an adaptationist account is certainly plausible. Yet its very plausibility may in fact undermine its credibility. The adaptiveness of creative behaviour seems so transparently true that this account may represent nothing more than another of those 'just so' stories that plague so many Darwinian explanations (Rose and Lauder, 1996). Indeed, this explanation may raise more questions than it answers. One problem concerns the personal attributes of highly creative individuals. As observed earlier, exceptional creators often feature characteristics that would seem to militate against the attainment of fitness in the sense of primary Darwinian selection. The most conspicuous instance is the disposition toward psychopathology (Eysenck, 1995). Whatever the assets of creativity, overt psychopathology does not appear to constitute adaptive behaviour. So why would natural selection favour its emergence? This question becomes even more problematic once it is recognized that creativity and psychopathology share a common genetic foundation, as is evident in the shared family pedigrees of these two characteristics

(Karlson, 1970; Richards *et al.*, 1988). Such a genetic link would seem to introduce selection pressures that would reduce the incidence of creativity in any human population.

One potential solution to this problem was offered by the distinguished evolutionists Julian Huxley and Ernst Mayr in collaboration with two psychiatrists who had special expertise in schizophrenia (Huxley *et al.*, 1964). After noting the strong evidence on behalf of the heritability of this common mental disorder, the authors observed that the gene for schizophrenia appears too frequently to be maintained by mutation alone. They accordingly examined the benefits and costs of possessing a genetic inclination toward schizophrenic disorder. For example, they discussed data showing that schizophrenics may be more physiologically robust than normal members of the population. The authors also suggested that the low fertility of schizophrenic males may be more than compensated by the high fertility by schizophrenic females. Therefore, Huxley, Mayr, and their coauthors concluded that the high incidence rate of schizophrenia could be 'the result of a balance between its selectively favourable and unfavourable properties' (p. 220).

Unfortunately, these authors focused on the biological repercussions of mental disorder. It could just as well be argued that psychopathological symptoms may have certain social consequences that contribute to the survival value of the corresponding genes. This very possibility was put forward by Hammer and Zubin (1968). They looked at psychopathology as a manifestation of a more general syndrome of what they styled 'the cultural unpredictability of behaviour'. Some individuals in a population inherit a certain tendency to do the unanticipated according to societal norms and role expectations. Although psychopathology is one manifestation of this genetic inclination, it is not by any means the only one. The innate proclivity for unpredictability may take the form of exceptional creativity, which can prove adaptive both to the individuals and to the culture that produces them. Whether this genetic endowment is positive or negative depends on certain cultural circumstances that channel the tendency in different directions. Hammer and Zubin directly compared this phenomenon with sickle-cell anaemia, which confers selective advantage in tropical climates where persons heterozygous on this trait gain increased resistance to malaria.

Whatever plausibility these arguments may have, they probably fall short of a complete Darwinian explanation. One basic objection concerns the presumed adaptiveness of creative behaviour. Although this assumption seems reasonable when it comes to technological domains—activities having to do with obtaining food, shelter, and self-protection—it is less apparent that creativity has evolutionary utility in fields that make no direct contribution to fitness. A clear-cut example is artistic creativity, an activity that is universal in the human species. To account for the evolution of these forms of creative activity, some other selection mechanism may be evoked.

Sexual selection

In 1859, when *Origin* first appeared, Darwin avoided discussing human evolution. But in 1871 he was willing to take on the subject, publishing *The Descent of Man and Selection in Relation to Sex*. In one respect, this title encompasses two separate books, one concerning the evolution of *Homo sapiens*, the other the evolutionary consequences of courtship and competition for mates. Yet in another, more profound respect, these two subjects are intimately connected. The connection stems from Darwin's belief that sexual selection played a major role in human evolution. In fact, he explicitly suggested that sexual selection might have driven the emergence of human creativity, including the 'nonadaptive' creative behaviour displayed by human beings. A peacock's brilliant tail display confers no advantage from the standpoint of natural selection, but does enable the male to procure a mate, and thus be favoured according to sexual selection. By the same token, singing, dancing, and other artistic activities do not help human beings to find food and shelter, but those activities may have a major part in obtaining suitable mates. For a long time, Darwin's ideas on sexual selection were neglected in favour of the concept of natural selection. However, after interest in sexual selection was revived, it became recognized as a potentially powerful evolutionary force in those species that engage in mate competition and selection (Andersson, 1994). Even more important from the perspective of this chapter, sexual selection can once more be seriously considered as a possible factor behind the evolution of human creativity.

Geoffrey Miller (1997, 1998, 2000) has extensively elaborated this very possibility. Miller (1997) began by arguing for the selective advantage of a personal characteristic quite similar to Hammer and Zubin's (1968) trait of cultural unpredictability. Miller's argument was founded on the notion of Machiavellian intelligence. According to the latter concept, human beings have had to evolve extremely complex cognitive and behavioural skills in order to survive the interpersonal politics of primate social systems (see also Byrne, Chapter 11). Such intricate systems require considerable acumen and dexterity to negotiate the elaborate web of cooperative and competitive activities that define an individual's place in the dominance hierarchies. Placement in the hierarchy is crucial insofar as it determines each individual's reproductive success. Because a premium is placed on being able to 'outsmart' rivals, the social primates have evolved a number of strategies that help prevent the disclosure of intentions. Among those strategies may be proteanism, that is, the capacity to be unpredictable when necessary in a given social situation (Driver and Humphries, 1988). Moreover, social proteanism would be useful in a diversity of circumstances besides domestic politics. In combat between rival males, for instance, the advantage often goes to the opponent whose moves cannot be anticipated.

One striking feature of Miller's (1997) theory of protean behaviour is that it provides an evolutionary explanation for the emergence of a mechanism that can generate quasi-randomness, the prerequisite for the production of genuinely unpredictable behaviour. Miller (1997) even made a specific connection between this intellectual capacity and Donald Campbell's (1960) blind-variation and selective-retention model. In doing so, Miller offered a primary Darwinian mechanism that would support the evolutionary emergence of the secondary Darwinian process of creativity. At the same time, selection for proteanism may also explain the connection between creativity and psychopathology (Simonton, 1999b). After all, mental disorder may be the unfortunate consequence of inheriting too much proclivity for proteanism. Presumably, the optimal amount of protean behaviour in the population would represent some equilibrium point between two maladaptive extremes on this trait—between psychosis and its opposite. At some point on the continuum between the optimally protean and the outright psychotic may then emerge the highly creative individual.

Even though the emergence of proteanism can help account for the evolution of the BVSR process hypothesized by secondary Darwinian theory of creativity, it cannot be the whole story. The range of human creativity is just far too vast. However, according to Miller (1998) this incipient capacity might have been considerably expanded once it became a criterion for mate selection. Both men and women could try to attract mates by demonstrating creative behaviours, whether it is through singing or dancing, tool-making or basket weaving. Even more crucial was the advent of the human capacity for language, for then courtship could take place on a more intellectual plane. Flirtatious displays of humour and wit, wisdom and creativity could be conveyed in verbal terms. Lovers could woo each other with poems and songs as well as gifts and adornments. In fact, Miller believes that sexual selection for these creative skills could have instigated a runaway evolutionary process that favoured the emergence of the human brain. This is a crucial possibility, because sexual selection is as strong a force for jungle dwellers as it is for denizens of the city. The desire to win an attractive mate is a cross-cultural and transhistorical universal for all peoples that have survived eons of natural selection.

Under this scenario, mate choice is working on both male and female simultaneously and equally, because these assets are of comparable value to the reproductive success of their offspring. So men and women concurrently evolve the capacity for creativity. Figuratively speaking, both males and females acquire brilliant intellectual plumage. Furthermore, the augmented intellectual abilities of both genders would quickly become preadaptations (or exaptations) that would be later coopted for purposes besides winning mates. The most prominent spin-off would be the rapid evolution of human culture, which provided a totally new basis for adaptations (Simonton, 1999b).

At present, it is not possible to determine conclusively the relative contributions of natural and sexual selection to the evolution of the human capacity for creativity. Some would argue that natural selection alone is sufficient to account for the emergence of this exceptional ability. Others are not so sure, believing that the hiatus between human creativity and non-human innovation is too vast, implicating more the runaway process of sexual selection than the more incremental process of natural selection.

Future developments

I devoted less space to primary Darwinism than to secondary Darwinism. Although this difference might simply reflect my training as a psychologist rather than as a biologist, the differential allotment may accurately reflect our current state of knowledge. Far more is known about the psychological foundations of human creativity than about the biological origins of that capacity. One cause for this contrast is that creative activity has not been a high-priority topic in evolutionary psychology (see, e.g. Barkow et al., 1992). For instance, a recent book on the evolution of cognition devotes discusses almost every topic imaginable—even gossip—except animal and human creativity (Heyes and Huber, 2000). What is obviously needed are more books like the current volume to provide the comparative basis for deciphering the processes underlying the evolution of creative behaviour.

Yet a complete comprehension of human creativity may require more than such extensive comparative inquiries. To an extent unprecedented in the evolution of life, each member of our species is the joint product of two interacting evolutionary forces, one biological and the other socio-cultural (Boyd and Richerson, 1985). Moreover, socio-cultural evolution has acquired sufficient power to override the dictates of biological evolution. As a result, human beings will often engage in behaviours that sacrifice their reproductive fitness for the sake of the socio-cultural system in which they live. Although altruistic behaviour represents the most discussed example, creative behaviour may well constitute another illustration. The potential selective disadvantage is apparent in the lives of those highly creative individuals, like Michelangelo, Newton, Descartes, and Beethoven, who left no biological progeny. In Dawkins' (1989) terms, these humans decided to contribute memes rather than genes to the human legacy.

A full scientific explanation for such oddities may thus require a complex model that specifies exactly how human creativity emerged through coevolutionary processes. Such a model would identify the essential theoretical nexus between primary and secondary BVSR mechanisms. The net result would be a Darwinian theory that will bridge the current chasm that separates the innovative behaviour of animals from the creative activities of humans.

Summary

Of all species that evolved on this planet, *Homo sapiens* is without doubt the one that acquired the greatest capacity for innovation. To appreciate the contrast between human creativity and animal innovation, this distinctive feature of the species should be placed in the context of Darwinian theory. This analysis has two parts. In the first part, human creative behaviour can be interpreted as a BVSR process analogous to what underlies biological evolution. After outlining the key features of this model, an overview of the supporting evidence is presented, with special focus on the cognitive processes, individual differences, developmental influences, creative careers, and socio-cultural phenomena associated with the behaviour's occurrence. There follows a brief discussion of the objections that have been raised against the model. In the second part, the evolution of *Homo sapiens* can be analysed in terms of the selection pressures that would support the emergence of this BVSR process in the human nervous system. These pressures include both natural and sexual selection, with the latter possibly exerting the most impact. Significantly, it was Charles Darwin himself who first suggested that sexual selection might have played a major role in the emergence of human creativity.

References

Andersson, M. (1994). *Sexual selection*. Princeton, NJ: Princeton University Press.

Barkow, J. H., Cosmides, L., and Tooby, J. (ed.) (1992). *The adapted mind: Evolutionary psychology and the generation of culture*. New York: Oxford University Press.

Basalla, G. (1988). *The evolution of technology*. Cambridge, England: Cambridge University Press.

Boden, M. A. (1991). *The creative mind: Myths & mechanisms*. New York: BasicBooks.

Boyd, R. and Richerson, P. J. (1985). *Culture and the evolutionary process*. Chicago, IL: University of Chicago Press.

Campbell, D. T. (1960). Blind variation and selective retention in creative thought as in other knowledge processes. *Psychological Review*, **67**, 380–400.

Campbell, D. T. (1965). Variation and selective retention in socio-cultural evolution. In *Social change in developing areas* (ed. H. R. Barringer, G. I. Blanksten, and R. W. Mack), pp. 19–49. Cambridge, MA: Schenkman.

Campbell, D. T. (1974). Evolutionary epistemology. In *The philosophy of Karl Popper* (ed. P. A. Schlipp), pp. 413–63. La Salle, IL: Open Court.

Cziko, G. (1995). *Without miracles: Universal Selection Theory and the second Darwinian revolution.* Cambridge, MA: MIT Press.

Cziko, G. (1998). From blind to creative. In defense of Donald Campbell's selectionist theory of human creativity. *Journal of Creative Behavior*, **32**, 192–209.

Dawkins, R. (1989). *The selfish gene* (rev. edn). Oxford, England: Oxford University Press.

Dennett, D. C. (1995). *Darwin's dangerous idea: Evolution and the meanings of life.* New York: Simon & Schuster.

Derks, P. and Hervas, D. (1988). Creativity in humor production: Quantity and quality in divergent thinking. *Bulletin of the Psychonomic Society*, **26**, 37–9.

Doty, R. M., Peterson, B. E., and Winter, D. G. (1991). Threat and authoritarianism in the United States, 1978–1987. *Journal of Personality and Social Psychology*, **61**, 629–40.

Driver, P. M. and Humphries, N. (1988). *Protean behavior: The biology of unpredictability.* Oxford, England: Oxford University Press.

Epstein, R. (1991). Skinner, creativity, and the problem of spontaneous behavior. *Psychological Science*, **2**, 362–70.

Eysenck, H. J. (1995). *Genius: The natural history of creativity.* Cambridge: Cambridge University Press.

Feist, G. J. (1998). A meta-analysis of personality in scientific and artistic creativity. *Personality and Social Psychology Review*, **2**, 290–309.

Fernberger, S. W. (1946). Scientific publication as affected by war and politics. *Science*, **104**, 175–7.

Finke, R. A., Ward, T. B., and Smith, S. M. (1992). *Creative cognition: Theory, research, applications.* Cambridge, MA: MIT Press.

Glover, J. A., Ronning, R. R., and Reynolds, C. R. (ed.) (1989). *Handbook of creativity.* New York: Plenum Press.

Goldberg, D. E. (1989). *Genetic algorithms in search, optimization, and machine learning.* Reading, MA: Addison-Wesley.

Guilford, J. P. (1967). *The nature of human intelligence.* New York: McGraw-Hill.

Hammer, M. and Zubin, J. (1968). Evolution, culture, and psychopathology. *Journal of General Psychology*, **78**, 151–64.

Harrison, R. G. (ed.) (1993). *Hybrid zones and the evolutionary process.* New York: Oxford University Press.

Heyes, C. and Huber, L. (ed.) (2000). *The evolution of cognition.* Cambridge, MA: MIT Press.

Huber, J. C. (2000). A statistical analysis of special cases of creativity. *Journal of Creative Behavior*, **34**, 203–25.

Huxley, J., Mayr, E., Osmond, H., and Hoffer, A. (1964). Schizophrenia as a genetic morphism. *Nature*, **204**, 220–1.

James, W. (1880). Great men, great thoughts, and the environment. *Atlantic Monthly*, **46**, 441–59.

Kantorovich, A. (1993). *Scientific discovery: Logic and tinkering.* Albany, NY: State University of New York Press.

Kantorovich, A. and **Ne'eman, Y.** (1989). Serendipity as a source of evolutionary progress in science. *Studies in History and Philosophy of Science*, **20**, 505–29.

Karlson, J. I. (1970). Genetic association of giftedness and creativity with schizophrenia. *Hereditas*, **66**, 177–82.

Kawai, M. (1965). Newly-acquired pre-cultural behavior in the natural troop of Japanese monkeys on Koshima Islet. *Primates*, **6**, 1–30.

Köhler, W. (1925). *The mentality of apes.* (E. Winter, Translator) New York: Harcourt, Brace.

Koza, J. R. (1992). *Genetic programming: On the programming of computers by means of natural selection.* Cambridge, MA: MIT Press.

Ludwig, A. M. (1995). *The price of greatness: Resolving the creativity and madness controversy.* New York: Guilford Press.

Mednick, S. A. (1962). The associative basis of the creative process. *Psychological Review*, **69**, 220–32.

Miller, G. F. (1997). Protean primates: The evolution of adaptive unpredictability in competition and courtship. In *Machiavellian intelligence II* (ed. A. Whiten and R. Byrne), pp. 312–40. Cambridge, England: Cambridge University Press.

Miller, G. F. (1998). How mate choice shaped human nature: A review of sexual selection and human evolution. In *Handbook of evolutionary psychology: Ideas, issues, and applications* (ed. C. B. Crawford and D. Krebs), pp. 87–129. Mahwah, NJ: Erlbaum.

Miller, G. (2000). *The mating mind: How sexual choice shaped the evolution of human nature.* New York: Doubleday.

Over, R. (1989). Age and scholarly impact. *Psychology and Aging*, **4**, 222–5.

Price, D. (1978). Ups and downs in the pulse of science and technology. In *The sociology of science* (ed. J. Gaston), pp. 162–71. San Francisco: Jossey-Bass.

Richards, R., Kinney, D. K., Lunde, I., Benet, M., and **Merzel, A. P. C.** (1988). Creativity in manic-depressives, cyclothymes, their normal relatives, and control subjects. *Journal of Abnormal Psychology*, **97**, 281–8.

Rose, M. R. and **Lauder, G. V.** (ed.) (1996). *Adaptation.* San Diego, CA: Academic Press.

Rothenberg, A. (1986). Artistic creation as stimulated by superimposed versus combined-composite visual images. *Journal of Personality and Social Psychology*, **50**, 370–81.

Runco, M. A. and **Pritzker, S.** (ed.) (1999). *Encyclopedia of creativity.* 2 Vols. San Diego: Academic Press.

Sales, S. M. (1973). Threat as a factor in authoritarianism: An analysis of archival data. *Journal of Personality and Social Psychology*, **28**, 44–57.

Simonton, D. K. (1975). Sociocultural context of individual creativity: A transhistorical time-series analysis. *Journal of Personality and Social Psychology*, **32**, 1119–33.

Simonton, D. K. (1976a). Biographical determinants of achieved eminence: A multivariate approach to the Cox data. *Journal of Personality and Social Psychology*, **33**, 218–26.

Simonton, D. K. (1976b). Do Sorokin's data support his theory? A study of generational fluctuations in philosophical beliefs. *Journal for the Scientific Study of Religion*, **15**, 187–98.

Simonton, D. K. (1976c). Ideological diversity and creativity: A re-evaluation of a hypothesis. *Social Behavior and Personality*, **4**, 203–7.

Simonton, D. K. (1976d). Philosophical eminence, beliefs, and zeitgeist: An individual-generational analysis. *Journal of Personality and Social Psychology*, **34**, 630–40.

Simonton, D. K. (1977a). Creative productivity, age, and stress: A biographical time-series analysis of 10 classical composers. *Journal of Personality and Social Psychology*, **35**, 791–804.

Simonton, D. K. (1977b). Eminence, creativity, and geographic marginality: A recursive structural equation model. *Journal of Personality and Social Psychology*, **35**, 805–16.

Simonton, D. K. (1980). Techno-scientific activity and war: A yearly time-series analysis, 1500–1903 A.D. *Scientometrics*, **2**, 251–5.

Simonton, D. K. (1983). Formal education, eminence, and dogmatism: The curvilinear relationship. *Journal of Creative Behavior*, **17**, 149–62.

Simonton, D. K. (1984a). Artistic creativity and interpersonal relationships across and within generations. *Journal of Personality and Social Psychology*, **46**, 1273–86.

Simonton, D. K. (1984b). *Genius, creativity, and leadership: Historiometric inquiries.* Cambridge, MA: Harvard University Press.

Simonton, D. K. (1984c). Is the marginality effect all that marginal? *Social Studies of Science*, **14**, 621–2.

Simonton, D. K. (1986). Biographical typicality, eminence, and achievement style. *Journal of Creative Behavior*, **20**, 14–22.

Simonton, D. K. (1987). Developmental antecedents of achieved eminence. *Annals of Child Development*, **5**, 131–69.

Simonton, D. K. (1988a). Age and outstanding achievement: What do we know after a century of research? *Psychological Bulletin*, **104**, 251–67.

Simonton, D. K. (1988b). Galtonian genius, Kroeberian configurations, and emulation: A generational time-series analysis of Chinese civilization. *Journal of Personality and Social Psychology*, **55**, 230–8.

Simonton, D. K. (1988c). *Scientific genius: A psychology of science.* Cambridge: Cambridge University Press.

Simonton, D. K. (1991a). Career landmarks in science: Individual differences and interdisciplinary contrasts. *Developmental Psychology*, **27**, 119–30.

Simonton, D. K. (1991b). Emergence and realization of genius: The lives and works of 120 classical composers. *Journal of Personality and Social Psychology*, **61**, 829–40.

Simonton, D. K. (1992). Leaders of American psychology, 1879–1967: Career development, creative output, and professional achievement. *Journal of Personality and Social Psychology*, **62**, 5–17.

Simonton, D. K. (1996). Creative expertise: A life-span developmental perspective. In *The road to expert performance: Empirical evidence from the arts and sciences, sports, and games* (ed. K. A. Ericsson), pp. 227–53. Mahwah, NJ: Erlbaum.

Simonton, D. K. (1997a). Creative productivity: A predictive and explanatory model of career trajectories and landmarks. *Psychological Review*, **104**, 66–89.

Simonton, D. K. (1997b). Foreign influence and national achievement: The impact of open milieus on Japanese civilization. *Journal of Personality and Social Psychology*, **72**, 86–97.

Simonton, D. K. (1999a). Creativity as blind variation and selective retention: Is the creative process Darwinian? *Psychological Inquiry*, **10**, 309–28.

Simonton, D. K. (1999b). *Origins of genius: Darwinian perspectives on creativity.* New York: Oxford University Press.

Simonton, D. K. (2000a). Creative development as acquired expertise: Theoretical issues and an empirical test. *Developmental Review*, **20**, 283–318.

Simonton, D. K. (2000b). Creativity: Cognitive, developmental, personal, and social aspects. *American Psychologist*, **55**, 151–8.

Simonton, D. K. (2002). *Great psychologists and their times: Scientific insights into psychology's history.* Washington, DC: American Psychological Association.

Simonton, D. K. (in press). Scientific creativity as a constrained stochastic behavior: The integration of product, person, and product perspectives. *Psychological Bulletin.*

Sternberg, R. J. (ed.) (1999). *Handbook of creativity.* Cambridge: Cambridge University Press.

Watson, J. D. (1968). *The double helix: A personal account of the discovery of the structure of DNA.* London: Weidenfeld & Nicolson.

West, M. A. and Rickards, T. (1999). Innovation. In *Encyclopedia of creativity* (ed. M. A. Runco and S. Pritzker), Vol. 2, pp. 45–55. San Diego, CA: Academic Press.

DISCUSSION

CHAPTER 15

TO INNOVATE OR NOT TO INNOVATE? THAT IS THE QUESTION

MARC D. HAUSER

Consider the yellow sticky note. No one, I assume, would contest the claim that this was truly an innovation. Though simple, it allowed people, ranging from the power suits of Wall Street to the harried mother of three children, to leave notes without gumming up the walls. When the standard square, yellow, sticky note mutated into different sizes, shapes, and colours, I doubt that anyone called these innovative changes. Although functionally useful (e.g. if you only have a few words to write, why waste it on a big sticky if a small one will do, and waste less paper?) these featural changes do not seem like the stuff of innovation. At least that seems like a reasonable starting intuition. But intuitions are somewhat unsatisfying in science, especially if they remain dormant and untested.

In this chapter, I would like to look back on the discussion of innovation provided in this volume, and develop three points. The first is historical, and simply explores closely related topics where a term used to describe human behaviour has been borrowed for the sake of describing animal behaviour. The second point concerns mechanism. Specifically, if we are going to take the problem of innovation in human and non-human animals seriously, then we need to understand the underlying mechanisms, and the extent to which they are shared across species. The third point is methodological. Having discussed several candidate mechanisms, or issues concerning mechanism, I conclude with a discussion of how future experiments on innovation might profitably proceed. In parallel with recent work on social learning, the study of innovation must begin to take the patterns of observation recorded from a variety of animals and design experiments that allow for more careful delineation of underlying mechanisms.

A matter of definition

Humans imitate, deceive, refer, laugh, think, plan, and remember. Nothing controversial here. My cat Cleo imitates, deceives, refers, laughs, thinks, plans, and remembers. For some, no controversy here either. For others, my claims about Cleo the cat are absurd. They are absurd either because there is no evidence for the capacity or because the evidence that has been mounted thus far falls far short of the capacity demonstrated in humans. This last position is the most interesting *vis-à-vis* the current discussion of innovation. To say that humans have a capacity X and that animals have a capacity that is *like* X is to make a claim about the underlying mechanism and its particular details. Similarity may arise at multiple levels, however. Capacity X may be similar in humans and non-human animals because it serves the same function; at this level of analysis, the underlying mechanism is not relevant. Capacity X might also be similar across species because the underlying neural mechanism is the same. Finally, capacity X might be similar across species because the genes that code for the mechanism are shared, even though the morphological structure itself is different. For example, the eye of an insect and a human are derived from homologous genes even though anatomically they look extremely different. At a functional level, however, both kinds of eyes have been designed to solve the same general problem: seeing. Although the resolution of human and insect eyes is different, both eye types have nevertheless evolved to solve the problem of detecting visually salient events. Using a particular term to describe a particular phenomenon therefore carries with it the burden of articulating what the underlying mechanisms are and how they are implemented. It also requires a clearly articulated statement about the level of analysis: functional, anatomical, or genetical.

Consider the use of the term 'referential' by biologists, anthropologists, and psychologists working on non-human animal vocalizations. When linguists state that words are referential they mean something like 'there is an arbitrary association between sound and meaning, such that the sound used picks out particular objects and events that are either imagined or real in the past, present or future'. When ethologists started to think about the possibility that human language evolved from an ancestral system, they examined the naturally produced vocalizations of animals for evidence of referential signals. In the first exploration, inspired by Struhsaker's (1967) field studies and Peter Marler's (1978) theoretical insights, Seyfarth and Cheney (Seyfarth et al., 1980; Cheney and Seyfarth, 1982) suggested that vervet monkey alarm calls and grunts were 'representational' signals (Cheney and Seyfarth, 1990). What they meant by this was that when listeners heard such calls, they evoked a representation of the object or event that the caller perceived. Thus, if the caller saw a leopard and gave a leopard alarm call, the listener would, upon hearing the call, have access to a mental representation of a leopard. Although this was one possible interpretation, there were others: rather than a label for the predator, the leopard alarm call represents a code of instruction such as 'run up into a tree'. Subsequent studies of comparable vocalizations invoked other terms, including 'symbolic' and most recently 'functionally referential' (Marler et al., 1992). The last term has had the most staying power, and deserves the most attention in the current context. When Marler and colleagues introduced this term, they inserted 'functional' to get away from the precise mechanism underlying the call, which was at the time of writing and up to the present, completely unclear (Hauser, 1996). A leopard alarm call is functionally referential in that it elicits from listeners a response that is functionally appropriate given the context at hand. Here, then, is a clear distinction between the use of words by human speakers of a language and the use of vocalizations by animals. Yes, some animal vocalizations are like human words in being functionally referential. But critically, by looking carefully at the causes of vocalizations as well as the responses, Cheney, Seyfarth, Zuberbühler, and others (Zuberbühler et al., 1997, 1999; Zuberbühler, 2000) have begun to elucidate all the ways in which non-human animal vocalizations are nothing like words, and some of the ways in which they are (Hauser et al., 2002). Critically, although early uses of words invented to describe human phenomenon are often applied a bit superficially to phenomenon that appear the same in humans, as time passes, and researchers begin to dig more deeply into the phenomenon, the terminology is tightened, as is the understanding of underlying mechanisms. 'Imitation' has undergone the same pattern of change and investigation (Galef, 1988; Whiten and Ham, 1992; Heyes and Galef, 1996; Byrne and Russon, 1998), and my guess is that 'innovation' will be next in line. Although some fear that the use of human terms to describe animal behaviour is doomed to failure because it ultimately falls victim to the cry of 'anthropomorphism' (Kennedy, 1992), I do not believe that this is necessarily the case. For those interested in the evolution of cognitive mechanisms, we must ultimately explore how a particular behaviour is mediated by the brain, and for this, words simply describe what we know or have yet to learn. When someone states that they have witnessed an innovative bird, monkey or human, it is only natural to ask 'Well, what do you mean by "innovative?"' Such probing questions are aimed at the problem of mechanism.

What drives innovation?

One of Gary Larson's little gems shows a blackboard filled with mathematical scribbles, two police detectives, and a few dead cats. The caption reads 'Notice all the computations, theoretical scribblings, and lab equipment Norm. Yes, curiosity killed these cats'. The cartoon is funny because it takes advantage of a cliché, and like many clichés, this one also turns out to be as false as 'Monkey see, monkey do' (Visalberghi and Fragaszy, 2001). In the same way that monkeys appear to be incapable of imitation *sensu strictu*, I do not believe animals are curious, at least in one sense. Although animals are curious in that they will explore and observe objects and events in their environment, they are not curious in the sense of seeking out the cause for particular objects and events. Cats might be curious about the scribbles on the blackboard, but they are not curious about the kinds of scribbles necessary to solve a problem. Let me explain by way of an example.

On Cayo Santiago, a small island off the coast of Puerto Rico that is home to some 1000 rhesus monkeys, garbage washes up on shore almost every day. The staff on Cayo picks up this trash and burns it on a regular basis. Sometimes, a coconut falls in the fire and explodes, providing vigilant rhesus with access to their most preferred food; moreover, since no rhesus monkey has ever figured out how to crack open a coconut, this explosion of coconut has showered the animals with a free dessert. The question is: given the number of rhesus monkeys who have observed this explosion, why have no rhesus monkeys ever figured out the trick? Why has some genius rhesus monkey, sitting around the campfire, not wondered 'If I take a coconut and place it in the fire, at some point it will explode as it did before'. It is kind of like playing the television and board game Jeopardy. THE ANSWER IS: Coconut explodes after it sits in the fire. QUESTION IS: How can you open a coconut? This is a no-brainer for any normal human. But in 60 years of watching these animals, no one has ever observed a rhesus monkey grabbing a coconut and placing it in the fire. Why not? My guess is that they lack a certain kind of curiosity, one that seeks out the relationship between cause and effect; this kind of curiosity may be necessary for certain kinds of innovation.

Let me be clear about what I am not claiming. I am not claiming that animals cannot perceive cause–effect relationships. A beautiful example of such comprehension comes from the work of David Premack (1986). In one set of experiments with his star chimpanzee Sara, Premack presented a series of cause–effect problems. In the basic design, Premack showed Sara two objects, one representing a transformation of the other. For example, the first object might be an apple and the second, two halves of an apple. Premack then showed Sara three objects (knife, sponge, tape dispenser), only one (knife) of which could cause the transformation from apple to apple halves. Sara consistently picked the appropriate causal agent (e.g. knife). Since it was possible for Sara to figure out some of these cases by simple association (e.g. apple and knife are frequently seen together), Premack ran a critical control, this time asking Sara to run the film backwards so to speak. When shown two halves of an apple and then an apple, the correct causal agent is the tape dispenser, not the knife. Sara selected the tape dispenser. This experiment shows that some animals, though not all (Visalberghi and Limongelli, 1994, 1996; Povinelli, 2000; Visalberghi and Fragaszy, 2001), can perceive cause–effect relationships and can pick out some of the forces that cause physical transformations to an object.

One of the things that makes Premack's apple halves-to-apple experiment so interesting is that it appears to demand a kind of reverse causality: how do you get a whole apple back if you have two apple halves? Although Sara seems to solve such problems, there might be a difference between perceiving the connection and acting in such a way as to cause the desired effect. To illustrate, consider a problem confronting chimpanzees living in the Kibale Forest of Uganda. Ugandans living on the forest edge set out snares within the forest in order to catch small game. The snare is quite simple, consisting of only a wire loop. When an animal such as a dik-dik walks into the snare, the wire loop closes and captures the animal. Unfortunately, though chimpanzees are not an intended target, sometimes chimpanzees are also caught by the snares. As a result, a large number of chimpanzees in this population have severed limbs. Although ideally one would prevent poachers from using snares in the forest, what is puzzling is that no chimpanzee has figured out how to remove the snares. It is a trivial problem: simply take one of the loose ends and push it so that the loop loosens and frees the captured limb. But the chimpanzee's natural tendency is to *pull* on one of the loose ends and this simply tightens the loop more, and ultimately, severs the limb. It is odd that this *solution* is not blocked by the pain each animal must feel as it tightens the wire around the limb. My main point here is that this is the kind of causal reasoning that chimpanzees apparently lack and this is surprising both because of the pain associated with a failed solution, and the great advantage that would come from a proper solution. Why have not any chimpanzees thought through this problem, reverse engineering the cause–effect relationship?

In contrast with the observations discussed above, studies of tool use show that animals can solve cause–effect problems, and this capacity may contribute to the fact that in primates at least, many of the innovations we observe arise in the domain of tool manufacture and use. But when animals use tools to solve problems, by what means do they come to understand the relationship between the tool and its role in solving the problem? Since most observations of tool use in nature occur after the innovation, or provide documentation of how naïve animals learn from experienced ones, we have little understanding of how the original innovation appeared. When the first brilliant chimpanzee invented termite fishing, did she look at the termite mound and wonder how she could obtain the termites efficiently, without breaking up the mound and without waiting for them to emerge one by one? Given this thought, did she then walk over, grab a branch, strip it of its lateral bits, and then insert it into the mound? Or did she happen to walk by a termite mound with a stick inside and notice termites climbing up? The bottom line is that this chimpanzee hit on a novel foraging technique. But in terms of the problem of innovation, it is important to understand how she came to acquire this technique. The first interpretation seems closer to what we mean by curiosity and innovation, the second closer to dumb luck. Dumb luck is not a bad thing, mind you. After all, sticky notes appear to have been discovered in much the same way.

The reason why curiosity and innovation of the kind I have been discussing might be difficult to evolve is because it potentially involves risk taking, breaking from the mould, trying to figureout ways of doing things differently, unconventionally. If you hit upon a solution accidentally, there will often be no risk. You can simply look at the consequences, evaluate whether it falls on the cost or benefit side of the economic pie chart, and then move forward. But if you are curious, and experiment with different solutions to a problem, you are potentially taking risks; risks may have high fitness payoffs, but they may also have high fitness costs. To be an innovator is to take risks, to block what is easy and common and try something new. Animals may only switch strategies when the current techniques fail, or when there are low costs associated with trying something different.

How might developmental and neurobiological studies inform our understanding of the problem of innovation? If innovation depends upon plasticity, upon being able to implement alternative responses, then we should expect to see innovations for phenotypes that are only weakly canalized in the Waddingtonian (1975) sense. Many of the examples described in this book seem to support this prediction. For instance, among birds, foraging innovations are extremely common. This is to be expected, especially among species that invade new habitats and thus, are confronted with novel foraging demands (see Chapter 3). Among primates, innovative foraging techniques are also common, as are social behaviours; since no social interaction is ever exactly the same, this also seems like an arena for plasticity (see Chapters 11–13). But in order for any of these innovative responses to emerge, remain stable within the individual, and pass on to others in the population, individuals must have the capacity to inhibit species-typical or common responses and select the new and uncommon response. Studies of primates in particular suggest that the prefrontal cortex plays an important role in inhibitory control. Although there are a wide variety of inhibitory problems, ranging from the control of emotions and motivational drives to the control of overlearned response biases, it is generally thought that prefrontal regions are recruited for these problems. In studies by the developmental psychologist Adele Diamond (Diamond and Goldman-Rakic, 1989; Diamond et al., 1989), results indicate that 2–4 month old rhesus, but not adults, have difficulty inhibiting a prepotent response to reach straight ahead for a food reward when the task requires reaching around a transparent barrier. If you lesion the dorso-lateral prefrontal cortex of an adult rhesus monkey, as opposed to other control regions, these individuals also perseverate, repeatedly reaching straight ahead for the food, and repeatedly failing to obtain it because of the transparent barrier. Diamond suggests that a mature dorso-lateral prefrontal cortex is necessary for inhibiting a pre-potent response bias. More recent studies with cotton-top tamarins (Santos *et al.*, 1999), however, suggest that even adults of this species have difficulty with Diamond's reaching task, indicating that there are not only developmental changes in the functioning of dorso-lateral prefrontal cortex, but species differences as well (Deacon, 1997; Rumbaugh, 1997); the latter may reveal differences in the selective pressures on brain evolution and design.

Finding the locus of inhibitory control will not, however, be easy. Although Diamond's work suggests a role for dorsolateral prefrontal cortex, considerable work on humans and non-human primates also suggests that the orbitofrontal cortex is important, as well as other regions (e.g. amygdala, anterior cingulate, ventral striatum). In particular, studies by Rolls, Watanabe, Damasio, and others suggest that the orbitofrontal cortex plays a role in weighting the emotional valence of a response (Damasio, 1994; Watanabe, 1996; Rolls, 1999; Roberts and Wallis, 2000). When humans and non-human primates make decisions, their choices are weighted according to positive and negative emotions or the odds of reward as opposed to punishment. An important question for future studies of the mechanisms underlying innovation in animals, therefore, is the extent to which innovative, novel, or risky behaviours are associated with positive or negative emotions. If trying something new is aversive, then there will be little in the way of emotional support for innovations, at least at the level of the individual innovator; Greenberg's chapter provides an interesting discussion of the relationship between innovation and neophobia.

If individuals are able to consider alternative solutions to current problems, then they will require a mechanism that allows two or more competing response alternatives to be held in mind at the same time in order to run simulations of their possible advantages and disadvantages. The prefrontal cortex is a likely candidate for such simulations since it plays a role in working memory, of being able to store something in mind in order to operate on it. Although we know that animals can generate expectations about future events (Galef, 1988; Shettleworth, 1998), and there is some evidence that chimpanzees have the rudiments of a theory of mind (Hare *et al.*, 2000, 2001) and thus can entertain what another individual knows, we do not know if animals sit around thinking about possible worlds or possible solutions to particular problems and it is difficult to work out how, methodologically, we might uncover whether animals run simulations, raising hypotheses about possible solutions to a problem and then running a feasibility test. But there are experiments we can run on innovation in animals and these will certainly advance our understanding (see below).

A final point on mechanism is raised by the comparative analyses presented in Chapter 2. Is the capacity for innovation a domain-specific or domain-general ability (Cosmides and Tooby, 1994; Hauser, 2000)? Is it more like memory, a capacity that is necessary for many different problems and is thus domain-general, or is it like language, a domain that recruits specific mechanisms (e.g. recursion, reference), that operate over domain-specific content (e.g. phonemes, words, sentences)? Although we do not yet know enough about this problem to say for sure, if non-human animals are anything like humans, innovation will turn out to be a domain-general capacity, one that can operate in the domain of feeding, social relationships, locomotion, and so on. Although I do not know of any quantitative literature on this in humans, history reveals numerous examples of brilliant artists who were equally accomplished writers (Egon Schiele), as well as scientists with gifts as essayists (Ed Wilson) or political analysts (Noam Chomsky). If the mechanisms underlying innovation in humans and other animals are similar, then innovative skills in one domain are likely to transfer to other domains, although the degree may vary depending upon opportunity.

Experiments on innovation

Many of the studies presented in this volume showcase the range of behaviour described as innovations observed in nature, and for primates and birds, the kinds of selection pressures that are most likely to lead to innovations. Some of these observations are now being used to fuel new tests, including comparative analyses (Lefebvre and Bolhuis, Reader and MacDonald, Sol) and experiments (Laland and van Bergen, Box, Greenberg, Lefebvre and Bolhuis). Given the fact that studies of innovation are closely tied to studies of social learning, it would seem that students of innovation could readily borrow from the past. Here, I briefly sketch some experiments that might be interesting to run. I divide these into studies of the (1) transmission of innovations and (2) causes of innovation.

What makes an innovation *good*, that is worthy of transmission to, or copying from, others? Presumably, a good innovation is one that works, that solves a problem that has either never been solved or if solved, provides an alternative and perhaps better solution (see discussions throughout this book, and also, in many recent treatments of memetics; Aunger, 2000). This suggests that innovators must have a receptive audience, individuals who recognize that a problem exists and that the innovator has hit upon a solution; if a solution exists, the audience must recognize that the innovator has hit upon an alternative that is either equally good or better than the previous solution. Not only must the audience recognize the innovation, they must be willing and able to try it. I may well recognize that Einstein's theory of relativity provides an innovation within the field of physics, but because of my limited command of physics, may not be able to use it. If Einstein lacked an audience, his innovation would have died with him. Presumably, many such innovations do in fact die because the community is unprepared for change, or unwilling. These loose speculations or comments suggest some interesting experiments.

Train an animal to solve a problem that no other animal in its group can solve, and now place what looks like an innovator back into its group; this is a technique that Stammbach (1988) used with long-tailed macaques, and that Lefebvre (1986) and Langen (1996) have used with birds. By understanding whether the innovative solution requires particular motor or conceptual abilities, and whether it depends on the qualities of the innovator (e.g. old vs young, female vs male, dominant vs subordinate), one should be able to make predictions about the diffusion of this solution through the population. For example, young individuals may not acquire the innovative technique because they lack the motor skills; thus, one might see them trying the right moves but failing due to dexterity. Similarly, one might see young animals trying the technique more than older animals due to their level of curiosity, in the same way that one might see males trying more than females because they are more impulsive, more risk-taking. And, innovations by older and more dominant individuals may be more likely to spread because these animals have clout, perhaps even tried and true 'wisdom'. A final twist, one that follows some of the elegant work by Galef (1996) on social learning in rats, would involve training innovators to perform actions that are functionally deleterious, neutral, or advantageous and then assessing whether the spread of this innovation is related to its fitness consequences. By systematically controlling what the trained innovator learns, one can systematically explore how and why certain innovations spread or fail to do so.

Although we cannot make animals innovate, we can certainly provide opportunities for innovation. This is in fact the kind of research program that Köhler began in his famous box and banana experiments on chimpanzees. By providing animals with various problems, and different kinds of tools to solve them, we should be able to explore the range of solutions generated, and the conditions under which they are most likely to be produced. For instance, are animals more likely to solve problems when they are alone, physically separated from their group? If animals are in view of each other, and working on the same problem, is there a collective benefit from several individuals attempting to solve the same problem? Once animals solve one problem, coming up with a novel solution, does this translate to other problems? Was it pure luck that the Japanese macaque Imo invented both the potato- and wheat-washing techniques or did her solution to one of these problems open the flood gates to all subsequent, and similarly looking problems?

In addition to designing experimental approaches to the problem of innovation, we also need some measure of innovation, both its function and degree of novelty. For example, in looking over the cases of innovation described in this volume, it appears that some innovations are extremely novel like the creation of a never before observed foraging technique, while other inventions represent small variations of an existing trait, such as the modification of song structure in birds by rearranging notes within the repertoire. Functionally, some innovations appear to be created for social purposes (e.g. birdsong; see Slater and Lachlan, Chapter 5) and others for the individual. If there is a social innovation, to the extent that social innovations confer a message they must be recognized and learned by other members of the community. Thus, for example, if an individual invents a new gesture or vocalization, this might count as an innovation, but it will die with the inventor unless it can be recognized by at least one other individual who can presumably also reproduce it. In selfish innovations, such as the many foraging techniques described for birds and primates, the immediate advantage is to the innovator, although clearly the innovation may spread to other group members. If the distinction between social and selfish innovations is meaningful, then there should be greater constraints on the design of social than selfish innovations. Once we make this distinction, it is also necessary to have an in-depth understanding of species-typical behaviour so that we can better assess the nature of apparently novel behaviours. For example, we need some way of distinguishing between the functionless stone handling of Japanese macaques, the virtuosity of a song bird running through what appears to be a Charlie Parker riff (Slater and Lachlan, Chapter 5), and the invention of novel feeding techniques and tools in birds and primates (Lefebvre and Bolhuis, Reader and MacDonald, Chapters 2 and 4). When Charlie Parker, John Coltrane, and Miles Davis brought in a new era of jazz, this was not simply a variation on a theme; it was not a rainbow coloured sticky note, but the sticky note itself. When some birds create new songs, they are innovative in that they are different from all other songs in the population. But are they rainbow coloured sticky notes or the first sticky note? My sense is that they are rainbow coloured. Making these distinctions is not only important at a descriptive level. It is essential if we are to understand the mechanisms driving innovations, and the pressures that favour their spread through a population, or equally, their dissolution.

Acknowledgements

For comments on earlier drafts of this manuscript, I thank Daniel Dennett, Jeff Galef, Kevin Laland, Louis Lefebvre, David Premack, and Simon Reader.

References

Aunger, R. (2000). *Darwinizing culture.* Oxford: Oxford University Press.

Byrne, R. and **Russon, A. E.** (1998). Learning by imitation: A hierarchical approach. *Behavioral and Brain Sciences,* **21,** 667–721.

Cheney, D. L. and **Seyfarth, R. M.** (1982). How vervet monkeys perceive their grunts: Field playback experiments. *Animal Behaviour,* **30,** 739–51.

Cheney, D. L. and **Seyfarth, R. M.** (1990). *How monkeys see the world: Inside the mind of another species.* Chicago, IL: Chicago University Press.

Cosmides, L. and **Tooby, J.** (1994). Origins of domain specificity: The evolution of functional organization. In *Mapping the mind: Domain specificity in cognition and culture* (ed. L. A. Hirschfeld and S. A. Gelman), pp. 85–116. Cambridge: Cambridge University Press.

Damasio, A. (1994). *Descartes' error.* Boston, MA: Norton.

Deacon, T. W. (1997). *The symbolic species: The co-evolution of language and the brain.*

Diamond, A. and **Goldman-Rakic, P. S.** (1989). Comparison of human infants and infant rhesus monkeys on Piaget's AB task: Evidence for dependence on dorsolateral prefrontal cortex. *Experimental Brain Research,* **74,** 24–40.

Diamond, A., Zola-Morgan, S., and **Squire, L. R.** (1989). Successful performance by monkeys with lesions of the hippocampal formation on AB and object retrieval, two tasks that mark developmental changes in human infants. *Behavioral Neuroscience,* **103,** 526–37.

Galef, B. G. Jr. (1988). Imitation in animals: History, definitions, and interpretation of data from the psychological laboratory. In *Social learning: Psychological and biological perspectives* (ed. T. Zentall and B. G. Galef), pp. 3–28. Hillsdale, NJ: Lawrence Erlbaum Associates.

Galef, B. G. Jr. (1996). Social enhancement of food preferences in Norway rats: A brief review. In *Social learning in animals: The roots of culture* (ed. C. M. Heyes and B. G. Galef), pp. 49–64. San Diego, CA: Academic Press.

Hare, B., Call, J., Agnetta, B., and **Tomasello, M.** (2000). Chimpanzees know what conspecifics do and do not see. *Animal Behaviour,* **59,** 771–85.

Hare, B., Call, J., and **Tomasello, M.** (2001). Do chimpanzees know what conspecifics know? *Animal Behaviour,* **61,** 139–51.

Hauser, M. D. (1996). *The evolution of communication.* Cambridge, MA: MIT Press.

Hauser, M. D. (2000). *Wild minds: What animals really think.* New York: Henry Holt.

Hauser, M. D., Chomsky, N., and **Fitch, W. T.** (2002). The faculty of language: What is it, who has it, and how did it evolve? *Science,* **298,** 1569–79.

Heyes, C. M. and **Galef, B. G. Jr.** (1996). *Social learning in animals: The roots of culture.* San Diego, CA: Academic Press.

Kennedy, J. S. (1992). *The new anthropomorphism.* Cambridge: Cambridge University Press.

Langen, T. A. (1996). Social learning of a novel foraging skill by white throated magpie jays (*Calocitta formosa,* Corvidae): A field experiment. *Ethology,* **102,** 157–66.

Lefebvre, L. (1986). Cultural diffusion of a novel food-finding behaviour in urban pigeons: An experimental field test. *Ethology*, **71**, 295–304.

Marler, P. (1978). Primate vocalizations: Affective or symbolic? In *Progress in ape research* (ed. G. Bourne), pp. 85–96. New York: Academic Press.

Marler, P., Evans, C. S., and **Hauser, M. D.** (1992). Animal signals? Reference, motivation or both? In *Nonverbal vocal communication: Comparative and developmental approaches* (ed. H. Papoucek, U. Jürgens, and M. Papoucek), pp. 66–86. Cambridge: Cambridge University Press.

Povinelli, D. (2000). *Folk physics for apes.* New York: Oxford University Press.

Premack, D. (1986). *Gavagai! or the future history of the animal language controversy.* Cambridge, MA: MIT Press.

Roberts, A. C. and **Wallis, J. D.** (2000). Inhibitory control and affective processing in the prefrontal cortex: Neuropsychological studies in the common marmoset. *Cerebral Cortex*, **10**, 252–62.

Rolls, E. T. (1999). *Brain and emotion.* Oxford: Oxford University Press.

Rumbaugh, D. (1997). Competence, cortex, and primate models: A comparative primate perspective. In *Development of the prefrontal cortex* (ed. N. Krasnegor, G. R. Lyon, and P. S. Goldman-Rakic), pp. 117–40. Baltimore: P.H. Brookes Publishing Company.

Santos, L. R., Ericson, B., and **Hauser, M. D.** (1999). Constraints on problem solving and inhibition: Object retrieval in cotton-top tamarins. *Journal of Comparative Psychology*, **113**, 1–8.

Seyfarth, R. M., Cheney, D. L., and **Marler, P.** (1980). Monkey responses to three different alarm calls: Evidence of predator classification and semantic communication. *Science*, **210**, 801–3.

Shettleworth, S. (1998). *Cognition, evolution and behavior.* New York: Oxford University Press.

Stammbach, E. (1988). Group responses to specially skilled individuals in a *Macaca fascicularis*. *Behaviour*, **107**, 241–66.

Struhsaker, T. T. (1967). Auditory communication among vervet monkeys (*Cercopithecus aethiops*). In *Social communication among primates* (ed. S. A. Altmann), pp. 281–324. Chicago, IL: Chicago University Press.

Visalberghi, E. and **Fragaszy, D. M.** (2001). Tool use in monkeys and apes. In *Frontiers of Life*. Vol. IV, part One: *The Biology of Behaviour* (ed. R. Dulbecco, D. Baltimore, F. Jacob and R. Levi-Montalcini). London: Academic Press.

Visalberghi, E. and **Limongelli, L.** (1994). Lack of comprehension of cause–effect relations in tool-using capuchin monkeys (*Cebus apella*). *Journal of Comparative Psychology*, **108**, 15–22.

Visalberghi, E. and **Limongelli, L.** (1996). Acting and understanding: Tool use revisited through the minds of capuchin monkeys. In *Reaching into thought: The minds of the great apes* (ed. A. E. Russon, K. A. Bard, and S. T. Parker), pp. 57–79. Cambridge: Cambridge University Press.

Waddington, C. H. (1975). *Evolution of an evolutionist.* Cornell: Cornell University Press.

Watanabe, M. (1996). Reward expectancy in primate prefrontal neurons. *Nature*, **382**, 629–32.

Whiten, A. and **Ham, R.** (1992). On the nature and evolution of imitation in the animal kingdom: Reappraisal of a century of research. *Advances in the Study of Behavior*, **21**, 239–83.

Zuberbühler, K. (2000). Causal knowledge of predators' behaviour in wild Diana monkeys. *Animal Behaviour*, **59**, 209–20.

Zuberbühler, K., Cheney, D. L., and **Seyfarth, R. M.** (1999). Conceptual semantics in a nonhuman primate. *Journal of Comparative Psychology*, **113**, 33–42.

Zuberbühler, K., Noe, R., and **Seyfarth, R. M.** (1997). Diana monkey long-distance calls: Messages for conspecifics and predators. *Animal Behaviour*, **53**, 589–604.

INDEX